Applied Mathematics Handbook

Applied Mathematics Handbook

Edited by **Frank West**

Willford Press
New York

Published by Willford Press,
118-35 Queens Blvd., Suite 400,
Forest Hills, NY 11375, USA
www.willfordpress.com

Applied Mathematics Handbook
Edited by Frank West

International Standard Book Number: 978-1-68285-223-1 (Hardback)

Printed in China.

Contents

Preface

The main aim of this book is to educate learners and enhance their research focus by presenting diverse topics covering this vast field. This is an advanced book which compiles significant studies by distinguished experts. This book addresses successive solutions to the challenges arising in the area of application, along with it; the book provides scope for future developments.

Mathematics has evolved over the period of time. In the modern era, it is not restricted to pure theory, infact its models and simulations have found their uses in various industries for problem solving. Applied mathematics focuses on the application of mathematical concepts across diverse disciplines of science, engineering, business, computer science, etc. The significant concepts of applied mathematics such as differential equations, approximation theory, probability, etc. have been discussed in detail within this text. It unravels the recent studies in this field. It will serve as a reference to a broad spectrum of readers and will also help those who are researching in this field.

It was a great honour to edit this book, though there were challenges, as it involved a lot of communication and networking between me and the editorial team. However, the end result was this all-inclusive book covering diverse themes in the field.

Finally, it is important to acknowledge the efforts of the contributors for their excellent chapters, through which a wide variety of issues have been addressed. I would also like to thank my colleagues for their valuable feedback during the making of this book.

Editor

1

Bernstein type inequalities on star-like domains in $\mathbb{R}^d$ with application to norming sets

András Kroó

Abstract Let P_n^d be the space of real algebraic polynomials of d variables and degree at most n, $K \subset \mathbb{R}^d$ a compact set, $||p||_K := \sup_{\mathbf{x}\in K} |p(\mathbf{x})|$ the usual supremum norm on K. Let $\varphi_K(\mathbf{x}) := \inf\{\alpha > 0 : \mathbf{x}/\alpha \in K\}$ denote the Minkowski functional of K. In this note we shall prove that if K is a star-like domain with Lipα boundary, that is $\varphi_K(\mathbf{x})$ satisfies the Lipα condition, $0 < \alpha \leq 1$ then the following Bernstein type inequality holds: for any $p \in P_n^d$, $\|p\|_K = 1$ and $\mathbf{x} \in \text{Int}K$

$$|\nabla p|(\mathbf{x}) \leq \frac{cn}{(1 - \varphi_K(\mathbf{x}))^{\frac{1}{\alpha}-\frac{1}{2}}},$$

where $|\nabla p|$ stands for the Euclidean length of the gradient of p. Furthermore, if $1 < \alpha \leq 2$ and K is a C^α star like-domain, that is $\nabla\varphi_K(\mathbf{x})$ has the Lip$(\alpha - 1)$ property, then the same inequality holds for the *tangential* derivatives of p. These new Bernstein type inequalities are applied for the study of cardinality of norming sets, or admissible meshes. The sequence of discrete sets $\mathbf{Y} = \{Y_n \subset K, n \in \mathbb{N}\}$ is called an *optimal admissible mesh* in K if there exist constants c_1, c_2 depending only on K such that

Communicated by S.K. Jain.

This paper was written during A. Kroó visit as **Chair of Excellence at the Carlos III University of Madrid**.

A. Kroó (✉)
Alfréd Rényi Institute of Mathematics, Hungarian Academy of Sciences, Budapest, Hungary
e-mail: kroo.andras@renyi.mta.hu

A. Kroó
Department of Analysis, Budapest University of Technology and Economics, Budapest, Hungary

$$||p||_K \le c_1 ||p||_{Y_n}, \quad p \in P_n^d, n \in \mathbb{N},$$

and $card(Y_n) \le c_2 n^d$, $n \in \mathbb{N}$. It was proved earlier that optimal admissible meshes exist in C^2 star-like domains. In this paper it will be shown that $C^{2-\frac{2}{d}}$ smoothness also suffices for their existence.

Keywords Multivariate polynomials · Bernstein type inequalities · Norming sets · Optimal meshes · Star-like sets

Mathematics Subject Classification (2000) 41A17

1 Introduction

Consider the space P_n^d of real algebraic polynomials of d variables and degree at most n. Let $K \subset \mathbb{R}^d, d \ge 2$, be any compact set and $\|p\|_K := \sup_{\mathbf{x}\in K} |p(\mathbf{x})|$ the usual supremum norm on K. Denote by $D_\mathbf{u} p$ the directional derivative of p in the direction $\mathbf{u} \in S^{d-1} := \{\mathbf{x} \in \mathbb{R}^d : |\mathbf{x}| = 1\}$, and $|\nabla p| := \max_{\mathbf{u}\in S^{d-1}} |D_\mathbf{u} p|$ the Euclidean length of its gradient. The classical Bernstein problem consists in estimating $|\nabla p|(\mathbf{x})$ for a given $p \in P_n^d$, $\|p\|_K = 1$ and $\mathbf{x} \in \mathrm{Int}K$. Typically, this estimate is given in terms of the degree n of the polynomials and the distance of point $\mathbf{x} \in \mathrm{Int}K$ to the boundary ∂K of the compact K. This problem goes back to Bernstein [3] who showed that when $d = 1$ and $K = [a, b]$ we have the sharp estimate

$$|p'(x)| \le \frac{n}{\sqrt{(x-a)(b-x)}} \|p\|_{[a,b]}, \quad x \in (a, b). \tag{1}$$

Let

$$\varphi_K(\mathbf{x}) := \inf\{\alpha > 0 : \mathbf{x}/\alpha \in K\} \tag{2}$$

denote the usual Minkowski functional of K. In case when $K \subset \mathbb{R}^d$ is a $\mathbf{0}$-symmetric convex body Sarantopoulos [13] established a complete analogue of (1)

$$|D_\mathbf{u} p|(\mathbf{x}) \le \frac{n\varphi_K(\mathbf{u})}{\sqrt{1-\varphi_K(\mathbf{x})^2}} \|p\|_K, \quad \mathbf{x} \in \mathrm{Int}K, \mathbf{u} \in S^{d-1}. \tag{3}$$

The above inequality was also independently verified by Baran [2]. Thus for Bernstein type inequalities on $\mathbf{0}$-symmetric convex bodies $\sqrt{1-\varphi_K(\mathbf{x})^2} \sim \sqrt{1-\varphi_K(\mathbf{x})}$ is the proper measure of distance from the given point to the boundary of the set. The problem of finding the correct measure of distance for Bernstein type inequalities on general compact sets in $\mathbb{R}^d$ was studied by Baran [1] and Totik [14] using potential theoretic methods. In [1, 14] this distance is given in terms of the so called equilibrium measure of the set, however this equilibrium measure can be rarely found explicitly.

In this note we shall provide an *explicit asymptotically sharp* measure of distance in Bernstein type inequalities for *star-like domains*. This will be accomplished for both general and *tangential* Bernstein type inequalities. We shall also apply these new inequalities in order to generalize some estimates for cardinality of the *norming sets*.

It should be noted that the star like property of the domains is used in this note for the sake of convenience, the methods applied can be extended to more general domains. A detailed examination of the proofs of the Bernstein type inequalities given by Theorems 1 and 2 reveals that they are based on embedding certain l_p balls into the domain, and technically this can be accomplished easier for a star-like domain. In particular, all results of the paper also hold for finite unions of star-like domains.

2 New results

Let $K \subset \mathbb{R}^d$ be a compact *star-like set* with respect to the origin, that is $\mathbf{0} \in \text{Int}K$ and for every $\mathbf{x} \in K$ we have that $[\mathbf{0}, \mathbf{x}) \subset \text{Int}K$. Note that the study of Bernstein type inequalities is shift independent hence assuming that the center of the star is in the origin does not restrict the generality of our considerations. Then the Minkowski functional $\varphi_K(\mathbf{x})$ of K given by (2) is a *homogeneous functional* satisfying the properties $\varphi_K(\mathbf{x}) < 1, \mathbf{x} \in \text{Int}K$, $\varphi_K(\mathbf{x}) = 1, \mathbf{x} \in \partial K$ and $\varphi_K(\mathbf{x}) > 1$ when $\mathbf{x}$ is not in K. Thus the quantity $1 - \varphi_K(\mathbf{x})$ measures the distance from the point to the boundary of the domain. Our goal is to show that in Bernstein type inequalities on star-like domains the proper measure of distance to the boundary is of the form $(1 - \varphi_K(\mathbf{x}))^\beta$ with some $\beta > 0$ which depends on the geometry of the boundary. In order to identify this parameter β let us introduce the Lip property of the boundary.

Definition 1 Let us say that the star-like set $K \subset \mathbb{R}^d$ has Lipα, $0 < \alpha \leq 1$ boundary if for some $M > 0$ depending on K we have

$$|\varphi_K(\mathbf{x}) - \varphi_K(\mathbf{x} + \mathbf{h})| \leq M|\mathbf{h}|^\alpha, \quad |\mathbf{h}| \leq 1, \ \mathbf{x} \in S^{d-1}. \tag{4}$$

With the above definition we have the next

Theorem 1 *If $K \subset \mathbb{R}^d$ is a star-like domain with Lipα, $0 < \alpha \leq 1$ boundary then for any $p \in P_n^d$, $\|p\|_K = 1$ and $\mathbf{x} \in \text{Int}K$*

$$|\nabla p|(\mathbf{x}) \leq \frac{c_K n}{(1 - \varphi_K(\mathbf{x}))^{\frac{1}{\alpha} - \frac{1}{2}}}, \tag{5}$$

where $c_K > 0$ depends only on K.

It is interesting to note in the above estimate that the measure of distance to the boundary of the set given by the quantity $(1 - \varphi_K(\mathbf{x}))^{\frac{1}{\alpha} - \frac{1}{2}}$ depends on the geometry of the set. It will be pointed out below that in general this is the sharp quantity measuring the distance to the boundary of Lipα, $0 < \alpha \leq 1$ domain. In the Lip 1 case this quantity reduces to $\sqrt{1 - \varphi_K(\mathbf{x})}$ which is essentially what is implied by (3) in case when K is convex.

Clearly, (5) yields an asymptotically optimal estimate in case when $\alpha = 1$ (or in particular K is convex). Nevertheless, a further improvement in the Bernstein inequality can be achieved when only *tangential derivatives* of polynomials are considered.

This approach was investigated in [9] were tangential Markov type inequalities were studied for smooth star-like domains.

Let us assume that the Minkowski functional is continuously differentiable on $\mathbb{R}^d \setminus \{\mathbf{0}\}$ and denote by $\nabla\varphi_K$ its gradient. For any $\mathbf{x} \in \partial K$ we have that $\nabla\varphi_K(\mathbf{x})$ gives the normal direction to the boundary at $\mathbf{x} \in \partial K$. In addition, let

$$T_K(\mathbf{x}) := \{\mathbf{u} \in S^{d-1} : \mathbf{u} \perp \nabla\varphi_K(\mathbf{x})\}$$

be the set of tangent unit vectors at $\mathbf{x} \in \partial K$. When the Minkowski functional is continuously differentiable on $\mathbb{R}^d \setminus \{\mathbf{0}\}$ we shall always assume in addition that the star-like domain K does not contain *critical points* on its boundary in the sense that tangent directions at any $\mathbf{x} \in \partial K$ are not collinear with $\mathbf{x}$.

Definition 2 Given $1 < \alpha \leq 2$, we say that the star-like domain $K \subset \mathbb{R}^d$ is C^α if for some $M > 0$ depending on K we have

$$|\nabla\varphi_K(\mathbf{x}) - \nabla\varphi_K(\mathbf{x} + \mathbf{h})| \leq M|\mathbf{h}|^{\alpha-1}, \quad \mathbf{x} \in S^{d-1}, \ |\mathbf{h}| \leq 1. \tag{6}$$

Theorem 2 *Let $K \subset \mathbb{R}^d$ be a C^α star-like domain with some $1 < \alpha \leq 2$. Then for any $p \in P_n^d$, $\|p\|_K = 1$, $\mathbf{x} \in \mathrm{Int}K \setminus \{\mathbf{0}\}$ and $\mathbf{u} \in T_K(\frac{\mathbf{x}}{\varphi_K(\mathbf{x})})$*

$$|D_{\mathbf{u}} p|(\mathbf{x}) \leq \frac{c_K n}{(1 - \varphi_K(\mathbf{x}))^{\frac{1}{\alpha} - \frac{1}{2}}}, \tag{7}$$

where $c_K > 0$ depends only on K.

Clearly, since $1 < \alpha \leq 2$ in (7) it provides a sharper Bernstein type inequality than (5), but only for tangential derivatives of the polynomial. It should be also noted that when $\alpha = 2$ (7) becomes a uniform Markov type estimate of order n.

Bernstein type inequalities are very useful, they are widely applied in various areas of analysis. One particular application that we intend to exhibit in this note is the study of the so called *norming sets* or *optimal meshes*, see e.g. [8] and [5], were these corresponding notions are introduced.

Definition 3 ([5]) A family of sets $\mathbf{Y} = \{Y_n \subset K, n \in \mathbb{N}\}$ is called an **admissible mesh** in K if there exist constants c_1, c_2 depending only on K such that

$$||p||_K \leq c_1 ||p||_{Y_n}, \quad p \in P_n^d, \ n \in \mathbb{N},$$

where the cardinality of Y_n grows at most polynomially, i.e. $card(Y_n) \leq c_2 n^m$, $n \in \mathbb{N}$, with some fixed $m \in \mathbb{N}$ depending only on K.

This definition of admissible meshes is similar to the notion of **norming sets** [8]. Admissible meshes are applied in various areas, for instance they are used for discrete least squares approximation, extracting discrete extremal sets of Fekete and Leja type, scattered data interpolation, etc. The study of admissible meshes has received lately a considerable attention, see for instance [4,6,12] were various applications and algorithms for the construction of admissible meshes is discussed.

Since $\dim P_n^d \sim n^d$ we clearly must have $m \geq d$ in the above definition (assuming, of course that no polynomial vanishes on K). Naturally, in optimal case we aim for a mesh with asymptotically minimal number of points in it, that is we would like to have $m = d$ in Definition 3. This leads to the following

Definition 4 We shall say that an admissible mesh (norming set) $\mathbf{Y} = \{Y_n \subset K, n \in \mathbb{N}\}$ in $K \subset \mathbb{R}^d$ is **optimal** if $card(Y_n) \leq cn^d, n \in \mathbb{N}$, with some $c > 0$ depending only on K.

The basic question in this respect consists in describing those sets $K \subset \mathbb{R}^d$ which possess optimal admissible meshes. Finding exact geometric properties characterizing sets with optimal admissible meshes appears to be a rather difficult problem. It was shown recently in [10] that any C^2 star-like domain possesses an optimal mesh. Applying the tangential Bernstein inequality (7) of Theorem 2 we can extend this result to C^α star-like domains with $2 > \alpha > 2 - \frac{2}{d}$. This is a substantial decrease in required smoothness of the star-like domain, especially for low dimensions.

Theorem 3 *Let* $1 \leq \alpha \leq 2$. *Assume that* $K \subset \mathbb{R}^d$ *is a star-like domain which is* C^α *if* $1 < \alpha \leq 2$ *and has Lip1 boundary if* $\alpha = 1$. *Then K possesses an admissible mesh* $\mathbf{Y} = \{Y_n \subset K, n \in \mathbb{N}\}$ *with* $card(Y_n) = O(n^d)$ *if* $\alpha > 2 - \frac{2}{d}$; $card(Y_n) = O(n^d \log n)$ *if* $\alpha = 2 - \frac{2}{d}$ *and* $card(Y_n) = O(n^{\frac{2d-2}{\alpha}})$ *if* $\alpha < 2 - \frac{2}{d}$ $(n \in \mathbb{N})$.

Hence by the above theorem when $\alpha > 2 - \frac{2}{d}$ the C^α star-like domains possess optimal meshes. This extends an earlier result given in [10], where the existence of optimal meshes was verified for C^2 star-like domains. Furthermore, when $\alpha = 2 - \frac{2}{d}$ the meshes are near optimal with only an extra log factor present. Finally, when $1 \leq \alpha < 2 - \frac{2}{d}$ the cardinality $O(n^{\frac{2d-2}{\alpha}})$ of admissible meshes provided by Theorem 3 is essentially better than the bound of order $O(n^{2d/\alpha})$ which can be deduced by a routine application of Markov type inequality $\|Dp\|_K = O(n^{2/\alpha})$, $\|p\|_K = 1$ (see [5] for details). In addition, when $\alpha = 1$ only estimates $card(Y_n) = O(n^{2d-2}), d > 2$ and $card(Y_n) = O(n^2 \log n), d = 2$ are applicable in the above theorem. In [10] these estimates for admissible meshes were given for the case of convex bodies in $K \subset \mathbb{R}^d$, Theorem 3 presents an extension of these bounds to the Lip1 star-like domains.

It should be also noted that since the method of proof of Theorem 3 is totally constructive it automatically provides an algorithm for explicit construction of optimal meshes in star-like domains.

3 Bernstein type inequalities for star-like domains

In this section we shall verify Theorems 1 and 2 which give Bernstein type bounds for star-like domains.

Proof of Theorem 1 Let $K \subset \mathbb{R}^d$ be a star-like domain with Lipα, $0 < \alpha \leq 1$ boundary, and consider any $p \in P_n^d$, $\|p\|_K = 1$ and $\mathbf{x} \in \text{Int}K$. Since $\mathbf{0} \in \text{Int}K$ a ball centered at the origin of some radius $c > 0$ is contained in K. We may assume without loss of generality that $|\mathbf{x}| > c/2$ since otherwise the point $\mathbf{x}$ is separated away from the boundary and the statement of the theorem easily follows from (3). Consider the univariate polynomial $g(t) := p(t\mathbf{x}), t \in \mathbb{R}$. Clearly, $t\mathbf{x} \in K$ for any $-c/|\mathbf{x}| \le t \le 1/\varphi_K(\mathbf{x})$. Thus $|g(t)| \le 1$ whenever $-c/|\mathbf{x}| \le t \le 1/\varphi_K(\mathbf{x})$. Hence using the univariate Bernstein inequality (1) for $g \in P_n^1$ with $a = -c/|\mathbf{x}|, b = 1/\varphi_K(\mathbf{x})$ easily yields setting $\mathbf{w} := \frac{\mathbf{x}}{|\mathbf{x}|}$

$$|\mathbf{x}||D_{\mathbf{w}}p|(\mathbf{x}) = |\langle \mathbf{x}, \nabla p(\mathbf{x})\rangle| = |g'(1)| \le \frac{n}{\sqrt{(1 + c/|\mathbf{x}|)(1/\varphi_K(\mathbf{x}) - 1)}}.$$

Since K contains a ball centered at the origin of radius $c > 0$ it follows that $c\varphi_K(\mathbf{x}) \le |\mathbf{x}|$ i.e., we obtain from the previous estimate that

$$|D_{\mathbf{w}}p|(\mathbf{x}) \le \frac{n}{\sqrt{(|\mathbf{x}| + c)(1 - \varphi_K(\mathbf{x}))}}\sqrt{\frac{\varphi_K(\mathbf{x})}{|\mathbf{x}|}} \le \frac{n}{c\sqrt{(1 - \varphi_K(\mathbf{x}))}} \tag{8}$$

This provides a bound for the derivative in radial direction which is clearly stronger than (5).

Now we proceed to the more difficult part of estimating the derivatives in any direction $\mathbf{u} \in S^{d-1}$ orthogonal to $\mathbf{x}$. Clearly we can restrict our considerations to the 2-dimensional plane containing $\mathbf{u}$ and $\mathbf{x}$, so without loss of generality we may assume that $d = 2, \mathbf{u} = (1, 0), \varphi_K(\mathbf{x}) = A < 1$ and $\mathbf{x} = (0, A)$ i.e., $(0, 1) = \mathbf{x}/\varphi_K(\mathbf{x}) \in \partial K, A > c/2$. Let $B > 0$ be the largest real number for which $[\mathbf{x}, \mathbf{z}] \subset K, \mathbf{z} := (B, A)$. Hence $\mathbf{z} \in \partial K$, i.e., $\varphi_K(\mathbf{z}) = 1$. Setting now $\mathbf{h} := (-B, 0)$ and using the Lipα property of K we obtain by (4)

$$1 - A = |\varphi_K(\mathbf{x}) - \varphi_K(\mathbf{z})| = |\varphi_K(\mathbf{z} + \mathbf{h}) - \varphi_K(\mathbf{z})| \le MB^\alpha.$$

This last relation immediately implies that

$$D := \{(x, y) \in \mathbb{R}^2 : 0 \le y \le 1 - M|x|^\alpha\} \subset K. \tag{9}$$

Now we shall inscribe into domain D a proper line passing through the point $\mathbf{x} = (0, A)$. Consider the line $L := \{y = A - M^{\frac{1}{\alpha}}(1 - A)^{\frac{\alpha-1}{\alpha}}x\}$. Since $A > c/2$ it follows by a routine calculation that

$$A - M^{\frac{1}{\alpha}}(1 - A)^{\frac{\alpha-1}{\alpha}}x \le 1 - Mx^\alpha$$

whenever $0 \le x \le x_1 := \frac{c}{2}M^{-\frac{1}{\alpha}}(1 - A)^{\frac{1-\alpha}{\alpha}}$, i.e., the line L passes inside $D \subset K$ for $0 \le x \le x_1$. Moreover, it is easy to see that this line intersects the curve $y = 1 - M(-x)^\alpha$ at the point $x_2 := -\eta(1 - A)^{\frac{1}{\alpha}}$ where $\eta > 0$ is the unique solution of the equation

$$1 = M\eta^\alpha + M^{\frac{1}{\alpha}}\eta.$$

Hence the line L is contained in the domain $D \subset K$ for $x \in [x_2, x_1]$. Now for the univariate polynomial $q(x) := p(x, A - M^{\frac{1}{\alpha}}(1-A)^{\frac{\alpha-1}{\alpha}}x)$ we have $|q(x)| \le 1$, $x \in [x_2, x_1]$. Hence applying again the univariate Bernstein inequality (1) with $a = x_2 = -\eta(1-A)^{\frac{1}{\alpha}}$, $b = x_1 = \frac{c}{2}M^{-\frac{1}{\alpha}}(1-A)^{\frac{1-\alpha}{\alpha}}$ and recalling that $A = \varphi_K(\mathbf{x})$ yields

$$|q'(0)| \le \frac{n}{\sqrt{-x_2 x_1}} \le \frac{c_K n}{(1-\varphi_K(\mathbf{x}))^{\frac{1}{\alpha}-\frac{1}{2}}}, \tag{10}$$

where c_K is a positive constant depending only on K. Furthermore,

$$|q'(0)| = \frac{\partial p}{\partial x}(\mathbf{x}) - M^{\frac{1}{\alpha}}(1-A)^{\frac{\alpha-1}{\alpha}}\frac{\partial p}{\partial y}(\mathbf{x}), \tag{11}$$

where by (8)

$$\left|\frac{\partial p}{\partial y}(\mathbf{x})\right| \le \frac{n}{c\sqrt{1-\varphi_K(\mathbf{x})}}. \tag{12}$$

Finally, combining relations (10)–(12) we arrive at

$$\begin{aligned}\left|\frac{\partial p}{\partial x}(\mathbf{x})\right| &\le |q'(0)| + \frac{n}{c\sqrt{1-\varphi_K(\mathbf{x})}}M^{\frac{1}{\alpha}}(1-\varphi_K(\mathbf{x}))^{\frac{\alpha-1}{\alpha}} \\ &\le \frac{c_K n}{(1-\varphi_K(\mathbf{x}))^{\frac{1}{\alpha}-\frac{1}{2}}} + \frac{M^{\frac{1}{\alpha}}n}{c(1-\varphi_K(\mathbf{x}))^{\frac{1}{\alpha}-\frac{1}{2}}} = \frac{c_K^* n}{(1-\varphi_K(\mathbf{x}))^{\frac{1}{\alpha}-\frac{1}{2}}}.\end{aligned}$$

This completes the proof of Theorem 1. □

Remark 1 Bernstein type inequality (5) implies a corresponding uniform Markov type estimate for star-like domains with Lipα boundary. Indeed, it is well known that univariate polynomials of degree at most n bounded by 1 on $[-1, 1]$ are also bounded by an absolute constant on the larger interval $[-1-1/n^2, 1+1/n^2]$, this easily follows, for instance, from the classical Remez inequality. Thus in case when $K \subset \mathbb{R}^d$ is a Lipα star-like domain a bound that holds for a $p \in P_n^d$ when $\varphi_K(\mathbf{x}) \le 1 - 1/n^2$ will also hold with another constant for every $\mathbf{x} \in K$. Since (5) implies a bound of magnitude $cn^{\frac{2}{\alpha}}$ if we put $\varphi_K(\mathbf{x}) = 1-1/n^2$ it follows that $\|Dp\|_K = O(n^{\frac{2}{\alpha}})$ whenever $p \in P_n^d$, $\|p\|_K = 1$. This Markov type estimate can be found in various papers (see e.g. [11] for corresponding references). The fact that a Markov type inequality of order $O(n^{\frac{2}{\alpha}})$ is sharp in general for Lipα domains goes back to an earlier paper by Goetgheluck [7]. Therefore Bernstein type inequality (5) must be sharp, as well, in the sense that the quantity $(1-\varphi_K(\mathbf{x}))^{\frac{1}{\alpha}-\frac{1}{2}}$ in (5) can not be replaced by another function tending slower to 0 when $\mathbf{x}$ approaches the boundary of the domain, since this would clearly yield a better Markov type inequality.

Proof of Theorem 2 Let $K \subset \mathbb{R}^d$ be a C^α star-like domain with some $1 < \alpha \le 2$. Consider any $\mathbf{x} \in \mathrm{Int}K \setminus \{\mathbf{0}\}$. Since the star-like domain K does not contain saddle points on its boundary, that is $\mathbf{x}/|\mathbf{x}|$ can not be a tangent direction to the boundary of K at any $\frac{\mathbf{x}}{\varphi_K(\mathbf{x})} \in \partial K$, a standard compactness argument yields that the angles between $\mathbf{x}$ and the tangents to ∂K at $\frac{\mathbf{x}}{\varphi_K(\mathbf{x})}$ are uniformly bounded away from zero. Hence without loss of generality we may assume that for the given $\mathbf{x} \in \text{Int}K\backslash\{\mathbf{0}\}$ any tangent $\mathbf{u}$ to ∂K at $\frac{\mathbf{x}}{\varphi_K(\mathbf{x})}$ is orthogonal to $\mathbf{x}$ (this can be always achieved by a uniformly bounded linear transformation of the space). In addition, in order to estimate $|D_{\mathbf{u}}p|(\mathbf{x})$ it suffices to work in the 2-dimensional plane containing $\mathbf{u}$ and $\mathbf{x}$.

Thus summarizing, we can assume that $d = 2, \mathbf{x} := (1-h, 0), h = 1 - \varphi_K(\mathbf{x})$, $(1,0) \in \partial K$ and $\mathbf{u} = (0,1)$ is the tangent to ∂K at $(1,0)$. Now choose an arbitrary $\mathbf{z} = (A, 0) \in \text{Int}K, 1 > A > 0$. Then $\varphi_K(\mathbf{z}) = A$. Let $B > 0$ be the largest real for which with $\mathbf{y} := (A, B)$ we have $[\mathbf{z}, \mathbf{y}] \subset K$. Thus $\mathbf{y} \in \partial K$. Furthermore

$$1 - A = \varphi_K(\mathbf{y}) - \varphi_K(\mathbf{z}) = \frac{\partial}{\partial y}\varphi_K(A, \xi)B \le c_1 B, \quad \xi \in [0, B].$$

Here and in what follows we denote by c_j positive constants depending only on K. Since $\mathbf{u} = (0,1)$ is the tangent to ∂K at $(1,0)$ we also have that $\frac{\partial}{\partial y}\varphi_K(1,0) = 0$. This and the previous relation together with the Lip$(\alpha - 1)$ property of $\nabla\varphi_K$ [see (6)] yield

$$1 - A = \frac{\partial}{\partial y}\varphi_K(A, \xi)B - \frac{\partial}{\partial y}\varphi_K(1,0)B \le c_2 B(1 - A + \xi)^{\alpha-1} \le c_3 B^{\alpha}.$$

This means that

$$D := \{(x, y) \in \mathbb{R}^2 : 0 \le x \le 1 - c_3|y|^{\alpha}\} \subset K. \tag{13}$$

We shall choose a proper $C > 1$ such that

$$Q := \{(x, y) \in \mathbb{R}^2 : 0 \le x \le 1 - h - Cy^2\} \subset D \subset K. \tag{14}$$

In view of (13) this will be accomplished provided that $c_3|y|^{\alpha} - Cy^2 \le h$ whenever $|y| \le (\frac{1-h}{C})^{1/2} < 1$. Clearly, if $\alpha = 2$ we simply can set $C = c_3$. So consider the case when $1 < \alpha < 2$. A routine calculation shows that

$$\max_{y\in[0,1]}(y^{\alpha} - Cy^2) = \beta C^{\frac{\alpha}{\alpha-2}}$$

with some $0 < \beta < 1$ depending only on α. Hence we can set

$$C := c_3^{\frac{2}{\alpha}}\left(\frac{h}{\beta}\right)^{\frac{\alpha-2}{\alpha}} = c_4 h^{\frac{\alpha-2}{\alpha}} \tag{15}$$

and with this choice of C inclusions (14) will hold.

Consider now the univariate trigonometric polynomial of degree $\le n$ defined by

$$t(\phi) := p\left(\frac{1-h}{2}(1 + \cos\phi), \frac{1}{2}\sqrt{\frac{1-h}{C}}\sin\phi\right),$$

where $p \in P_n^2$, $\|p\|_K = 1$. Setting $x = \frac{1-h}{2}(1+\cos\phi)$, $y = \frac{1}{2}\sqrt{\frac{1-h}{C}}\sin\phi$ it follows that

$$1 - h - Cy^2 - x = 1 - h - \frac{1-h}{4}\sin^2\phi - \frac{1-h}{2}(1+\cos\phi)$$
$$= \frac{1-h}{4}(1+\cos^2\phi - 2\cos\phi) \geq 0, \quad \phi \in [-\pi, \pi].$$

Thus in view of inclusions (14) and $\|p\|_K = 1$ we obtain that $|t(\phi)| \leq 1$ $\phi \in [-\pi, \pi]$.

Hence using the Bernstein inequality for trigonometric polynomials of degree $\leq n$ we arrive at

$$n \geq |t'(0)| = \frac{1}{2}\sqrt{\frac{1-h}{C}}\left|\frac{\partial p}{\partial y}(1-h, 0)\right|,$$

i.e., also using (15)

$$\left|\frac{\partial p}{\partial y}(\mathbf{x})\right| \leq \frac{c_5 n h^{\frac{\alpha-2}{2\alpha}}}{\sqrt{1-h}}.$$

Recall that K contains a ball centered at the origin and radius c, thus the statement of the theorem holds trivially if $|\mathbf{x}| < c/2$ i.e., $1 - h < c/2$. On the other hand when $1 - h \geq c/2$ we obtain from the last estimate that

$$\left|\frac{\partial p}{\partial y}(\mathbf{x})\right| \leq \frac{c_6 n}{h^{\frac{1}{\alpha}-\frac{1}{2}}}.$$

Since $h = 1 - \varphi_K(\mathbf{x})$ this completes the proof of the theorem. □

Remark 2 Again similar to the Remark 1 given after the proof of Theorem 1 the Bernstein type inequality of Theorem 2 yields a uniform Markov type upper bound for tangential derivatives of polynomials on C^α star-like domains when $1 < \alpha \leq 2$. Namely it implies that for $p \in P_n^d$, $\|p\|_K = 1$

$$|D_{\mathbf{u}}p(\mathbf{x})| = O(n^{\frac{2}{\alpha}}), \quad \mathbf{u} \in T_K(\mathbf{x}),$$

uniformly for $\mathbf{x} \in \partial K$. Since the above tangential Markov type inequality is known to be sharp, in general (see [9] for details), it follows that the tangential Bernstein type inequality given by Theorem 2 also provides the sharp measure of distance to the boundary of the domain.

Proof of Theorem 3 First let us point out how the Bernstein type inequalities of Theorems 1 and 2 will be used in the proof. Given two points $\mathbf{x}, \mathbf{y} \in K_\rho := \{\mathbf{z} \in K : \varphi_K(\mathbf{z}) = \rho\}, 0 < \rho < 1$ and any $p \in P_n^d$, $\|p\|_K = 1$ we shall need a proper estimate for $|p(\mathbf{x}) - p(\mathbf{y})|$ in terms of $|\mathbf{x} - \mathbf{y}|$. Clearly we can pass to the 2-dimensional plane containing the origin and $\mathbf{x}, \mathbf{y}$, i.e., we may assume

$d = 2$ and consider the polar coordinate representation for K_ρ given by the polar curves $K_\rho = (r(t)\cos t, r(t)\sin t), t \in [0, 2\pi]$ where the radial functions $r(t)$ depend on ρ and are Lip1 ($\alpha = 1$) or $C^\alpha (1 < \alpha \leq 2)$, respectively. Then setting $Q(t) := p(r(t)\cos t, r(t)\sin t)$ easily yields that with some positive constant c_K depending only on K

$$|p(\mathbf{x}) - p(\mathbf{y})| = |Q(t_1) - Q(t_2)| = \left| \int_{t_1}^{t_2} Q'(t)\, dt \right| \leq c_K |\mathbf{x} - \mathbf{y}| D_T p$$

where $D_T p$ denotes the largest magnitude of tangential derivative (when $1 < \alpha \leq 2$) or the largest gradient (when $\alpha = 1$) of p on the level curve K_ρ, respectively. But in both of these cases Theorems 1 and 2 give an upper bound of magnitude $\frac{n}{(1-\rho)^{\frac{1}{\alpha}-\frac{1}{2}}}$ for these derivatives. Hence we obtain from the above estimate that

$$|p(\mathbf{x}) - p(\mathbf{y})| \leq c_K \frac{n|\mathbf{x} - \mathbf{y}|}{(1-\rho)^{\frac{1}{\alpha}-\frac{1}{2}}}, \quad \mathbf{x}, \mathbf{y} \in K_\rho. \tag{16}$$

As before, let c be the radius of the ball centered at the origin which is contained in K. Set $a = c^{5/2}/50$ and choose an integer q such that $1 - c/2 \leq a^2 q^2 \leq 1 - c/4$ (such an integer obviously exists). Set $m := qn$ and

$$\rho_j := 1 - \frac{a^2 j^2}{n^2}; \quad K_j := K_{\rho_j} = \{\mathbf{x} \in \mathbb{R}^d : \varphi_K(\mathbf{x}) = \rho_j\},$$
$$1 \leq j \leq m = qn, \ n \in \mathbb{N}. \tag{17}$$

We may assume without loss of generality that K is also contained in the unit ball centered at the origin, i.e.,

$$c\varphi_K(\mathbf{x}) \leq |\mathbf{x}| \leq \varphi_K(\mathbf{x}), \quad \mathbf{x} \in \mathbb{R}^d.$$

Consider first $\mathbf{x} \in K$ such that $\rho_1 \geq \varphi_K(\mathbf{x}) \geq \rho_m = 1 - a^2 q^2$ that is $\rho_{j+1} \leq \varphi_K(\mathbf{x}) \leq \rho_j$ for some $1 \leq j \leq m-1$. Set $t := \frac{\varphi_K(\mathbf{x})}{\rho_{j+1}}$, $\mathbf{y} := \mathbf{x}/t$. Then $\mathbf{y} \in K_{j+1}$, $\mathbf{x} = t\mathbf{y}$, $t \geq 1$. Hence using that $c/2 \geq \rho_m = 1 - a^2 q^2 \geq c/4$ we have

$$|\mathbf{x} - \mathbf{y}| = (t-1)|\mathbf{y}| \leq t - 1 \leq \frac{\rho_j - \rho_{j+1}}{\rho_{j+1}} \leq \frac{12a^2 j}{cn^2};$$
$$\left| \frac{\mathbf{x}}{\varphi_K(\mathbf{x})} - \mathbf{x} \right| = \frac{|\mathbf{x}|(1-\rho_j)}{\varphi_K(\mathbf{x})} \geq c(1 - \rho_j).$$

By these estimates we have for any $p \in P_n^d$, $\|p\|_K = 1$ using the univariate Bernstein inequality (1) for $p(r\mathbf{x}), 0 \leq r \leq 1//\varphi_K(\mathbf{x})$

$$|p(\mathbf{x}) - p(\mathbf{y})| \leq \frac{n|\mathbf{x} - \mathbf{y}|}{\sqrt{|\mathbf{y}||\mathbf{x}/\varphi_K(\mathbf{x}) - \mathbf{x}|}} \leq \frac{12a^2 j}{c^2 n\sqrt{(1-\rho_j)\rho_{j+1}}} \leq \frac{24a}{c^{5/2}} \leq \frac{1}{2}. \tag{18}$$

If $\rho_1 \le \varphi_K(\mathbf{x})$ we can use the univariate Markov inequality

$$\|q'\|_{[a,b]} \le \frac{2n^2}{b-a}\|q\|_{[a,b]}, \quad q \in P_n^1$$

yielding for $p(r\mathbf{x}), 0 \le r \le 1//\varphi_K(\mathbf{x})$ with some $\mathbf{y} = t\mathbf{x} \in K_1$

$$|p(\mathbf{x}) - p(\mathbf{y})| \le 2n^2\frac{\varphi_K(\mathbf{x})}{|\mathbf{x}|}|\mathbf{x}-\mathbf{y}| \le \frac{2n^2}{c}\frac{a^2}{\rho_1 n^2} \le \frac{1}{2},$$

which is the same as (18).

Thus for every $\mathbf{x} \in K$ such that $1 - a^2q^2 \le \varphi_K(\mathbf{x})$ there exists some $\mathbf{y} \in K_j$, $1 \le j \le n$ for which (18) holds.

Let us denote again by c_j positive constants depending only on K.

In what follows for the sets A, B in $\mathbb{R}^d$ we shall denote by

$$d(A, B) := \sup_{\mathbf{x}\in A} \inf_{\mathbf{y}\in B} |\mathbf{x}-\mathbf{y}|$$

the density of the set B in the set A. Evidently, given $h > 0$ we can choose a discrete set $\Omega \subset S^{d-1}$ so that $d(\Omega, S^{d-1}) \le h$ and card$\Omega \le c(d)h^{1-d}$ with some constant $c(d)$ depending only on d. Also note that since the boundary of K is Lip1 for any $\mathbf{x}, \mathbf{y} \in \partial K$ with $|\frac{\mathbf{x}}{|\mathbf{x}|} - \frac{\mathbf{y}}{|\mathbf{y}|}| \le h$ we have $|\mathbf{x}-\mathbf{y}| \le c_2 h$.

Based on these remarks and utilizing the constant c_K from (16) we can set

$$h_j := \frac{1}{4c_K n}\left(\frac{a^2 j^2}{n^2}\right)^{\frac{1}{\alpha}-\frac{1}{2}}, \quad 1 \le j \le m \tag{19}$$

and choose $N_j \le c_1 h_j^{1-d}$ points $\{\mathbf{y}_{i,j} \in K_j, 1 \le i \le N_j\} := Y_{n,j}$ on K_j so that

$$d(K_j, Y_{n,j}) \le h_j, \quad 1 \le j \le m. \tag{20}$$

Set $K^* := \{\mathbf{x} \in \mathbb{R}^d : \varphi_K(\mathbf{x}) \le 1 - a^2q^2\}$. Since for $\mathbf{x} \in K^*$ we have that

$$|\mathbf{x}| \le \varphi_K(\mathbf{x}) \le 1 - a^2q^2 \le c/2$$

it follows that K^* is contained in the ball of radius $c/2$ centered at the origin. Recalling that ball of radius c centered at the origin is contained in K we obtain using (3) for this ball

$$|p(\mathbf{x}) - p(\mathbf{y})| \le \frac{2n}{c}|\mathbf{x}-\mathbf{y}|, \quad \mathbf{x}, \mathbf{y} \in K^*. \tag{21}$$

Furthermore, we can choose a discrete set $Y_n^* \subset K^*$ with card$(Y_n^*) \le c_2 n^d$ so that

$$d(K^*, Y_n^*) \le \frac{c}{4n}. \tag{22}$$

Set now $\mathbf{Y}_n := \cup_{1\le j\le m} Y_{n,j} \cup Y_n^*$. Using (19) the cardinality of this set can be estimated as follows

$$\operatorname{card}\mathbf{Y}_n \le c_5 \left(n^d + n^{\frac{2(d-1)}{\alpha}} \sum_{j=1}^{m} j^{(d-1)(1-2/\alpha)} \right).$$

If $\alpha > 2 - \frac{2}{d}$ the the sum above is of magnitude $n^{d-\frac{2(d-1)}{\alpha}}$ yielding that $\operatorname{card}(\mathbf{Y}_n) = O(n^d)$. When $\alpha = 2-\frac{2}{d}$ the above sum is of magnitude $\log n$ and we obtain $\operatorname{card}(\mathbf{Y}_n) = O(n^d \log n)$. Finally, if $\alpha < 2 - \frac{2}{d}$ the sum is convergent and $d < \frac{2(d-1)}{\alpha}$ so we end up with $\operatorname{card}(\mathbf{Y}_n) = O(n^{\frac{2(d-1)}{\alpha}})$. Thus the cardinality of $\mathbf{Y}_n$ is of magnitude stated in Theorem 3.

It remains now to show that $\mathbf{Y}_n$ is an admissible mesh. Choose any $p \in P_n^d$, $\|p\|_K = 1$ and $\mathbf{x} \in K$, $p(\mathbf{x}) = 1$. Assume first that $\mathbf{x} \in K^*$. Then by (22) we can find $\mathbf{y} \in Y_n^* \subset K^*$, $|\,\mathbf{x} - \mathbf{y}\,| \le \frac{c}{4n}$. Thus using (21) we obtain

$$1 - p(\mathbf{y}) = p(\mathbf{x}) - p(\mathbf{y}) \le \frac{2n}{c}|\mathbf{x} - \mathbf{y}| \le \frac{1}{2},$$

i.e., $||p||_{\mathbf{Y}_n} \ge 1/2$ in this case.

Now let $\varphi_K(\mathbf{x}) \ge 1 - a^2q^2$. Then as it was shown above there exists some $\mathbf{y} \in K_j, 1 \le j \le n$ for which (18) holds. Moreover, since $p(\mathbf{x}) = 1$ we have $|p(\mathbf{y})| \ge 1/2$. Furthermore, by (19) and (20) we can find $\mathbf{y}_{i,j} \in Y_{n,j} \subset K_j$ such that

$$|\mathbf{y} - \mathbf{y}_{i,j}| \le h_j = \frac{1}{4c_K n}\left(\frac{a^2 j^2}{n^2}\right)^{\frac{1}{\alpha}-\frac{1}{2}}.$$

Thus using again (16) with $\rho = \rho_j$ defined by (17) we obtain

$$|p(\mathbf{y}) - p(\mathbf{y}_{i,j})| \le c_K |\mathbf{y} - \mathbf{y}_{i,j}| \frac{n}{\left(\frac{a^2 j^2}{n^2}\right)^{\frac{1}{\alpha}-\frac{1}{2}}} \le \frac{1}{4}.$$

Since $|p(\mathbf{y})| \ge 1/2$ the last estimate implies that $|p(\mathbf{y}_{i,j})| \ge 1/4$ which means that $||p||_{\mathbf{Y}_n} \ge 1/4$ in this case, as well.

The proof of Theorem 3 is completed. □

References

1. Baran, M.: Complex equilibrium measure and Bernstein type inequality for compact sets in $\mathbb{R}^n$. Proc. Am. Math. Soc. **123**, 485–494 (1995)

2. Baran, M.: Bernstein type theorems for compact sets in $\mathbb{R}^n$ revisited. J. Approx. Theory **79**, 190–198 (1994)
3. Bernstein, S.N.: Sur l'ordre de la meilleure approximation des fonctions continues par les polynômes de degré donné. Mem. Cl. Sci. Acad. R. Belg. **4**, 1–103 (1912)
4. Bos, L., De Marchi, S., Sommariva, A., Vianello, M.: Weakly admissible meshes and discrete extremal sets. Numer. Math. Theory Methods Appl. **4**, 1–12 (2011)
5. Calvi, J.P., Levenberg, N.: Uniform approximation by discrete least squares polynomials. J. Approx. Theory **152**, 82–100 (2008)
6. De Marchi, S., Marchioro, M., Sommariva, A.: Polynomial approximation and cubature at approximate Fekete and Leja points of the cylinder. Appl. Math. Comput. **218**, 10617–10629 (2012)
7. Goetgheluck, P.: Inegalité de Markov dans les ensembles effilés. J. Approx. Theory **30**, 149–154 (1980)
8. Jetter, K., Stöckler, J., Ward, J.D.: Error estimates for scattered data interpolation. Math. Comput. **68**, 733–747 (1999)
9. Kroó, A.: Markov type inequalities for surface gradients of multivariate polynomials. J. Approx. Theory **118**, 235–245 (2002)
10. Kroó, A.: On optimal polynomial meshes. J. Approx. Theory **163**, 1107–1124 (2011)
11. Kroó, A., Szabados, J.: Markov–Bernstein type inequalities for multivariate polynomials on sets with cusps. J. Approx. Theory **99**, 134–152 (1999)
12. Piazzon, F., Vianello, M.: Analytic transformation of admissible meshes. East J. Approx. **16**, 313–322 (2010)
13. Sarantopoulos, Y.: Bounds on the derivatives of polynomials on Banach spaces. Math. Proc. Cambr. Philos. Soc. **110**, 307–312 (1991)
14. Totik, V.: Polynomial inverse images and polynomial inequalities. Acta Math. **187**, 139–160 (2001)

2

Hopf algebras in non-associative Lie theory

J. Mostovoy · J. M. Perez-Izquierdo ·
I. P. Shestakov

Abstract We review the developments in the Lie theory for non-associative products from 2000 to date and describe the current understanding of the subject in view of the recent works, many of which use non-associative Hopf algebras as the main tool.

1 Introduction: Lie groups, Lie algebras and Hopf algebras

Non-associative Lie theory, that is, Lie theory for non-associative products, appeared as a a subject of its own in the works of Malcev who constructed the tangent structures corresponding to Moufang loops. Its general development has been slow; nevertheless, by now many of the basic features of the theory have been understood, with much of the progress happening in the last 10 years or so. In the present paper we outline the non-associative Lie theory in general and review the recent developments.

Communicated by Efim Zelmanov.

J. Mostovoy (✉)
Departamento de Matemáticas, CINVESTAV-IPN, Apartado Postal 14-740,
07000 México D.F., Mexico
e-mail: jacob@math.cinvestav.mx

J. M. Perez-Izquierdo
Departamento de Matemáticas y Computación, Universidad de La Rioja, 26004 Logroño, Spain
e-mail: jm.perez@unirioja.es

I. P. Shestakov
Instituto de Matemática e Estatística, Universidade de São Paulo,
Caixa Postal 66281, São Paulo, SP 05311-970, Brazil

I. P. Shestakov
Sobolev Institute of Mathematics, Novosibirsk 630090, Russia
e-mail: shestak@ime.usp.br

The history of the subject before 2000 has been summarized in the paper of Sabinin [51]. Probably, the single most important advance to that date has been the definition of a Sabinin algebra by Mikheev and Sabinin in 1986 (they use the term *hyperalgebra*). Sabinin algebras are the structures tangent to general non-associative unital local products; under some convergency conditions one can recover the product from a Sabinin algebra. Lie, Malcev, Bol, Lie–Yamaguti algebras and Lie triple systems are all particular instances of Sabinin algebras.

Classical Lie theory relates three kinds of objects: Lie groups, Lie algebras and cocommutative Hopf algebras. Until recently, however, Hopf algebras had little, if any, role in the non-associative Lie theory. To a certain extent, this gap has now been filled, and here we shall give an outline of the theory of local loops and Sabinin algebras from the point of view of non-associative Hopf algebras. We shall be guided by the analogy with the theory of Lie groups and Lie algebras, so first we recall the basics of the classical theory.

At the heart of Lie theory lies the equivalence of the following three categories:

- simply-connected finite-dimensional Lie groups;
- finite-dimensional Lie algebras;
- irreducible cocommutative finitely generated Hopf algebras.

Let us describe explicitly the functors that establish this equivalence. In what follows all the vector spaces, Lie algebras etc will be defined over a field of characteristic zero.

1.1 Lie groups → Lie algebras

There are several ways to produce the Lie algebra of a Lie group. Let G be a Lie group with the unit element e. The most common construction identifies the tangent space T_eG with the vector space of left-invariant vector fields on G. The commutator of two such fields (considered as derivations acting on functions) is again left-invariant, and defines the Lie bracket on T_eG.

This definition can be stated in different terms. On the tangent bundle TG there exists a canonical flat connection ∇. It can be defined by the associated parallel transport: for any path between $a, b \in G$ it sends T_aG to T_bG by the differential of the left multiplication by ba^{-1}. The curvature tensor of ∇ is identically zero and the torsion tensor T is covariantly constant, that is $\nabla \mathrm{T} = 0$ on G. This implies that the torsion tensor T is defined completely by its value on T_eG, which is a linear map

$$T_eG \otimes T_eG \to T_eG.$$

This map coincides, up to sign, with the Lie bracket on T_eG.

1.2 Lie algebras → Lie groups

One way to obtain a Lie group from a Lie algebra is via the Baker–Campbell–Hausdorff formula which expresses the formal power series

$$\log(\exp x \exp y) \tag{1}$$

in terms of iterated commutators in x and y. For finite-dimensional Lie algebras, the Baker–Campbell–Hausdorff series always has a non-zero radius of convergence and, hence, defines a local Lie group.

It can be shown that finite-dimensional local Lie groups can always be extended to global Lie groups, but the proof of this fact is somewhat mysterious and uses the facts that seem to come from outside of Lie theory, such as the Ado theorem. The Ado theorem claims that every finite-dimensional Lie algebra is isomorphic to a Lie subalgebra of the algebra of the matrices. Therefore, it integrates to a local Lie group which is embedded as a local subgroup into the group of invertible matrices. Then it can be shown that this local subgroup generates a subset in the group of matrices which is an image of a smooth injective Lie group homomorphism. This last step is far from being trivial and uses in a crucial way the associativity of the matrix multiplication.

1.3 Lie algebras → Hopf algebras

Each Lie algebra $\mathfrak{g}$ can be embedded into an associative algebra A in such a way that the bracket in $\mathfrak{g}$ is induced by the commutator in A. Among all such embedding one is universal. The corresponding algebra (called universal enveloping algebra) is constructed as the quotient of the unital tensor algebra on $\mathfrak{g}$ by the relations

$$x \otimes y - y \otimes x - [x, y]$$

for all $x, y \in \mathfrak{g}$ and is denoted by $U(\mathfrak{g})$. The Lie algebra $\mathfrak{g}$ is naturally embedded into the tensor algebra on $\mathfrak{g}$ and this embedding descends to the embedding

$$\mathfrak{g} \to U(\mathfrak{g})$$

into the universal enveloping algebra.

The algebra $U(\mathfrak{g})$ can be given a coproduct

$$\delta : U(\mathfrak{g}) \to U(\mathfrak{g}) \otimes U(\mathfrak{g})$$

which turns it into a Hopf algebra. It is defined by declaring all the elements of $\mathfrak{g} \subset U(\mathfrak{g})$ to be primitive:

$$\delta(x) = x \otimes 1 + 1 \otimes x.$$

Since the coproduct in a Hopf algebra is an algebra homomorphism, this is sufficient to determine δ completely. According to the Poincaré–Birkhoff–Witt theorem, as a coalgebra $U(\mathfrak{g})$ is isomorphic to the symmetric algebra $k[\mathfrak{g}]$. This latter can be thought of as the universal enveloping algebra of the abelian Lie algebra which coincides with $\mathfrak{g}$ as a vector space and whose Lie bracket is identically zero.

It is immediate that as a coalgebra $U(\mathfrak{g})$ is cocommutative. Also, by construction, it is primitively generated (that is, generated by primitive elements). The antipode S on $U(\mathfrak{g})$ is uniquely determined; in particular, on each $x \in \mathfrak{g}$ we have $S(x) = -x$.

1.4 Hopf algebras → Lie algebras

A straightforward calculation shows that the subspace of primitive elements in a Hopf algebra is always closed under the commutator. In particular, there is a functor that assigns to each Hopf algebra the Lie algebra of its primitive elements. It is not hard to prove that it gives an equivalence of the category of cocommutative primitively generated Hopf algebras to that of Lie algebras (of not necessarily finite dimension).

1.5 Lie groups → Hopf algebras

Let D_eG be the space of distributions (that is, linear functionals on functions) on the Lie group G, supported at the unit e. In other words, D_eG is the space spanned by the delta function supported at e and all of its derivatives. The multiplication on G gives rise to a convolution product on D_eG; the diagonal map $G \to G \times G$ induces a coproduct on D_eG. It can be seen that D_eG is a Hopf algebra; moreover, it is precisely the universal enveloping algebra of the Lie algebra of G.

1.6 Hopf algebras → Lie groups

To recover a Lie group from a Hopf algebra, one uses the fact that any cocommutative Hopf algebra generated by its primitive elements is a universal enveloping algebra of some Lie algebra. In particular, by the Poincaré–Birkhoff–Witt theorem, it can be thought of as the symmetric algebra $k[V]$ on some vector space V over the base field k with a modified product

$$\mu : k[V] \otimes k[V] \to k[V].$$

Consider the composition of this product with the projection $k[V] \to V$ onto the space of primitive elements. The resulting map

$$F_\mu : k[V] \otimes k[V] \to V$$

can be interpreted as follows. Choose a basis $e_1, \ldots, e_n$ in V. Then $k[V]$ is the algebra of the polynomials in the e_i. Each coordinate of the map F_μ is a formal power series in x_i and y_j, where $x_1, \ldots, x_n$ and $y_1, \ldots, y_n$ are the coordinates dual to the e_i in the first and the second copy of V respectively. It turns out that for any Hopf algebra such that V is of finite dimension these formal power series all converge in a neighbourhood U of the origin in $V \times V$ and thus give an analytic map

$$V \times V \supseteq U \to V.$$

This map is, in fact, a finite-dimensional local Lie group and these are always locally equivalent to simply-connected Lie groups.

1.7 The Ado theorem

The fundamental triangle of Lie theory is remarkably robust: it can be generalized to a variety of situations. For instance, finite-dimensional Lie algebras can be replaced by infinite-dimensional Lie algebras, Lie algebras in tensor categories other than vector spaces, or, as we shall see, can be substituted for more general kinds of tangent algebras. There is only one construction that does not always survive into the more general context.

Note that while recovering a Lie group either from its Lie algebra or from its Hopf algebra of distributions we first construct a formal Lie group, that is, a collection of power series which a priori may not converge. In order to see that this formal Lie group comes from an actual Lie group we need the Ado theorem, which states that every finite-dimensional Lie algebra has a faithful finite-dimensional representation. In particular, it follows that any finite-dimensional formal Lie group can be obtained from a local subgroup in a matrix group. A further argument is then needed to see that every such local subgroup extends to an injective homomorphism of a Lie group to the group of invertible matrices.

This proof breaks down already in the case of infinite-dimensional Lie groups. It also fails in the case of general finite-dimensional loops. Nevertheless, there are situations (Moufang loops, nilpotent loops) where each formal multiplication extends to a global loop. In these situations we have an analogue of the Ado theorem, though we should point out that this theorem alone does not guarantee the globalizability of formal products.

1.8 Nilpotent groups

Lie algebras and Hopf algebras also appear in the theory of discrete nilpotent groups.

Let G be a discrete group. Write $\mathbb{Q}G$ for the group algebra of G over the rationals and let $\Delta \subset \mathbb{Q}G$ be the augmentation ideal. The ith dimension subgroup D_iG (or $D_i(G, \mathbb{Q})$) consists of all those $g \in G$ for which

$$g - 1 \in \Delta^i.$$

The dimension subgroups form a descending central series which is closely related to the lower central series $\gamma_i G$ defined inductively by $\gamma_1 G = G$ and $\gamma_i G = [\gamma_{i-1} G, G]$ for $i > 1$. Namely, D_iG consists of all $g \in G$ for which there exists $n \geq 1$ with the property that $g^n \in \gamma_i G$.

The graded abelian group

$$\mathcal{L}G = \bigoplus_{i>0} D_iG/D_{i+1}G$$

associated with the dimension series is actually a Lie ring with the Lie bracket induced by the group commutator

$$[a, b] = a^{-1}b^{-1}ab.$$

Tensored with the rational numbers, this Lie ring becomes a Lie algebra and its universal enveloping algebra can be explicitly identified with the graded algebra

$$\bigoplus_{i\geq 0} \Delta^i/\Delta^{i+1}$$

associated with the filtration on $\mathbb{Q}G$ by the powers of the augmentation ideal.

In general, a discrete group G cannot be recovered from the Lie algebra $\mathcal{L}G \otimes \mathbb{Q}$ coming from the dimension series (equivalently, from the lower central series). However, if G is a finitely generated nilpotent group, $\mathcal{L}G \otimes \mathbb{Q}$ is finite-dimensional and can be integrated to a Lie group into which G embeds as a discrete subgroup if it is torsion-free.

One group for which the dimension series and the lower central series can be described explicitly is the free group F_n on n generators $x_1, \dots, x_n$. Let $\mathbb{Z}\langle\langle X_1, \dots, X_n\rangle\rangle$ be the ring of formal power series in n non-commuting variables. The group of invertible elements in this ring consists of power series which start with ± 1. We denote this group by $\mathbb{Z}\langle\langle X_1, \dots, X_n\rangle\rangle^*$. There is a homomorphism

$$\begin{aligned} \mathcal{M} : F_n &\to \mathbb{Z}\langle\langle X_1, \dots, X_n\rangle\rangle^*, \\ x_i &\mapsto 1 + X_i, \end{aligned} \tag{2}$$

which is, in fact, injective. The mth terms of the dimension and of the lower central series of F_n coincide and consist of those elements which map under $\mathcal{M}$ to the power series of the form

$$1 + \text{terms of degree at least } m.$$

This, together with the injectivity of $\mathcal{M}$, shows that the free group F_n is residually nilpotent.

For the results on dimension subgroups and powers of the augmentation ideal we refer the reader to [44] (Chapter 3, Section 4) and to [31].

The non-associative versions of the above constructions and statements are the subject of the present paper. There are fundamental (and vast) parts of Lie theory which have not been understood yet in the non-associative context, most notably the representation theory. By no means we want to suggest that there are inherent obstructions to this, apart from time and effort.

2 Local, infinitesimal and formal loops

The non-associative Lie theory deals with the following three equivalent categories:

- formal loops;
- Sabinin algebras;
- irreducible cocommutative and coassociative non-associative Hopf algebras.

In contrast to the usual finite-dimensional Lie theory, there is no procedure to extend formal loops to global multiplications on manifolds. This, however, should not be seen as a problem since many non-associative products are inherently of local nature. The true analogue of a Lie group in the non-associative Lie theory is not a smooth loop, but, rather, a local loop, or a germ of a local loop (We shall use the term "infinitesimal loops" for germs of local loops). A formal loop produces a local loop whenever the power series that define it converge in some neighbourhood of the origin; verifying this is a problem from analysis and we shall not discuss it here.

2.1 Local loops and smooth loops

Let M be a smooth finite-dimensional manifold.[1] A local multiplication on M (or, more precisely, on a non-empty open subset $U \subseteq M$) is a smooth map

$$F : U \times U \to M.$$

If there exists $e \in U$ with the property that

$$F|_{e\times U} = \mathrm{Id}(U) = F|_{U\times e},$$

the local multiplication F is called unital, or a local loop. The point e is referred to as the unit; if the unit exists, it is necessarily unique. Often, when working with a fixed local loop F, we shall write $F(x, y)$ simply as $x \cdot y$ or xy.

For any local loop there exist two local multiplications $V \times V \to M$ with $V \subseteq U$, denoted by x/y and $y\backslash x$. They are defined by

$$x/y \cdot y = x$$

and

$$y \cdot y\backslash x = x$$

for all $x, y \in V$, and, for obvious reasons, are called the right and the left division, respectively. The existence of both divisions follows from the fact that the right and left multiplication maps

$$R_y = F|_{U\times y} : U \to M$$

and

$$L_y = F|_{y\times U} : U \to M$$

are the identity when $y = e$. We take V to be the largest neighbourhood on which both divisions are defined.

[1] we assume that $\dim M > 0$ is part of the definition of a smooth manifold.

In general, there is no reason to expect that the right and the left divisions would be expressed as multiplications by an inverse of any kind.

A local loop is called a smooth loop if $U = V = M$ in the above definitions. Recall that a discrete loop, or, simply, a loop, is a set M with a globally defined product for which there exists a unit element, and such that the left and the right multiplications by a fixed element of M are bijections. In particular, for each loop there are globally defined right and left divisions. In these terms, a smooth loop can be defined as a loop which, at the same time, is a smooth manifold, with the multiplication and the divisions being smooth maps.

2.2 Example: invertible elements in algebras

Call an element a of a unital non-associative algebra invertible if both equations $ax = 1$ and $ya = 1$ have a unique solution. The invertible elements of a finite-dimensional algebra over the real numbers form a local loop.

This local loop is not necessarily a smooth loop. Consider, for instance, the Cayley–Dickson algebras $\mathbb{A}_n$ on $\mathbb{R}^{2^n}$. When $n > 3$ there exist pairs of invertible elements in $\mathbb{A}_n$ whose product is not invertible.

2.3 Example: homogeneous spaces

Recall that a smooth manifold M is a homogeneous space for a Lie group G if G acts transitively on M and the action is smooth. Choose a point $e \in M$. Then we have a smooth map

$$p : G \to M,$$

which sends $g \in G$ to $g(e) \in M$. This map identifies M, as a smooth manifold, with the left coset space G/G_e, where G_e is the stabilizer of e. Conversely, if G is a Lie group and H is a closed subgroup, the set G/H of left cosets of G by H is a smooth manifold and a homogenous space for G.

Let M be a homogeneous space for G and $U \subseteq M$ a neighbourhood of a point $e \in M$. Assume that we are given a section of p over U, that is, a smooth map $i : U \to G$ such that $i(e)$ is the unit in G and $p \circ i = \mathrm{Id}(U)$. Then M is a local loop, with the multiplication $U \times U \to M$ defined as

$$(x, y) \mapsto p\,(i(x)i(y))\,.$$

When p is actually a homomorphism of Lie groups, that is, when G_e is a normal subgroup in G, this local loop structure is the same thing as the product on M restricted to $U \times U$.

There are many important examples of homogeneous spaces, among them spheres, hyperbolic spaces and Grassmannians.

2.4 Infinitesimal loops

Let F and F' be two local loops on $U \subseteq M$ and $U' \subseteq M'$ respectively. A morphism $F \to F'$ is a smooth map $f : M \to M'$ such that $f(U) \subseteq U'$ and such that

$$F'(f(x), f(y)) = f(F(x, y))$$

for all $x, y \in U$. With this notion of a morphism local loops form a category.

If a morphism f from F to F' is an open embedding, we shall say that F is a restriction of F'. In this situation, the local loops F and F' may be thought of as being "locally the same" near the unit. Certainly, they should not be distinguished by any infinitesimal algebraic structure at the unit, such as those arising in the classical Lie theory (think of formal groups or Lie algebras). With this in mind, we define local equivalence as the equivalence relation on the category of local loops generated by open embeddings. An equivalence class of local loops is called an infinitesimal loop.

Strictly speaking, it is the infinitesimal loops and not the local loops that are the main subject in the non-associative Lie theory. However, for the sake of simplicity, we shall speak of local loops rather than infinitesimal loops, wherever possible. This should not lead to confusion.

2.5 Analytic local loops and formal loops

Examples of local loops are very easy to construct via power series, since the only condition a multiplication of a local loop has to satisfy is the existence of the unit. Consider an n-tuple of power series

$$F_i \in \mathbb{R}[[x_1, \ldots, x_n, y_1, \ldots, y_n]]$$

where $1 \leq i \leq n$, and assume that all of them converge in some neighbourhood of the origin in $\mathbb{R}^{2n}$. The map

$$(x_1, \ldots, x_n, y_1, \ldots, y_n) \mapsto (F_1(x_j, y_k), \ldots, F_n(x_j, y_k))$$

defines a local loop on $\mathbb{R}^n$, with the origin as the unit, if and only if for all i

$$F_i = x_i + y_i + \sum_{k,l=1}^{n} x_k y_l \cdot P_i^{kl} \tag{3}$$

with $P_i^{kl} \in \mathbb{R}[[x_1, \ldots, x_n, y_1, \ldots, y_n]]$.

A local loop on an analytic manifold whose multiplication can be written in this form in some coordinate chart is called analytic. Similarly, an infinitesimal loop is analytic if it has an analytic representative.

The convergence of the power series F_i that specify an analytic loop is often irrelevant. For instance, a power series with no constant term can be substituted into another power series instead of a variable, and thus we do not need convergence in order to speak of the algebraic identities, such as associativity or Moufang identities, satisfied by an analytic loop. In what follows we shall refer to an n-tuple of formal power series F_i of the form (3) as a formal loop.

Let us state this definition in the form that does not make reference to any explicit basis in $\mathbb{R}^n$, nor to the fact that n is finite. For a vector space V and a field k write $k[V]$ for the symmetric algebra of V over k. A formal map from V to a vector space W is a linear map $k[V] \to W$ which annihilates the constants. A formal multiplication is a formal map from $V \times V$ to V. Given that $k[V \times V]$ is canonically isomorphic to $k[V] \otimes k[V]$, this is the same as a linear map

$$k[V] \otimes k[V] \to V.$$

A formal multiplication F is a formal loop if

$$F|_{1 \otimes k[V]} = \pi_V = F|_{k[V] \otimes 1},$$

where $\pi_V : k[V] \to V$ is the projection of a polynomial onto its linear part.

Since $k[V]^*$ is the algebra of formal power series in V, a choice of coordinates in V specifies a formal loop as an n-tuple of formal power series in $2n$ variables, where $n = \dim V$ is not necessarily finite.

2.6 The canonical connection and geodesic loops

Let F be a local loop on $U \subseteq M$ and let $V \subseteq U$ be a neighbourhood on which the left division is well-defined. The canonical connection ∇ on the tangent bundle TV has virtually the same definition as for the Lie groups (apart from the fact that it is not defined on the whole of M): for $a, b \in G$ the corresponding parallel transport sends T_aG to T_bG by the differential of the smooth map

$$x \mapsto b(a\backslash x).$$

It is clear that ∇ is flat. There is no reason, however, for the torsion tensor of ∇ to be constant.

It can be seen that each flat connection on a neighbourhood of a point comes from a local loop, called the geodesic loop of the connection. Indeed, a connection ∇ on a manifold M gives rise to the exponential map

$$\exp_a : T_aM \supseteq U_a \to M,$$

defined on a neighbourhood U_a of the origin in T_aM, for each $a \in M$. If U_a is chosen to be small enough, this map has an inverse, denoted by $\log_a$. Fix a point $e \in M$. Then there is a neighbourhood $U \subseteq M$ of e such that

$$a \cdot b = \exp_a (a \log_e b), \tag{4}$$

is well defined for all $a, b \in U$. Here $a \log_e b$ stands for the parallel transport of the vector $\log_e b \in T_e M$ to $T_a M$. The operation $\cdot$ defines a local loop on $U \subseteq M$ and it is a straightforward check that ∇ is its canonical connection. This loop is called the geodesic loop of ∇. In order to be precise, we should consider the geodesic loop as an infinitesimal rather than local loop since there is a choice involved in the construction of the neighbourhood U.

It can be shown that the necessary condition for an infinitesimal loop (which is also sufficient in the case of analytic loops) to be a geodesic loop of a flat connection is that it should satisfy the right alternative property. Namely, it should be represented by a local loop such that

$$(a \cdot b) \cdot b = a \cdot (b \cdot b)$$

whenever both sides are defined. In particular, each local loop F gives rise to a right alternative infinitesimal loop, which is the geodesic loop of the canonical connection of F. One can recover a local loop F from the corresponding geodesic loop with the help of the additional operation Φ defined by

$$a \cdot b = a \times \Phi(a, b). \tag{5}$$

Here $\cdot$ denotes the multiplication in F and $\times$ is the product in the geodesic loop. Φ here can be any function of a and b that satisfies $\Phi(e, b) = b$ and $\Phi(a, e) = e$.

The construction of the geodesic loop associated with a flat connection is due to Kikkawa [24] and Sabinin [49].

There is also a corresponding formal notion of a canonical connection. Given a formal loop F on a vector space V, the formal canonical connection of F is the restriction of F to the subspace

$$k[V] \otimes V \subset k[V] \otimes k[V].$$

Among all the formal loops that give rise to the same formal connection there is exactly one right alternative formal loop.

The notion of a geodesic loop is of crucial importance in non-associative Lie theory. Indeed, there are two constructions of the tangent structure to a local (infinitesimal, formal) loop and both are based on the fact that there is a flat connection associated with a local loop, see the next section. In fact, in the approach to the Lie theory taken by Sabinin in [52] the role of non-associative analogues of Lie groups is played not by loops but by affinely connected manifolds. This involves additional algebraic structures such as odules; at the moment of the writing of the present paper the theory of odules has not noticeably advanced beyond the results of [52].

3 Sabinin algebras

Many of the well-known generalizations of Lie algebras involve only one or two operations: Malcev algebras have one binary bracket, Lie triple systems one ternary bracket, Bol, Lie-Yamaguti and Akivis algebras one binary and one ternary bracket. In contrast, Sabinin algebras which are the most general form of the tangent structure for loops, have an infinite set of independent operations, and, as a consequence, admit an infinite number of axiomatic definitions. At the moment there are three different natural constructions for the set of operations in a Sabinin algebra. Two of these constructions are due to Sabinin and Mikheev and we shall review them in this section. The third set of operations, which appeared for the first time in [57] in the study of the primitive elements in non-associative bialgebras, will be considered in the next section. Finding a complete set of relations satisfied by this third set of operations is presently an open problem.

Remark 1 The fact that the definition of Mikheev and Sabinin involved an infinite number of operations may have contributed to its slow acceptance. For some time Akivis algebras were considered as possible analogues of Lie algebras. The definition of an Akivis algebra involves two operations: an antisymmetric binary bracket and a ternary bracket, with only one identity that relates the two operations and generalizes the Jacobi identity. In spite of the elegance and simplicity of this definition, the category of Akivis algebras is not equivalent to that of formal loops and, hence, is not suitable as a basis for non-associative Lie theory. We shall consider the Akivis algebras in more detail in Sect. 4.7.

Both ways of deriving the structure of a Sabinin algebra from a non-associative product that were proposed by Mikheev and Sabinin consist of two steps. First, a local loop is replaced by the corresponding geodesic loop (which, in the analytic case is the same as a right alternative local loop), and then the tangent operations for a geodesic loop are extracted from its canonical connection. This results in two infinite sets of completely independent operations. Mikheev and Sabinin used the term "hyperalgebra" for the algebraic structure tangent to a geodesic loop and "hyperalgebra with multioperators" for the general tangent structure. In our terminology hyperalgebras with multioperators will be called Sabinin algebras, and hyperalgebras (which are the same as hyperalgebras with the trivial multioperator) will be called flat Sabinin algebras. The adjective "flat" here is supposed to reflect the fact that these are the tangent structures to general flat connections.

3.1 Flat Sabinin algebras from the torsion tensor of a connection

Let k be a field of characteristic zero. A flat Sabinin algebra is a vector space V over k together with a set of maps

$$V^{\otimes n+2} \to V$$
$$X_1 \otimes \cdots \otimes X_n \otimes Y \otimes Z \mapsto \langle X_1, \ldots, X_n; Y, Z\rangle$$

for all integers $n \geq 0$, satisfying the following identities:

$$\langle X_1, \ldots, X_n; Y, Z\rangle = -\langle X_1, \ldots, X_n; Z, Y\rangle,$$

$$\langle X_1, \ldots, X_r, A, B, X_{r+1}, \ldots, X_n; Y, Z\rangle - \langle X_1, \ldots, X_r, B, A, X_{r+1}, \ldots, X_n; Y, Z\rangle$$
$$+ \sum_{k=0}^{r} \sum_{\alpha} \langle X_{\alpha_1}, \ldots, X_{\alpha_k}, \langle X_{\alpha_{k+1}}, \ldots, X_{\alpha_r}; A, B\rangle, X_{r+1}, \ldots, X_n; Y, Z\rangle = 0,$$

$$\sigma_{X,Y,Z}\Bigg(\langle X_1, \ldots, X_r, X; Y, Z\rangle$$
$$+ \sum_{k=0}^{r} \sum_{\alpha} \langle X_{\alpha_1}, \ldots, X_{\alpha_k}; \langle X_{\alpha_{k+1}}, \ldots, X_{\alpha_r}; Y, Z\rangle, X\rangle \Bigg) = 0.$$

Here α varies over the set of all bijections $\{1, \ldots r\} \to \{1, \ldots r\}, i \to \alpha_i$ such that $\alpha_1 < \alpha_2 < \cdots < \alpha_k, \alpha_{k+1} < \cdots < \alpha_r, k = 0, 1, \ldots, r, r \geq 0$, and $\sigma_{X,Y,Z}$ denotes the sum over all cyclic permutations of X, Y, Z.

The above definition may seem complex, but it has the following geometric meaning.

Consider an affine connection ∇ at a point e of a manifold M. It is characterized by two tensors: the curvature tensor and the torsion tensor which are related by the Bianchi identities. If ∇ is flat (as it is in the case of a canonical connection) the curvature tensor is identically zero. If we assume that ∇ is analytic, the torsion tensor T is uniquely determined in a neighbourhood of e by its value and the values of all its covariant derivatives at this point. These derivatives are multilinear operations on T_eM. Set

$$\langle Y, Z\rangle = \mathrm{T}(Y, Z)$$

and

$$\langle X_1, \ldots, X_n; Y, Z\rangle = \nabla_{X_1} \ldots \nabla_{X_n} \mathrm{T}(Y, Z)$$

for all $n > 0$. Then the brackets defined in this way are antisymmetric in Y and Z, since the torsion tensor is antisymmetric, and satisfy the identities which come from the Bianchi identities and their derivatives, and from the fact that the curvature is zero. These identities are precisely those that appear in the definition of a flat Sabinin algebra.

From this construction it is clear that given a flat Sabinin algebra $\mathfrak{l}$ one can reconstruct, locally, the corresponding flat connection, and, hence, the infinitesimal loop, provided that the operations of $\mathfrak{l}$ define converging power series. The convergence conditions are explicitly stated by Mikheev and Sabinin in [34,50]. In general, even if the convergence conditions are not satisfied, the geometric reasoning can be applied so as to produce a formal loop from any flat Sabinin algebra.

3.2 Flat Sabinin algebras from pairs of Lie algebras

An alternative approach to flat Sabinin algebras is to define them as the algebraic structure that exists on a complement to a subalgebra in a Lie algebra.

Start with a direct sum decomposition of vector spaces

$$\mathfrak{g} = \mathfrak{h} \oplus \mathfrak{l}, \tag{6}$$

where $\mathfrak{g}$ is a Lie algebra and $\mathfrak{h}$ is a subalgebra. Let $\pi_{\mathfrak{l}}$ be the linear projection map $\pi_{\mathfrak{l}} : \mathfrak{g} \to \mathfrak{l}$ with $\mathfrak{h}$ as the kernel. It will be convenient to introduce a simplified notation for the right-normed iterated bracket in $\mathfrak{g}$:

$$[X_1, X_2, \ldots, X_n] = [X_1, [X_2, [\ldots, X_n] \ldots]].$$

Define a sequence of multilinear operations[2] on $\mathfrak{l}$ by setting for each $n \geq 2$ and $X_1, \ldots, X_n \in \mathfrak{l}$

$$(X_1, \ldots, X_n) := \pi_{\mathfrak{l}}[X_1, \ldots, X_n].$$

There are two kinds of relations satisfied by these operations. The relations of the first kind reflect the fact that they come from Lie brackets. Namely, the anti-symmetry of the Lie bracket gives

$$(X_1, X_2) + (X_2, X_1) = 0, \tag{7}$$

and the Jacobi identity translates into the following identity:

$$(X_1, X_2, X_3) + (X_2, X_3, X_1) + (X_3, X_1, X_2) = 0. \tag{8}$$

The identities of the second kind express the fact that $\mathfrak{h}$ is a subalgebra and not just a vector subspace. If $A, B \in \mathfrak{g}$ are arbitrary elements, then

$$\pi_{\mathfrak{l}}[\pi_{\mathfrak{l}}A - A, \pi_{\mathfrak{l}}B - B] = 0.$$

Setting $A = [X_1, \ldots, X_n]$ and $B = [Y_1, \ldots, Y_m]$ we get

$$\begin{aligned} &\pi_{\mathfrak{l}}[[X_1, \ldots, X_n], [Y_1, \ldots, Y_m]] + ((X_1, \ldots, X_n), (Y_1, \ldots, Y_m)) \\ &\quad = ((X_1, \ldots, X_n), Y_1, \ldots, Y_m) - ((Y_1, \ldots, Y_m), X_1, \ldots, X_n). \end{aligned}$$

Now, the first term in this expression can be rewritten iteratively using the Jacobi identity in terms of the right-normed brackets so that the last relation takes the form

$$\begin{aligned} &-((X_1, \ldots, X_n), (Y_1, \ldots, Y_m)) + ((X_1, \ldots, X_n), Y_1, \ldots, Y_m) \\ &\quad -((Y_1, \ldots, Y_m), X_1, \ldots, X_n) = \sum_{\alpha} (-1)^{h(\alpha)} (X_{\alpha_1}, \ldots X_{\alpha_n}, Y_1, \ldots, Y_m), \end{aligned} \tag{9}$$

where the summation is taken over all permutations α of the set $\{1, \ldots, n\}$ for which there exists s such that $\alpha_i < \alpha_{i+1}$ for $i < s$, and $\alpha_i > \alpha_{i+1}$ for $i \geq s$, and $h(\alpha) = n - s$ (Clearly, for any such α we have that $\alpha_s = n$).

[2] Our notation may be potentially confusing where it concerns the trilinear operation $(\cdot, \cdot, \cdot)$. Here this is not the associator.

We can now define a flat Sabinin algebra as a vector space with a sequence of multilinear operations $(X_1, \ldots, X_n)$ for all $n \geq 2$ satisfying the relations (7–9). It can be shown that if $\mathfrak{l}$ is a flat Sabinin algebra in this sense, there exists a decomposition (6) which gives rise to the operations on $\mathfrak{l}$, [40].

The relation between flat Sabinin algebras in this sense and germs of flat affine connections is quite straightforward. Given such a connection ∇ in the neighbourhood of a point e on a manifold M, we can identify the tangent space T_eM with the space of all ∇-parallel vector fields. Set $\mathfrak{g}$ to be the Lie subalgebra generated by the ∇-parallel vector fields inside the Lie algebra of all vector fields, and $\mathfrak{h}$ to be the subalgebra of $\mathfrak{g}$ consisting of the vector fields that vanish at e. Then we have the direct sum decomposition

$$\mathfrak{g} = \mathfrak{h} \oplus T_eM,$$

and T_eM acquires the structure of a flat Sabinin algebra.

One can, in fact, write down explicit formulae relating the two sets of operations in Sabinin algebras, see [34,50].

3.3 Multioperators

Let F be a right alternative local loop F on $U \subseteq M$ and

$$\Phi : U \times U \to M$$

a smooth function such that $\Phi(e, b) = b$, $\Phi(a, e) = e$ and such that the Jacobian matrix of $\Phi(a, b)$ with respect to b is the identity when $b = e$. Define a new local loop F_Φ by setting

$$F_\Phi(a, b) = F(a, \Phi(a, b)).$$

The canonical connection of F_Φ coincides with that of F; in particular, F is the geodesic loop associated with F_Φ (Strictly speaking, F_Φ is defined on a smaller neighbourhood than U. As usual, one can replace local loops here by infinitesimal loops).

If Φ is analytic, one can write

$$\Phi(a, b) = b + \sum_{m \geq 1, n \geq 2} \Phi_{m,n}(a, \ldots, a; b, \ldots, b), \tag{10}$$

where $\Phi_{m,n}(X_1, \ldots, X_m; Y_1, \ldots, Y_n)$ is linear in each X_j and Y_j and invariant with respect to all the permutations of the X_i and of the Y_j:

$$\Phi_{m,n}(X_1, \ldots, X_m; Y_1, \ldots, Y_n) = \Phi_{m,n}(X_{\sigma(1)}, \ldots, X_{\sigma(m)}; Y_{\tau(1)}, \ldots, Y_{\tau(n)}) \tag{11}$$

for all $\sigma \in \Sigma_m$, $\tau \in \Sigma_n$.

This motivates the following definition. A Sabinin algebra is a flat Sabinin algebra $\mathfrak{l}$ together with a set of operations

$$\Phi_{m,n} : \mathfrak{l}^{\otimes m+n} \to \mathfrak{l}$$

for $m \geq 1, n \geq 2$, satisfying the symmetry conditions (11). Given all that has been said about the flat Sabinin algebras, it is not hard to show that Sabinin algebras form a category which is equivalent to the category of all formal loops.

3.4 Free Sabinin algebras

The free Sabinin algebra on a set of generators S is the universal Sabinin algebra generated by S. It can be constructed as the vector space spanned by the symbols corresponding to all the possible operations $\langle \cdot, \dots, \cdot; \cdot, \cdot \rangle$ and $\Phi_{m,n}(\cdot, \dots, \cdot; \cdot, \dots, \cdot)$ and their compositions, whose arguments are elements of S, with the relations of a Sabinin algebra imposed. Another way to construct the free Sabinin algebras is by using the techniques described in the next section. Namely, the free Sabinin algebra on a set of generators S is the set of primitive elements in the free non-associative algebra on the same set of generators.

Several types of bases in free Lie algebras are known: Hall bases, Lyndon–Shirshov bases [48], right normed bases of Chibrikov [11]. The construction of Lyndon–Shirshov bases was extended by Chibrikov [12] to free Sabinin algebras. The dimensions of the graded summands of a free Sabinin algebra on n generators were calculated in [6].

Free Sabinin algebras enjoy many other properties characteristic of free Lie algebras [12]. In particular, the variety of Sabinin algebras is Schreier (every subalgebra of a free Sabinin algebra is free). This implies that all automorphisms of finitely generated free Sabinin algebras are tame, and that the occurrence problem for free Sabinin algebras is decidable. Also, finitely generated subalgebras of free Sabinin algebras are residually finite and the word problem is decidable for the variety of Sabinin algebras.

3.5 Further remarks

Just as in the associative case, there exists a Baker–Campbell–Hausdorff formula for Sabinin algebras. We shall discuss it in the next section.

The two sets of multilinear operations in a Sabinin algebra are completely independent and can be considered as Taylor expansions of two non-linear operations. Sabinin in [52] takes this approach pointing out that it can be applied to non-analytic loops, and branding the constructions of [34] and [50] as obsolete. The book [52] was written before it was discovered that Sabinin algebras are important in the theory of non-associative bialgebras; in this latter context non-linear operations are not easy to deal with.

4 Non-associative Hopf algebras

It would be fair to say that most of the constructions and results on Sabinin algebras mentioned in the previous section were the product of a quest for a Lie theory of flat connections. The subject of non-associative Hopf algebras has been developed in the context of the long-standing problem of finding an alternative enveloping algebra for an arbitrary Malcev algebra. While the existence of such an alternative envelope is still an open question, it turned out [42] that Malcev algebras have universal enveloping algebras which are very similar in their properties to usual cocommutative Hopf algebras. Later, it turned out that a similar construction can be carried out for Bol algebras [45] and, more generally, all Sabinin algebras [46].

Another motivation for the development of the machinery of non-associative Hopf algebras was the question of whether the commutator and the associator are the only primitive operations in a non-associative bialgebra. It appeared as a conjecture in [20]; if it were true, it would imply an important role for the Akivis algebras in non-associative Lie theory. This conjecture was answered in the negative in [57]: in non-associative bialgebras, apart from the commutator and the associator, there exists an infinite series of independent primitive operations. These operations, called the Shestakov–Umirbaev operations, are obtained from the associator by means of a procedure resembling linearization; they are closely related to the associator deviations in the nilpotency theory of loops (see Sect. 8.1).

The role of the non-associative Hopf algebras in the fundamental questions of Lie theory such as integration was clarified in [39]. Hopf algebraic techniques are also relevant in other problems, such as the Ado theorem, which will be discussed in Sect. 7.4.

4.1 The definition and basic properties

In what follows we shall consider algebras and coalgebras over a field k of characteristic zero.

A unital bialgebra is a unital algebra which, at the same time, is a counital coalgebra in such a way that the coproduct is an algebra homomorphism (or, equivalently, such that the product is a coalgebra morphism). It will be convenient to use Sweedler's notation in which the coproduct is written as

$$\delta(x) = \sum x_{(1)} \otimes x_{(2)}.$$

A unital bialgebra H with the coproduct δ and the counit ϵ is called a (non-associative) Hopf algebra if is endowed with two additional bilinear operations, the right and the left division

$$\begin{aligned} / : H \times H \to H, \qquad & \backslash : H \times H \to H, \\ (x, y) \mapsto x/y, \qquad & (x, y) \mapsto x\backslash y, \end{aligned}$$

such that

$$\sum (yx_{(1)})/x_{(2)} = \epsilon(x)y = \sum (y/x_{(1)})x_{(2)}$$

and

$$\sum x_{(1)}\backslash(x_{(2)}y) = \epsilon(x)y = \sum x_{(1)}(x_{(2)}\backslash y).$$

When the product in a Hopf algebra is associative and the coproduct is coassociative we get the usual notion of a Hopf algebra. In this case the antipode S can be defined as $S(x) = 1/x$ and, generally, we have

$$x\backslash y = S(x)y \quad \text{and} \quad x/y = xS(y)$$

for all x, y.

The subspace of a Hopf algebra generated by the unit is a simple subcoalgebra. If this subspace is the only simple subcoalgebra, the Hopf algebra is called irreducible. In irreducible Hopf algebras the right and left division are uniquely determined.

While non-associativity is an essential property of Hopf algebras that appear in Lie theory, the coproduct in many interesting cases is, actually, coassociative. For instance, the loop algebra kL of a loop L is always a coassociative Hopf algebra. In what follows by a Hopf algebra we shall understand a cocommutative and coassociative Hopf algebra, unless explicitly stated otherwise.

4.2 Primitive elements and Shestakov–Umirbaev operations

In any associative bialgebra the set of primitive elements is closed under the algebra commutator. This can be expressed by saying that the commutator is a primitive operation. Any other primitive operation can be written in terms of the commutators, in the following sense: if an associative bialgebra H is generated by a set $\{x_i\}$ where each x_i is primitive, then each primitive element in H is a linear combination of iterated commutators in the x_i. We express this by saying that the commutator forms a complete set of primitive operations.

In non-associative bialgebras there are other primitive operations, most notably the algebra associator

$$(a, b, c) = (ab)c - a(bc).$$

As discovered by Shestakov and Umirbaev in [57], the set of primitive operations obtained from commutators and associators is not complete. For instance, in the free unital non-associative algebra on one primitive generator x the element

$$(x^2, x, x) - 2x(x, x, x) = (x^2x)x - x^2x^2 - 2x(x^2x) + 2x(xx^2)$$

is primitive, yet cannot be obtained from x by taking linear combinations and compositions of commutators and associators.

For $m, n \geq 1$ let $\underline{u} = x_1, \ldots, x_m$ and $\underline{v} = y_1, \ldots, y_n$ be sequences of primitive elements in a bialgebra and write $u = ((x_1x_2)\ldots)x_m$ and $v = ((y_1y_2)\ldots)y_n$

for the corresponding left-normed products. The Shestakov–Umirbaev operations $p(x_1, \dots, x_m; y_1, \dots, y_n; z)$ are defined inductively by

$$(u, v, z) = \sum u_{(1)}v_{(1)} \cdot p(\underline{u}_{(2)}; \underline{v}_{(2)}; z),$$

where (x, y, z) denotes the associator and z is primitive. Here Sweedler's notation is extended so as to mean that the sum is taken over all decompositions of the sequences $\underline{u}$ and $\underline{v}$ into pairs of subsequences $\underline{u}_{(1)}, \underline{u}_{(2)}$ and $\underline{v}_{(1)}, \underline{v}_{(2)}$; the expressions $u_{(1)}$ and $v_{(1)}$ are the left-normed products of the elements of $\underline{u}_{(1)}$ and $\underline{v}_{(1)}$, respectively. The subsequences $\underline{u}_{(i)}, \underline{v}_{(i)}$ are allowed to be empty, in which case we write $\underline{u}_{(i)} = 1$ and $\underline{v}_{(i)} = 1$; and, by definition, we have

$$p(1; \underline{v}; z) = p(\underline{u}; 1; z) = 0.$$

For instance, the operation which corresponds to $m = n = 1$ is just the associator. The operations corresponding to $m = 2, n = 1$ and $m = 1, n = 2$ are

$$p(x_1, x_2; y; z) = (x_1x_2, y, z) - x_1(x_2, y, z) - x_2(x_1, y, z)$$

and

$$p(x; y_1, y_2; z) = (x, y_1y_2, z) - y_1(x, y_2, z) - y_2(x, y_1, z),$$

respectively.

When the bialgebra in question is a Hopf algebra, the Shestakov–Umirbaev operations can be written with the help of the left division:

$$p(x_1, \dots, x_m; y_1, \dots, y_n; z) = \sum (u_{(1)}v_{(1)}) \backslash (u_{(2)}, v_{(2)}, z).$$

This form of the definition may be preferable since it expresses each operation directly via the associator.

In [57] it is shown that all Shestakov–Umirbaev operations are primitive, and that together with the commutator they form a complete set of primitive operations.

In associative algebras the commutator satisfies the Jacobi identity; in particular, the primitive elements of a bialgebra A form a Lie subalgebra of the commutator algebra of A. In the non-associative case, the Lie algebras are replaced by Sabinin algebras. Indeed, set

$$\begin{aligned}
&\langle 1; y, z\rangle = \langle y, z\rangle = -[y, z] = -yz + zy \\
&\langle x_1, \dots, x_m; y, z\rangle = \langle \underline{u}; y, z\rangle = -p(\underline{u}; y; z) + p(\underline{u}; z; y) \\
&\Phi^{SU}(x_1, \dots, x_m; y_1, \dots, y_n) = \\
&\quad \frac{1}{m!}\frac{1}{n!} \sum_{\tau\in\Sigma_m, \sigma\in\Sigma_n} p(x_{\tau(1)}, \dots, x_{\tau(m)}; y_{\sigma(1)}, \dots, y_{\sigma(n-1)}; y_{\sigma(n)})
\end{aligned}$$

with $u = ((x_1 x_2) \dots) x_m$, Σ_m the symmetric group on m letters and $m \geq 1, n \geq 2$. Then the identities of a Sabinin algebra are satisfied for all x_i, y_j, z in an arbitrary non-associative algebra, [57].

In particular, we get a functor from non-associative algebras to Sabinin algebras

$$A \mapsto \text{УШ}(A)$$

that assigns to an algebra A the Sabinin algebra $\text{УШ}(A)$ on the same vector space. In the case when A is a bialgebra, the primitive elements of A form a Sabinin subalgebra

$$\mathrm{Prim}(A) \subset \text{УШ}(A),$$

and we have a functor

$$A \to \mathrm{Prim}(A)$$

generalizing the corresponding functor from associative bialgebras (or Hopf algebras) to Lie algebras.

4.3 Universal enveloping algebras for Sabinin algebras

It was proved in [46] that each Sabinin algebra $\mathfrak{l}$ can be realized as the subspace of primitive elements in a certain non-associative irreducible Hopf algebra $U(\mathfrak{l})$. Moreover, the correspondence

$$\mathfrak{l} \to U(\mathfrak{l})$$

is functorial.

When $\mathfrak{l}$ is a Lie algebra, the Hopf algebra $U(\mathfrak{l})$ is the usual universal enveloping algebra of $\mathfrak{l}$. The properties of $U(\mathfrak{l})$ also closely mirror those of the associative enveloping algebras. There is a version of the Poincaré–Birkhoff–Witt theorem which says that $U(\mathfrak{l})$, as a coalgebra, is isomorphic to the symmetric algebra $k[\mathfrak{l}]$. As a corollary, if $\mathfrak{l}$ is given a basis (e_i), the algebra $U(\mathfrak{l})$ is additively spanned by the left normed products $(\dots(e_{i_1} e_{i_2}) \dots) e_{i_n}$ with $i_k \leq i_{k+1}$. As a consequence, Hopf algebras $U(\mathfrak{l})$ are right and left Noetherian for finite-dimensional $\mathfrak{l}$ and have no zero divisors.

Most importantly, we have the Milnor–Moore theorem which states that each irreducible Hopf algebra A is the universal enveloping algebra of the Sabinin algebra of its primitive elements:

$$A = U(\mathrm{Prim}(A)).$$

This implies that the universal enveloping algebra functor is an equivalence of the category of Sabinin algebras with that of irreducible Hopf algebras.

An explicit construction of $U(\mathfrak{l})$ can be found in [46] or in Section 4.3 of [39].

It seems plausible that for a flat Sabinin algebra $\mathfrak{l}$ represented as complement to a Lie sub algebra $\mathfrak{h}$ in a Lie algebra $\mathfrak{g}$ the algebra $U(\mathfrak{l})$ should have an explicit description in terms of $U(\mathfrak{g})$ and $U(\mathfrak{h})$. At the moment we only have a description of this kind for Malcev algebras and Lie triple systems (Sect. 6).

4.4 Loops and Hopf algebras of distributions

A distribution on a manifold M is a linear functional on the functions on M. A distribution is said to be supported at a point if its value on a function only depends on the germ of the function at this point. Write D_eM be the space of distributions on a manifold M supported at the point $e \in M$. Under some reasonable conditions, which we shall assume to be satisfied, D_eM as a vector space is spanned by the Dirac delta function at e and all of its (higher) derivatives.

The space D_eM has the natural structure of a coalgebra with the coproduct induced by the diagonal map $M \to M \times M$. As a coalgebra, it is isomorphic to the symmetric algebra $k[T_eM]$, and the space of primitive elements in D_eM consists of all the derivatives along some vector in T_eM. In particular, it is cocommutative, coassociative and irreducible, since all these properties hold in $k[T_eM]$.

A local loop structure $U \times U \to M$ in a neighbourhood of e induces a product

$$D_eU \otimes D_eU \to D_eM = D_eU.$$

This product is readily seen to be a coalgebra homomorphism, and, hence, $D_eU = D_eM$ is a Hopf algebra. If L is a local loop on U, the notation D_eL will stand for D_eU with this Hopf algebra structure.

In the case when a local loop L is analytic, the product on D_eL can be expressed in terms of the coefficients of the power series that define L. In particular, this makes it possible to define the Hopf algebra of distributions for any formal loop, and the explicit formulae in the formal case are quite straightforward.

Indeed, any formal map

$$\theta : k[V] \to W$$

can be lifted to a unique coalgebra morphism

$$\theta' : k[V] \to k[W],$$

which is defined as

$$\theta'(\mu) = \sum_{n=0}^{\infty} \frac{1}{n!} \theta(\mu_{(1)}) \cdots \theta(\mu_{(n)}) = \epsilon(\mu)1 + \theta(\mu) + \cdots .$$

Note that this is always a finite sum since the terms with n greater than the degree of μ vanish.

In particular, any formal loop

$$F : k[V \times V] = k[V] \otimes k[V] \to V$$

lifts to a product

$$F' : k[V] \otimes k[V] \to k[V],$$

which gives $k[V]$ the structure of a non-associative Hopf algebra. If the series that define F converge, this Hopf algebra coincides with the Hopf algebra of the distributions $D_e F$.

By the Milnor–Moore and the Poincaré–Birkhoff–Witt theorems, each irreducible Hopf algebra can be considered as a product on $k[V]$ for some vector space V. Taking the primitive part of this product we get a formal loop:

$$k[V] \otimes k[V] \to k[V] \to \mathrm{Prim}(k[V]) = V.$$

This establishes an equivalence between the category of formal loops and that of irreducible Hopf algebras.

4.5 The two structures of a Sabinin algebra on the tangent space to a loop

Given a local loop L on a manifold M, there are two ways to obtain the operations of a Sabinin algebra on $T_e M$. The first way was described by Sabinin and Mikheev: the brackets are the derivatives of the torsion tensor of the canonical connection of L and the multioperator Φ measures the failure of L to be right alternative. The second possibility is to use the Sabinin algebra operations, as defined by Shestakov and Umirbaev, on the space of primitive elements in the Hopf algebra $D_e L$.

It turns out that these two Sabinin algebra structures are almost, though not quite, the same. Namely, the brackets in both Sabinin algebras coincide, while the Mikheev–Sabinin multioperator Φ is different from the Shestakov–Umirbaev multioperator Φ^{SU}. At the moment it is not known how to express the Mikheev–Sabinin multioperator in terms of the Shestakov–Umirbaev primitive operations.

4.6 Exponentials, logarithms and the Baker–Campbell–Hausdorff formula

In the associative case, the exponential map is the unique, up to a rescaling, power series that sends the primitive elements of a complete Hopf algebra bijectively onto its group-like elements. Since the group-like elements in a Hopf algebra form a group, the Baker–Campbell–Hausdorff formula (1) endows the Lie algebra of the primitive elements with the structure of a group. This construction can be applied to any Lie algebra so as to obtain the corresponding formal Lie group.

Complete Hopf algebras can be defined in the non-associative context as well; the exponential and logarithmic power series in this situation are series in one non-associative variable. In contrast with the associative case, however, the exponential and the logarithm are not defined uniquely by the condition that they interchange the primitive and the group-like elements. In fact, this condition is satisfied by an infinite-dimensional affine family of power series all of which can be used to define a functor from Sabinin algebras to formal loops via a formula of Baker–Campbell–Hausdorff type. For instance, there exists a unique non-associative exponential series with the additional property that

$$\exp 2x = \exp x \exp x$$

and whose linear term is x. The first few terms of the corresponding Baker–Campbell–Hausdorff formula were calculated explicitly by Gerritzen and Holtkamp in [14]. They do not speak of Sabinin algebras and, in particular, do not discuss whether their formula provides any kind of integration.

A different exponential series was considered in [39]. Replace in the usual exponential series the term x^n with the left-normed product of n copies of a non-associative variable x:

$$\exp x = 1 + x + \frac{1}{2!}x^2 + \frac{1}{3!}x^2 x + \frac{1}{4!}(x^2 x)x + \cdots$$

The exponential defined in this way is, essentially, the exponential map of the canonical connection on the loop of formal power series in one non-associative variable and non-zero constant term.

The non-associative logarithm, which is defined as the inverse of the exponential, can be calculated explicitly and its coefficients involve Bernoulli numbers. Let $f : \mathbb{N} \to R$ be a sequence of elements of some ring R, indexed by the natural numbers. It can be extended to an R-valued function defined on all non-associative monomials in x. Given a non-associative monomial τ define f_τ inductively as follows. For $\tau = x$ set $f_\tau = 1$. If $\tau \neq x$, there is only one way of writing τ as a product $(\ldots((x\tau_1)\tau_2)\ldots)\tau_k$. Set

$$f_\tau = f_k \cdot f_{\tau_1} \cdot \cdots \cdot f_{\tau_k}.$$

With this notation we have

$$\log(1 + x) = \sum_\tau \frac{B_\tau}{\tau!} \cdot \tau,$$

where τ varies over the set of all non-associative monomials in x, and B_τ and $\tau!$ are the extensions of the Bernoulli numbers and the factorial, respectively, to the set of non-associative monomials [39].

Rather than use (1), we define the formal power series $\mathcal{L}(x, y)$ in two non-associative variables x and y by the formula

$$\mathcal{L}(x, y) = \log\left(\sum_{m,n\geq 0} \frac{1}{m!n!} x^m y^n\right),$$

where all the products inside the logarithm are assumed to be left-normed. It can be seen that this series can be written as an infinite linear combination of primitive operations in x and y. If applied to a Sabinin algebra $\mathfrak{l}$, it gives a formal loop whose tangent algebra has the same brackets as $\mathfrak{l}$ but whose multioperator is zero. In particular, it provides formal integration for flat Sabinin algebras [40].

If $\mathfrak{l}$ is not flat, one defines the Baker–Campbell–Hausdorff series to be

$$\mathcal{L}(x, \Phi(x, y)),$$

where $\Phi(x, y)$ is the power series defined by the multioperator as in (10). This formula provides formal integration for arbitrary Sabinin algebras.

4.7 Akivis algebras and their envelopes

An Akivis algebra is a vector space with one bilinear and one trilinear operation, denoted by $[\cdot, \cdot]$ and $(\cdot, \cdot, \cdot)$ respectively, such that

$$[x, y] = -[y, x]$$

and

$$[[x, y], z] + [[y, z], x] + [[z, x], y] = (x, y, z) + (y, z, x) + (z, x, y) \\ -(x, z, y) - (z, y, x) - (y, x, z)$$

for all x, y, z. Any algebra is an Akivis algebra, with $[\cdot, \cdot]$ being the commutator and $(\cdot, \cdot, \cdot)$ the associator. If A is an algebra, denote by $Ak(A)$ the algebra A considered as an Akivis algebra. Each Sabinin algebra is also automatically an Akivis algebra since the commutator and the associators are both primitive operations.

It was proved in [55] that for each Akivis algebra $\mathfrak{a}$ there is a Hopf algebra $U_{Ak}(\mathfrak{a})$ together with an injective morphism of Akivis algebras $\mathfrak{a} \hookrightarrow U_{Ak}(\mathfrak{a})$ which is universal in the sense that any Akivis algebra morphism $\mathfrak{a} \to Ak(A)$ lifts to a unique algebra homomorphism $U_{Ak}(\mathfrak{a}) \to A$.

If $\mathfrak{a}$ is actually a Sabinin algebra, $U_{Ak}(\mathfrak{a})$ is, in general, different from $U(\mathfrak{a})$. In particular, the image of $\mathfrak{a}$ in $U_{Ak}(\mathfrak{a})$ is contained in the subspace of primitive elements, but not every primitive element of $U_{Ak}(\mathfrak{a})$ comes from $\mathfrak{a}$.

4.8 Dual Hopf algebras

Inverting the arrows in the categorical definition of an associative Hopf algebra one arrives to the definition of the same structure. In the non-associative case this is no longer true.

A unital bialgebra H with the coproduct δ and the counit ϵ is called a dual Hopf algebra if is endowed with two additional linear operations $H \to H \otimes H$, called the right and the left codivision

$$x \mapsto \sum x_{/1/} \otimes x_{/2/}, \quad x \mapsto \sum x_{\backslash 1\backslash} \otimes x_{\backslash 2\backslash},$$

such that

$$\sum x_{/1/(1)} \otimes x_{/1/(2)} x_{/2/} = \sum x_{(1)} \otimes \epsilon(x_{(2)}) = \sum x_{(1)/1/} \otimes x_{(1)/2/} x_{(2)}$$

and

$$\sum x_{\backslash 1\backslash} x_{\backslash 2\backslash(1)} \otimes x_{\backslash 2\backslash(2)} = \sum \epsilon(x_{(1)}) \otimes x_{(2)} = \sum x_{(1)} x_{(2)\backslash 1\backslash} \otimes x_{(2)\backslash 2\backslash}.$$

The dual of a finite-dimensional Hopf algebra is, rather tautologically, a dual Hopf algebra. In particular, the space of functions on a finite loop is a dual Hopf algebra, with the codivisions defined as

$$\sum f_{/1/}(x) \otimes f_{/2/}(y) = f(x/y) \quad \text{and} \quad \sum f_{\backslash 1\backslash}(x) \otimes f_{\backslash 2\backslash}(y) = f(y\backslash x).$$

It is dual to the loop algebra, which is a cocommutative and and coassociative Hopf algebra.

A more general example is the space of functions on an algebraic loop.

An (affine) algebraic loop is an affine variety with the structure of a loop for which the multiplication and both divisions are morphisms (regular functions). Since the space $\mathcal{O}(X)$ of regular functions on any affine variety X has the property that

$$\mathcal{O}(X \times X) = \mathcal{O}(X) \otimes \mathcal{O}(X),$$

it follows that for an affine algebraic loop X the space $\mathcal{O}(X)$ is a commutative and associative dual Hopf algebra, with the codivisions defined in the same fashion as for finite loops. Moreover, over an algebraically closed field k the category of affine algebraic loops is equivalent to that of finitely generated commutative and associative dual Hopf algebras (The details are, essentially, the same as for affine algebraic groups).

Remark 2 In principle, one may consider bialgebras with two divisions and two codivisions satisfying the identities that hold in Hopf algebras and in dual Hopf algebras. However, it can be shown that, in general, there is no way to introduce codivisions on the Hopf algebras that are of main interest here, namely, the universal enveloping algebras of Sabinin algebras.

5 Identities and special classes of loops

5.1 Identities in Hopf algebras coming from identities in loops

Informally speaking, a non-associative word is an expression formed from several indeterminates by applying the multiplication and the right and the left divisions (In particular, a monomial is a non-associative word formed using multiplication only). If w is such a word, a local loop on M is said to satisfy the identity $w = e$ if for all the values of the indeterminates in M for which w is defined, the result of performing all the operations in w is the unit. More generally, one can speak of the identities of the form $w_1 = w_2$ where w_1 and w_2 are two words. As we already pointed out, identities also make sense in formal loops.

An identity satisfied by a local (formal) loop can be lifted to an identity in its algebra of distributions. If we start with a formal loop F on a vector space V, then a non-associative word $\boldsymbol{w}$ in n variables gives rise to a formal map $w : k[V]^{\otimes n} \to V$ which can be lifted to a coalgebra map $w' : k[V]^{\otimes n} \to k[V]$. Then the identity

$$\boldsymbol{w}_1 = \boldsymbol{w}_2 \tag{12}$$

holds in F if and only if we have

$$w'_1 = w'_2 \tag{13}$$

with any choice of arguments in $k[V]$. The identity (13) in the algebra of distributions is referred to as the linearization of the identity (12).

The simplest example is the associativity in a formal loop which is necessary and sufficient for the associativity of its algebra of distributions. An analogous statement holds for the commutativity. A less trivial example is the left Moufang identity

$$x(y(xz)) = ((xy)x)z.$$

It translates into the identity

$$\sum \mu_{(1)}(\nu(\mu_{(2)}\eta)) = \sum((\mu_{(1)}\nu)\mu_{(2)})\eta \tag{14}$$

in the corresponding algebra of distributions. Non-associative Hopf algebras in which (14) is satisfied are called Moufang–Hopf algebras.

Geodesic loops are right alternative; their algebras of distributions satisfy

$$\sum \mu(\nu_{(1)}\nu_{(2)}) = \sum(\mu\nu_{(1)})\nu_{(2)}. \tag{15}$$

5.2 Identities in Sabinin algebras coming from identities in loops

Identities in loops also produce identities in Sabinin algebras. In principle, there is a straightforward way to find all the identities in a Sabinin algebra that are implied by a given loop identity. Namely, the Baker–Campbell–Hausdorff formula expresses the loop product in terms of the operations in the corresponding Sabinin algebra; imposing an identity on a loop, therefore, imposes an infinite set of identities on its Sabinin algebra. This set can be hugely redundant and choosing a subset consisting of independent identities is difficult in practice. Nevertheless, this is not necessarily an impossible task: Kuzmin [26] used this method to show that a Malcev algebra integrates to a local Moufang loop.

Since loop identities translate into identities in Hopf algebras and in Sabinin algebras one can state various specializations of the general Lie theory for loops that satisfy a given set of identities. The most fundamental example of this is the Lie theory for right alternative loops which relates them to flat Sabinin algebras and Hopf algebras satisfying (15). There are many other such Lie theories for loops with identities; we describe some of them below.

5.3 Moufang loops and Malcev algebras

Historically, the first generalization of the Lie theory to non-associative products was the theory of Moufang loops and Malcev algebras. Recall that a Malcev algebra is a vector space with an antisymmetric bracket that satisfies

$$[J(a,b,c),a] = J(a,b,[a,c])$$

where

$$J(a,b,c) = [[a,b],c] + [[b,c],a] + [[c,a],b].$$

In this setup, we have the following three equivalent categories: simply-connected finite-dimensional Moufang loops, finite-dimensional Malcev algebras, and finitely generated irreducible Moufang-Hopf algebras. The equivalence of first two categories was established in the works of Malcev [30], Kuzmin [26] and Kerdman [23], while the third category was introduced only recently in [42].

An important feature of this case is the fact that finite-dimensional Malcev algebras always integrate to globally defined Moufang loops. While the original proof of this fact [23] is rather inaccessible, the modern argument uses the theory of groups and algebras with triality and consists in translating the problem into the context of Lie groups and Lie algebras, see Sect. 6 for more details. The Ado theorem also holds for finite-dimensional Malcev algebras [42], see more in Sect. 7.

Moufang loops provide an example of how one can pass from loop identities to Sabinin algebra identities via the identities in the bialgebra of distributions. Given a Moufang loop M, the Moufang–Hopf identity (14), which holds in D_eM, implies that the tangent space $T_eM = \mathrm{Prim}(D_eM)$ belongs to the generalized alternative nucleus of D_eM. For an algebra A the generalized alternative nucleus is defined as

$$\mathrm{N_{alt}}(A) = \{a \in A \mid (a,x,y) = -(x,a,y) = (x,y,a) \text{ for all } x,y \in A\}.$$

It is a Malcev algebra with the bracket given by $[a,b] = ab - ba$. The rest of the Sabinin algebra operations in T_eM can be expressed via the Malcev bracket [46]; as a consequence, the Malcev algebra structure is sufficient to reconstruct a local Moufang loop.

A long-standing open problem proposed by Kuzmin [26] is whether all Malcev algebras are special, that is, whether each Malcev algebra can be embedded into the commutator algebra of an alternative algebra. The formal integration procedure gives an embedding of a Malcev algebra into the generalized alternative nucleus of its universal enveloping algebra. However, the universal enveloping algebra of a Malcev algebra is not alternative, in general. One can define the universal alternative enveloping algebra of a Malcev algebra $\mathfrak{m}$ as the maximal alternative quotient of $U(\mathfrak{m})$. Explicit calculations of both kinds of enveloping algebras for specific Malcev algebras can be found in [7,9,59].

5.4 Bol loops and Bol algebras

The variety of left Bol loops is defined by the left Bol identity

$$x(y(xz)) = (x(yx))z.$$

Its linearization is the identity

$$\sum \mu_{(1)}(\nu(\mu_{(2)}\eta)) = \sum (\mu_{(1)}(\nu\mu_{(2)}))\eta. \tag{16}$$

A (left) Bol algebra is a vector space with one bilinear antisymmetric bracket and one trilinear bracket that satisfy the following identities:

$$[a, a, b] = 0,$$
$$[a, b, c] + [b, c, a] + [c, a, b] = 0,$$
$$[x, y, [a, b, c]] = [[x, y, a], b, c] + [a, [x, y, b], c] + [a, b, [x, y, c]]$$

and

$$[a, b, [x, y]] = [[a, b, x], y] + [x, [a, b, y]] + [x, y, [a, b]] + [[a, b], [x, y]].$$

The category of infinitesimal analytic finite-dimensional Bol loops is equivalent to the category of finite-dimensional left Bol algebras and to the category of irreducible Hopf algebras satisfying (16). While the correspondence between Bol loops and Bol algebras is well-studied (see [34,41,52]), the Hopf algebras in this context were first considered in [45].

Similarly to the case of Moufang loops, one can arrive to the Bol algebra identities via the distribution bialgebras. The identity (16) satisfied in the bialgebra of distributions shows that the primitive elements always belong to the left alternative nucleus of the distribution bialgebra, where the left alternative nucleus of an algebra A is defined as

$$\mathrm{LN_{alt}}(A) = \{a \in A \mid (a, x, y) = -(x, a, y) \text{ for all } x, y \in A\}.$$

Any subspace of $\mathrm{LN_{alt}}(A)$ which is closed under the operations

$$[a, b] = ab - ba$$

and

$$[a, b, c] = a(bc) - b(ac) - c[a, b]$$

is a (left) Bol algebra. This is the case, in particular, for T_eM inside $\mathrm{LN}_{\mathrm{alt}}(D_eM)$, where M is a Bol loop [45]. All the other Sabinin algebra operations can be expressed in terms of these two brackets and, thus the structure of a Bol algebra is sufficient to reconstruct a local left Bol loop.

The speciality problem for Malcev algebras has its generalization to Bol algebras. Namely, one may ask whether each Bol algebra is contained in a left alternative algebra as a Bol subalgebra. Hentzel and Peresi [19] have shown that this is not the case. Using computer-aided computations, they found an identity of degree eight which is satisfied in the Bol algebra of any left alternative algebra, but not in the free Bol algebra.

An important property that holds for Malcev algebras but not for general Bol algebras is the Ado theorem: it was proved in [47] that it fails for Lie triple systems (which are Bol algebras with the trivial binary bracket).

5.5 Nilpotent loops and nilpotent Sabinin algebras

Let us say that a loop L is nilpotent of class n if the $n + 1$st term of the commutator-associator filtration (see Sect. 8.1) of L is trivial:

$$\gamma_{n+1}L = \{e\}.$$

This is not the standard definition of nilpotency in loops; however, there are strong indications that it is the correct one [18,35,36,40]. Since the commutator-associator filtration is defined in terms of words (namely, commutators, associators and associator deviations, which are defined in Sect. 8.1, and their compositions) nilpotent loops of class n form a variety.

The corresponding Sabinin algebras, unsurprisingly, are the nilpotent Sabinin algebras of class n. Define a bracket of weight k in a Sabinin algebra to be a multilinear operation in k arguments formed by composing the brackets and the operations $\Phi_{m,n}$. Then a Sabinin algebra is nilpotent of class n if all the brackets of weight greater than n vanish in it. For a flat Sabinin algebra, nilpotency is the same as the nilpotency of the Lie algebra in which it is a summand according to the construction of Sect. 3.2, see [40].

The Lie theory of nilpotent loops and nilpotent Sabinin algebras exhibits various features typical of Lie and Malcev algebras. In particular, a nilpotent Sabinin algebra always integrates to a globally defined simply-connected nilpotent loop of the same class. This is due to the fact that the Baker–Campbell–Hausdorff series in this case consists of a finite number of terms and, hence, always converges. Also, the Ado theorem holds, both for Sabinin algebras and (globally) for simply-connected nilpotent loops (see Sect. 7.4 for more details on the Ado theorem in the non-associative setting). While in the case of Malcev algebras one may think that these good properties are a consequence of Malcev algebras being "close" to Lie algebras, nilpotent Sabinin algebras may have an arbitrarily big (though finite) number of independent operations.

5.6 Connections with quasigroups

In various contexts in geometry where non-associative algebra proves to be useful, it is quasigroups rather than loops that appear naturally. While we do not intend to discuss this subject in detail, there are two points that we should mention here.

One important application of quasigroups is the theory of symmetric spaces as developed by Loos [27]; this is where bialgebras first appeared within non-associative Lie theory. A symmetric space can be defined algebraically as a manifold M with a product $\circ$ that satisfies, for all $x, y \in M$, the following identities:

$$x \circ x = x,$$
$$x \circ (x \circ y) = y,$$
$$x \circ (y \circ z) = (x \circ y) \circ (x \circ z)$$

and such that the map $S_x : y \mapsto x \circ y$ has the unique fixed point $y = x$ in some neighbourhood of x. Geometrically, the product $x \circ y$ is the reflection of the point y about the point x. The fact that each $e \in M$ is idempotent is sufficient to define a bialgebra structure on D_eM for each e and linearize the identities in M to obtain identities in D_eM. For instance, if M is a symmetric space, in each D_eM we have:

$$\sum \mu_{(1)}\mu_{(2)} = \mu, \ \sum \mu_{(1)}(\mu_{(2)}\nu) = \epsilon(\mu)\nu, \ \mu(\eta\nu) = \sum (\mu_{(1)}\eta)(\mu_{(2)}\nu). \quad (17)$$

The local uniqueness of the fixed point for the multiplication map S_e implies that the right multiplication in D_eM by the only group-like element

$$e : f \mapsto f(e)$$

is bijective. This, together with the identities (17), can be used to deduce algebraically that the tangent space $T_eM = \mathrm{Prim}(D_eM)$ is a Lie triple system with the product $[a, b, c] = a(bc) - b(ac)$.

The second point is related to the terminology concerning the identities in loops. Given a local quasigroup, that is, a locally invertible map $M \times M \to M$ defined in a neighbourhood of a point (a, b), one can define a local loop on a neighbourhood of $a \cdot b \in M$ by

$$xy = (x/b) \cdot (a\backslash y),$$

where $\cdot$, $\backslash$ and $/$ are the product and the two divisions in the quasigroup. This construction associates with a local quasigroup a family of local loops, and it may be useful to compare the identities that hold in each local loop. It turns out that the Moufang identity is satisfied at the same time in each of its local loops; in particular, it is sufficient to verify this identity in one of the local loops in order to conclude that it holds in all of them. In contrast, the monoassociativity identity

$$(xx)x = x(xx) \tag{18}$$

may be satisfied in one of the local loops but not in the others. A monoassociative quasigroup is defined as a local quasigroup all of whose local loops satisfy (18) (This definition is of importance in web theory where monoassociative quasigroups correspond to hexagonal three-webs). As a consequence, a monoassociative local loop is sometimes defined as a local loop of a monoassociative quasigroup [1,8]. This is a much stronger condition on a loop than just the identity (18). In particular, each right alternative local loop satisfies (18); note that flat Sabinin algebras have an infinite number of independent operations. On the other hand, Sabinin algebras tangent to monoassociative loops have three independent operations: one bilinear, one trilinear and one quadrilinear [33,54]. The complete set of relations for these operations is not known.

6 Constructions involving associative Hopf algebras

Since each Sabinin algebra can be thought of a subspace of a Lie algebra, it is not surprising that non-associative Hopf algebras can be interpreted in terms of associative Hopf algebras. The construction of the universal enveloping algebra of a Sabinin algebra given in [46] is based on such an interpretation. In general, this construction does not produce closed formulae; however, there are important special cases when it can be significantly simplified. This happens, for instance, for Moufang-Hopf algebras and for universal enveloping algebras of Bol algebras. More generally, it can be done whenever we are given a local product on a homogeneous space G/H together with its lifting to the product in G by means of a local section; in this case the non-associative algebra of distributions $D_e(G/H)$ can be expressed in terms of the associative Hopf algebra D_eG.

6.1 Groups with triality

One important step in the study of Malcev algebras was to establish its relation with groups with triality [13, 15, 32]. Each self-map d of the octonion algebra, antisymmetric with respect to its standard quadratic form, gives rise to two antisymmetric maps d', d'' uniquely defined by the relation

$$d(xy) = d'(x)y + xd''(y).$$

This phenomenon is known as the local triality principle and, in general, it is often related in one or another way with exceptional behaviour such as that of exceptional Lie algebras or Jordan algebras [21]. Behind it, there is a certain representation of the symmetric group Σ_3 on three letters, acting on the Dynkin diagram of the multiplication Lie algebra of the split octonions, which is a central simple Lie algebra of type D_4

[53]. This symmetry lifts to the corresponding group Spin(8) and gives it the structure of a group with triality. More generally, an abstract group with triality is a group G together with two automorphisms σ, ρ such that

$$\sigma^2 = \rho^3 = \mathrm{id}, \quad \sigma\rho\sigma = \rho^2,$$

and

$$P(g)\rho(P(g))\rho^2(P(g)) = 1,$$

where $P(g) = g\sigma(g)^{-1}$. The automorphisms σ and ρ define a representation of Σ_3, which is independent on a particular choice of the generators σ y ρ of orders 2 and 3, respectively. If $H \subseteq G$ is the subgroup fixed by σ, the quotient G/H is a Moufang loop with the product

$$g_1H * g_2H = \rho(P(g_1))^{-1}g_2H.$$

Let $\mathfrak{g}$ and $\mathfrak{h}$ be the Lie algebras of G and H, respectively; the group Σ_3 acts on $\mathfrak{g}$ with $\mathfrak{h}$ being fixed by σ. The Malcev algebra $\mathfrak{m}$ tangent to G/H is the complement to $\mathfrak{h}$ in $\mathfrak{g}$ consisting of the eigenvectors of σ with eigenvalue -1. The bracket in $\mathfrak{m}$ is defined by

$$[x, y]_{\mathfrak{m}} = [\rho^2(x) - \rho(x), y].$$

This is the construction of a Lie algebra with triality. Each Malcev algebra arises as the -1-eigenspace of σ in such a Lie algebra.

In order to construct the universal enveloping algebra $U(\mathfrak{m})$, embed G/H into G by means of the map

$$gH \mapsto \rho(P(g))^{-1},$$

which is a global section of the principal fibration $G \to G/H$. The image of this embedding is the set $\rho(\mathcal{M}(G))$ where

$$\mathcal{M}(G) = \{P(g) \mid g \in G\}.$$

The Moufang product on $\mathcal{M}(G)$ which comes from G/H can be expressed directly in terms of the product on G as

$$x * y = \rho(x)^{-1}y\rho^2(x)^{-1},$$

see [16]. Linearizing this formula we get that the subcoalgebra

$$\mathcal{M}(U(\mathfrak{g})) = \left\{\sum x_{(1)}\sigma(S(x_{(2)})) \mid x \in U(\mathfrak{g})\right\},$$

where S is the antipode of $U(\mathfrak{g})$, with the product

$$x * y = \sum \rho(S(x_{(1)}))y\rho^2(S(x_{(2)}))$$

is isomorphic to $U(\mathfrak{m})$, see [3].

One can speak of the categories of Lie groups with triality, Lie algebras with triality and Hopf algebras with triality. The constructions of this paragraph establish their equivalence to the categories of analytic Moufang loops, Malcev algebras and Moufang-Hopf algebras, respectively. In this way, for instance, the global integrability of finite-dimensional Lie algebras implies global integrability of finite-dimensional Malcev algebras.

6.2 Bruck loops

A similar construction works for Bruck loops. These are left Bol loops that satisfy the automorphic inverse identity

$$(xy)^{-1} = x^{-1}y^{-1},$$

where x^{-1} is shorthand for e/x. The binary operation in the corresponding Bol algebra in this case is trivial, and the structure of a Bol algebra reduces to that of a Lie triple system. This shows that Bruck loops are related to symmetric spaces. Indeed, the product

$$x \circ y = x(y^{-1}x)$$

endows an arbitrary local Bruck loop with the structure of a locally symmetric space. Conversely, given a point e in a symmetric space the Bruck loop product is recovered by means of the relation

$$xy \circ e = x \circ (e \circ (y \circ e)).$$

While the product of a locally symmetric space can be globalized, this is not true for the corresponding local loop since the map $x \mapsto x \circ e$ may be only locally, and not globally, bijective.

A symmetric space can be considered as a quotient G/H, where G is a Lie group with an involutive automorphism σ and H is an open subgroup in the fixed set G^σ. Locally, the quotient G/H can be embedded into G with the help of the map

$$gH \mapsto P(g) = g\sigma(g)^{-1},$$

which is a local section of the quotient map $G \to G/H$. The image of this map coincides, locally, with the set

$$\mathcal{M}(G) = \{P(g) \mid g \in G\}.$$

It can be shown that in a neighbourhood of the unit in $\mathcal{M}(G)$ each element has a unique square root. The product of a local Bruck loop on $\mathcal{M}(G)$ can be written then as

$$x * y = \sqrt{x}\, y \sqrt{x}$$

for $x, y \in \mathcal{M}(G)$.

Similarly, a Lie triple system can be considered as the -1-eigenspace of an involutive automorphism in a Lie algebra $\mathfrak{g}$. As a consequence, the universal enveloping algebra of a Lie triple system, viewed as a Sabinin algebra, can be defined in terms of the Hopf algebra $U(\mathfrak{g})$ and the automorphism σ as the coalgebra

$$\mathcal{M}(U(\mathfrak{g})) = \left\{ \sum x_{(1)}\sigma(S(x_{(2)})) \mid x \in U(\mathfrak{g}) \right\},$$

with the product

$$x * y = \sum r(x_{(1)}) y r(x_{(2)})$$

where r is the map corresponding to the linearization of the square root, see [38].

7 Representation theory

Compared to the representation theory of Lie groups and Lie algebras, the representation theory of loops and Sabinin algebras is still in its infancy. It is, probably, too early to give a coherent picture of the field; here we touch only on a few topics.

The theory of loop representations has been put on firm foundations by Smith [58]; the same approach works for Sabinin algebras and non-associative Hopf algebras. It rests on the observation of Beck [2] that the concept of a split-null extension has a natural definition in the categorical setting: if X is an object in a category $\mathcal{C}$, an X-module is an abelian group in the comma category $\mathcal{C} \downarrow X$. The objects of this category are morphisms $Y \xrightarrow{\pi} X$ and the morphisms are commutative diagrams of the form

$$\begin{array}{ccc} Y_1 & \longrightarrow & Y_2 \\ & {\scriptstyle \pi_1}\searrow \quad \swarrow{\scriptstyle \pi_2} & \\ & X & \end{array}$$

This categorical description encompasses all reasonable definitions of modules; in particular, that of a group representation and that of a module over a Lie algebra. It has an advantage of producing immediate bijections between the representations of a formal group, of its Lie algebra and of the universal enveloping algebra of the latter. The same is true in the non-associative context: the equivalence of the category of formal loops to that of Sabinin algebras and to the category of irreducible Hopf algebras identifies that the corresponding sets of representations. Let us spell out what Beck's definition means in practice.

7.1 Loop representations

Let **Loops** be the category of all loops and let L be an arbitrary loop. Introduce the following notation:

- $L[X]$ — the free product of L with the free loop on one generator X;
- L_a and R_a — the operators of left and right multiplication, respectively, by $a \in L$ in $L[X]$;
- $\mathrm{U}(L)$ — the group of self-maps of $L[X]$ generated by the L_a and R_a for all $a \in L$;
- $\mathrm{U}(L)_e$ — the stabilizer of the identity in $L[X]$.

Abelian groups in the category $\mathbf{Loops} \downarrow L$ are in one-to-one correspondence with the representations of the group $\mathrm{U}(L)_e$. Namely, given a $\mathrm{U}(L)_e$-module E_e, define

$$E = E_e \times L$$

as a set, endowed with the product

$$(x, a)(y, b) = (r_{a,b}(x) + s_{a,b}(y), ab), \tag{19}$$

where

$$r_{a,b} = R_{ab}^{-1} R_b R_a \quad \text{and} \quad s_{a,b} = R_{ab}^{-1} L_a R_b.$$

Then the projection of the loop E onto L is an abelian group in $\mathbf{Loops} \downarrow L$. The sum, inverse and the neutral element that define on $E \to L$ the structure of an abelian group are induced by the corresponding operations in E_e. Each abelian group in $\mathbf{Loops} \downarrow L$ arises in this way, with the module E_e defined uniquely up to an isomorphism.

If L belongs to a variety of loops $\mathbf{V}$, the category **Loops** in this construction can be replaced by $\mathbf{V}$. The free loop and the free product in this case should be taken in $\mathbf{V}$; instead of the group $\mathrm{U}(L)_e$ one obtains a group $\mathrm{U}(L; \mathbf{V})_e$. The abelian groups in $\mathbf{V} \downarrow L$ are then in correspondence with certain representations of $\mathrm{U}(L; \mathbf{V})_e$. For instance, if $\mathbf{V} = \mathbf{Groups}$ is the variety of groups, we have

$$\mathrm{U}(L; \mathbf{Groups})_e \simeq L.$$

Given a representation

$$\rho : L \to \mathrm{Aut}(V)$$

the product (19) takes the form

$$(x, a)(y, b) = (x + \rho_a(y), ab).$$

See [58] for further details.

7.2 Hopf and Sabinin algebra representations

Let $\mathcal{H}$ be the category of all irreducible Hopf algebras and H an object in $\mathcal{H}$. Beck's definition translates to this context in the manner similar to the case of loops.

An abelian group $A \xrightarrow{\pi} H$ in the category $\mathcal{H} \downarrow H$ is a morphism such that the Hopf algebra A factorizes as

$$A \cong k[V] \otimes H,$$

where $k[V]$ is the symmetric algebra of a vector space V and the product in A is given by

$$(p \otimes a)(q \otimes b) = \sum (r_{a_{(1)},b_{(1)}}(p))(s_{a_{(2)},b_{(2)}}(q)) \otimes a_{(3)}b_{(3)}$$

in terms of the action of some operators $r_{a,b}$, $s_{a,b}$ on V whose action is extended to $k[V]$ as

$$r_{a,b}(pq) = \sum r_{a_{(1)},b_{(1)}}(p)\, r_{a_{(2)},b_{(2)}}(q)$$

and, similarly,

$$s_{a,b}(pq) = \sum s_{a_{(1)},b_{(1)}}(p)\, s_{a_{(2)},b_{(2)}}(q).$$

The subalgebra $k[V]$ appears naturally as the equalizer of the coalgebra morphisms $\pi \colon A \to H$ and

$$\begin{aligned} \epsilon 1 \colon A &\to H \\ x &\mapsto \epsilon(x)1. \end{aligned}$$

The sum, inverse and the neutral element in $A \xrightarrow{\pi} H$ are given by

$$\begin{array}{ccc} \sum(p \otimes a_{(1)}) \otimes (q \otimes a_{(2)}) & \xrightarrow{\quad + \quad} & pq \otimes a \\ & {\scriptstyle \pi} \searrow \quad \swarrow {\scriptstyle \pi} & \\ & \epsilon(p)\epsilon(q)a & \end{array}$$

$$\begin{array}{ccc} (p, a) & \xrightarrow{\quad - \quad} & (S(p), a) \\ & {\scriptstyle \pi} \searrow \quad \swarrow {\scriptstyle \pi} & \\ & \epsilon(p)a & \end{array} \qquad \begin{array}{ccc} a & \xrightarrow{\quad 0 \quad} & 1 \otimes a \\ & {\scriptstyle \pi} \searrow \quad \swarrow {\scriptstyle \pi} & \\ & a & \end{array}$$

where S is the antipode in $k[V]$. The category $\mathcal{H} \downarrow H$ has finite products: The product $A_1 \otimes_H A_2$ of $A_1 \xrightarrow{\pi_1} H$ and $A_2 \xrightarrow{\pi_2} H$ is the equalizer of the morphisms $\pi_1 \otimes \epsilon$ and $\epsilon \otimes \pi_2$.

Usually, in the associative context it is the vector space V that is called an H-module and the abelian group $A \xrightarrow{\pi} H$ is not given a special name. In this case the action of $r_{a,b}$ is trivial, that is, $r_{a,b}(p) = \epsilon(a)\epsilon(b)p$, while $s_{a,b}(q) = \epsilon(b)s_{a,1}(q)$. If we write $a \cdot q = s_{a,1}(q)$, the formula for the product in A becomes

$$(p \otimes a)(q \otimes b) = \sum p(a_{(1)} \cdot q) \otimes a_{(2)}b,$$

which is nothing but the smash product of the H-module algebra $k[V]$ with H, where the H-module algebra structure on $k[V]$ is induced by the H-module structure on V.

Modules over Sabinin algebras are somewhat easier to describe than modules over loops and Hopf algebras. Beck's notion of a module for a Sabinin algebra $\mathfrak{l}$ coincides with the usual idea of a split-null extension. In other words, a module over $\mathfrak{l}$ is a Sabinin algebra $V \oplus \mathfrak{l}$ such that $\mathfrak{l}$ is a subalgebra and such that each multilinear operation vanishes if at least two of its arguments belong to V.

7.3 An extended concept of a representation

Sometimes there is rationale for enlarging the class of the extensions considered in the definition of the modules over a loop (Hopf algebra, Sabinin algebra). For instance, the requirement that the extension $\pi : E \to L$ lies within the variety $\mathcal{M}$ of Moufang loops leads to conditions on $r_{a,b}$ and $s_{a,b} \in \mathrm{U}(L; \mathcal{M})_e$; these conditions are expressed as the vanishing of certain elements of the group algebra of $\mathrm{U}(L; \mathcal{M})_e$ while acting on $\pi^{-1}(e)$. These elements may fail to vanish on a tensor product of $\mathrm{U}(L; \mathcal{M})_e$-modules. As a consequence, in contrast to the case of groups, a tensor product of L-modules in general will not be a L-module. In the same fashion, a tensor product of Sabinin algebra representations cannot be expected to be a representation.

The failure of the tensor product of two representations to be a representation is a basic reflection of non-associativity. In case of the Moufang loops this problem can be overcome by studying abelian groups $E \to L$ such that E is not necessarily a Moufang loop, but requiring only that the image of the zero section $L \to E$ is contained in the set of Moufang elements of E. These are the elements a such that

$$a(x(ay)) = ((ax)a))y \quad \text{and} \quad ((xa)y)a = x(a(ya))$$

for all $x, y \in E$; Moufang elements in any loop form a Moufang loop. In the context of Hopf algebras this corresponds to studying abelian groups $A \xrightarrow{\pi} H$ such that in the decomposition $A \cong k[V] \otimes H$ the subspace $\mathrm{Prim}(H)$ (but not V) is contained in the generalized alternative nucleus of A. As for Malcev algebras, representations of this kind correspond to split-null extensions $(V \oplus \mathfrak{m}, [\ ,\ ])$ of the Malcev algebra $\mathfrak{m}$ such that the bracket $[\ ,\ ]$ is anticommutative and the adjoint map

$$r_a\colon x \mapsto r_a(x) = [x, a]$$

satisfies

$$[[r_a, r_b], r_c] = -[r_{[a,b]}, r_c] + r_{[[a,b],c]+[[a,c],b]+[a,[b,c]]}. \tag{20}$$

Not only the resulting representation theory behaves well with respect to tensor products but also among its other good properties is the existence of faithful finite-dimensional representations for finite-dimensional Malcev algebras, see [42].

In general, no clear criterion is known which would determine the adequate class of abelian groups $E \xrightarrow{\pi} L$ for a given loop L. Note that once such class is chosen, methods of non-associative Lie theory provide us with the classes of extensions for the corresponding Hopf and Sabinin algebras.

7.4 The Ado theorem

The Ado theorem is usually stated as a result about representations. While this may be appropriate in the context of Lie algebras, in the more general setting of Sabinin algebras it is more convenient to use a different wording which does not mention representations at all.

Each Sabinin algebra $\mathfrak{l}$ appears as a Sabinin subalgebra of $\mathit{УШ}(A)$ for some non-associative algebra A; for instance, one can take A to be the universal enveloping algebra $U(\mathfrak{l})$. Let us say that a variety of Sabinin algebras satisfies the Ado theorem if for each finite-dimensional Sabinin algebra $\mathfrak{l}$ in that variety A can be chosen to be finite-dimensional. By the universal property of $U(\mathfrak{l})$, this is the same as to say that the $U(\mathfrak{l})$ has an ideal of finite codimension which contains no non-trivial primitive elements.

We have mentioned before that, while Lie, Malcev and nilpotent Sabinin algebras satisfy the Ado theorem, this is not the case for all Sabinin algebras, Lie triple systems being a counterexample. The universal enveloping algebra of a central simple Lie triple system of finite dimension has no proper right ideals apart from the augmentation ideal [38,47]. It would be interesting to find criteria for a variety of loops to satisfy the Ado theorem.

The Ado theorem can be also considered as a statement about loops rather than their tangent algebras. Let us say that a local loop Q is locally linear if can be locally embedded as a subloop into the local loop of invertible elements in some finite-dimensional algebra A. This is equivalent to saying that the Sabinin algebra of Q is isomorphic to a subalgebra of $\mathit{УШ}(A)$, which is the property described by the Ado theorem. Since the Ado theorem fails for Lie triple systems, local Bruck loops are not locally linear in general.

We should point out that local linearity is a weaker concept than global linearity. This has nothing to do with non-associativity since the difference already exists for groups. For Moufang loops there also are examples which illustrate this phenomenon. For instance, the seven-dimensional projective plane is a Moufang loop which locally embeds into the invertible elements of the octonions. This local embedding, however, cannot be globalized [56].

8 Discrete loops and Sabinin algebras

The definition of a nilpotent loop, and, generally, of the lower central series of a loop, was given by Bruck in [10]. Since then, the theory of nilpotent groups has made significant advances and it became increasingly clear that Bruck's definition does not provide an adequate analogue of the associative lower central series. Namely, one should expect that

- the successive quotients of the lower central series for a finitely generated loop are finitely generated abelian groups;
- the graded abelian group associated with the lower central series carries an algebraic structure similar to that of a Lie ring;
- there exists a close relation to the dimension series.

A lower central series for loops satisfying all the above properties has been defined and studied in [35–37] and we review this construction in this section. A closely related definition was also given by Lemieux et al. in [28]. Essentially, the argument of [28] consists in defining the dimension series for the free loops and then pushing it to arbitrary loops using the universal property of the free loops. Recently, Hartl and Loiseau [17,18] defined higher commutators in arbitrary semi-abelian categories which lead in the case of loops to the same definition as the one discussed here.

8.1 The commutator–associator series

Let L be a loop. The commutator–associator filtration on L is defined in terms of commutators, associators and associator deviations. The commutator of two elements a, b of L is

$$[a, b] = (ab)/(ba)$$

and the associator of a, b and c is defined by

$$(a, b, c) = ((ab)c)/(a(bc)).$$

There is an infinite number of associator deviations. These are functions $L^{l+3} \to L$ characterized by a non-negative number l, called *level* of the deviation, and l indices $\alpha_1, \ldots, \alpha_l$ with $0 < \alpha_i \leq i + 2$. The deviations of level one are

$$\begin{aligned}(a, b, c, d)_1 &= (ab, c, d)/((a, c, d)(b, c, d)),\\ (a, b, c, d)_2 &= (a, bc, d)/((a, b, d)(a, c, d)),\\ (a, b, c, d)_3 &= (a, b, cd)/((a, b, c)(a, b, d)).\end{aligned}$$

By definition, the deviation $(a_1, \ldots, a_{l+3})_{\alpha_1,\ldots,\alpha_l}$ of level l is equal to

$$A(a_{\alpha_l} a_{\alpha_l+1})/(A(a_{\alpha_l})A(a_{\alpha_l+1})),$$

where $A(x)$ stands for the deviation $(a_1, \ldots, a_{\alpha_l-1}, x, a_{\alpha_l+2}, \ldots, a_{l+3})_{\alpha_1,\ldots,\alpha_{l-1}}$ of level $l - 1$. The associator is thought of as the associator deviation of level zero.

Now, set $\gamma_1 L = L$ and for $n > 1$ define $\gamma_n L$ to be the minimal normal subloop of L containing

- $[\gamma_p L, \gamma_q L]$ with $p+q \geq n$;
- $(\gamma_p L, \gamma_q L, \gamma_r L)$ with $p+q+r \geq n$;
- $(\gamma_{p_1} L, \dots, \gamma_{p_{l+3}} L)_{\alpha_1,\dots,\alpha_l}$ with $p_1 + \dots + p_{l+3} \geq n$.

The subloop $\gamma_n L$ is called the nth commutator–associator subloop of L. For a group G the subgroup $\gamma_n G$ is the nth term of the lower central series of G.

The commutator–associator subloops of a loop L are normal in L. Moreover, they are fully invariant, that is, are preserved by all automorphisms of L. If L is finitely generated, each quotient $\gamma_i L/\gamma_{i+1} L$ is a finitely generated abelian group. The crucial property of the commutator–associator filtration is that for an arbitrary loop L the commutator, the associator and the associator deviations induce multilinear operations on the graded abelian group

$$\mathcal{L}_\gamma L = \bigoplus \gamma_i L/\gamma_{i+1} L;$$

these operations respect the grading. It can be shown that the algebraic structure given by these multilinear operations on $\mathcal{L}_\gamma L \otimes \mathbb{Q}$ is precisely that of a Sabinin algebra. This, however, is quite non-trivial and requires an understanding of the relation between the commutator–associator filtration and the dimension series.

8.2 The dimension series

Let L be a discrete loop and $\mathbb{Q}L$ its loop algebra over the rational numbers. Denote by $\Delta \subset \mathbb{Q}L$ the augmentation ideal and by Δ^n its nth power, that is, the submodule of $\mathbb{Q}L$ spanned over $\mathbb{Q}$ by all products of at least n elements of Δ with any arrangement of the brackets. The loop L can be thought of a subset of $\mathbb{Q}L$ and we define the nth dimension subloop of L as

$$D_n L = L \cap (1 + \Delta^n).$$

It can be shown that $D_n L$ is, indeed, a series of fully invariant subloops of L. The successive quotients $D_i L/D_{i+1} L$ are torsion-free abelian groups, and it turns out that the commutator, the associator and the associator deviations respect the filtration by the $D_n L$. The induced operations on the associated graded group can be identified as follows.

The loop algebra $\mathbb{Q}L$ is a bialgebra, with the coproduct defined as $\delta(g) = g \otimes g$ for $g \in L$. The associated graded algebra

$$\mathcal{D}L = \bigoplus_{i \geq 0} \Delta^i/\Delta^{i+1}$$

is then a cocommutative non-associative primitively generated Hopf algebra. If we set

$$\mathcal{L}L = \bigoplus_{i > 0} D_i L/D_{i+1} L,$$

then $\mathcal{L}L \otimes \mathbb{Q} \subset \mathcal{D}L$ and the primitive elements of $\mathcal{D}L$ are precisely the elements of $\mathcal{L}L \otimes \mathbb{Q}$. Therefore, $\mathcal{L}L \otimes \mathbb{Q}$ is a Sabinin algebra. It can be shown that the Shestakov-Umirbaev operations $p_{m,n}$ on $\mathcal{D}L$ coincide with the operations induced by certain associator deviations on $\mathcal{L}L \otimes \mathbb{Q}$ (See [37] for a more precise statement).

These results can be applied to the study of the commutator–associator filtration. For each n the subloop D_nL contains the commutator–associator subloop $\gamma_n L$. Moreover, the analogue of the Jennings theorem [22] is true:

$$D_nL = \sqrt{\gamma_n L},$$

that is, D_nL consists of all those elements of L whose kth power, for at least one k and at least one arrangement of the parentheses, lies in $\gamma_n L$. In particular, we have that

$$\mathcal{L}L \otimes \mathbb{Q} = \mathcal{L}_\gamma L \otimes \mathbb{Q}.$$

Remark 3 The dimension subloops can be defined with the help of the group algebra with coefficients in any ring, and the result may depend on this ring. A particularly interesting case is that of the integer coefficients. For many years it was conjectured that the integer dimension subgroups of a group coincide with its lower central series. The eventual counterexample published by E. Rips in 1972 was very involved, and while by now there are more counterexamples to this conjecture, the nature of the difference between the integer dimension series and the lower central series is still absolutely mysterious. It might be interesting to find a loop where the difference between the integer dimension series and the lower central series is due to the non-associative effects.

8.3 The Magnus map and the dimension series of a free loop

Let $\mathbb{Z}\{\{X_1, \ldots, X_n\}\}$ be the non-associative ring of formal power series in n non-commutative and non-associative variables. The invertible elements in this ring are the power series which start with ± 1. They form a loop, which we denote by $\mathbb{Z}\{\{X_1, \ldots, X_n\}\}^*$. Let $F_{\{n\}}$ the free loop on n generators $x_1, \ldots, x_n$. The homomorphism

$$\begin{aligned} \mathcal{M} : F_{\{n\}} &\to \mathbb{Z}\{\{X_1, \ldots, X_n\}\}^*, \\ x_i &\mapsto 1 + X_i, \end{aligned}$$

is the non-associative version of the map (2). As in the associative case, the ith term of the dimension series of $F_{\{n\}}$ consists of those elements which are sent by $\mathcal{M}$ to the power series of the form

$$1 + \text{terms of degree at least } i.$$

This, however, is not enough to conclude that the loop $F_{\{n\}}$ is residually nilpotent since it is not known at the moment whether $\mathcal{M}$ is injective, even in the case of the free loop on one generator.

A similar construction can be performed for the free commutative loop. It not known if the corresponding Magnus map is injective in this case, either.

The Magnus map identifies the completion of the Sabinin algebra $\mathcal{L}F_{\{n\}} \otimes \mathbb{Q}$ with the primitive elements in $\mathbb{Z}\{\{X_1, \ldots, X_n\}\}$. In particular, this Sabinin algebra is the completion of the free Sabinin algebra on n generators.

9 Quantum loops

An analogy with associative Hopf algebras suggests that there might exist deformations of the universal enveloping algebras for Sabinin algebras which are not necessarily coassociative or cocommutative. So far, no interesting examples of this kind have been found. Let us indicate, nevertheless, what one may be looking for.

In view of the fact that the Hopf algebra of distributions supported at the identity of a loop naturally satisfies the linearizations of the identities satisfied by the loop, one may assume (rightly or wrongly) that its possible deformations should satisfy the same linearized identities. Let us illustrate this logic with the variety of Moufang loops.

The left and right Moufang identities can be written as

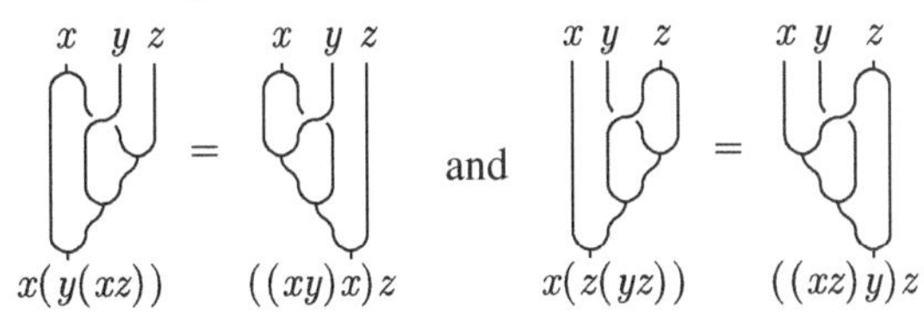

where

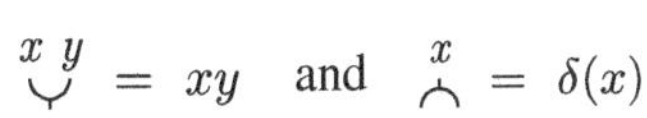

with $\delta(x) = (x, x)$. The same diagrams represent the Moufang-Hopf identities satisfied in the corresponding algebra of distributions, with the difference that in this case $\delta(x) = \sum x_{(1)} \otimes x_{(2)}$. One may suspect that the universal enveloping algebra of the central simple exceptional Malcev algebra of traceless octonions $\mathbb{M}(\alpha, \beta, \gamma)$ has non-cocommutative deformations that satisfy the Moufang–Hopf identities. Given that Hopf algebras are, in spirit, self-dual objects, it is natural to require that these deformations satisfy not coassociativity but, rather, the identities dual to Moufang–Hopf identities. These latter can be represented as

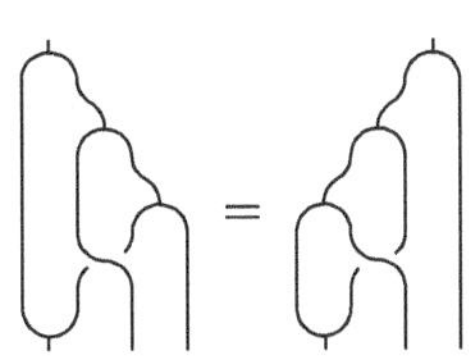

(left co-Moufang–Hopf)

and

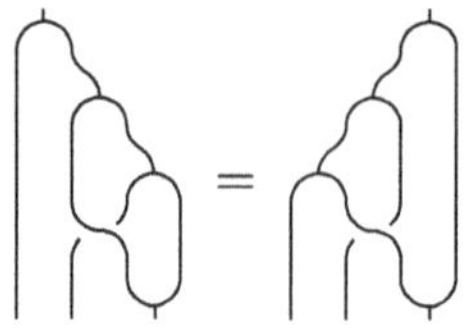

(right co-Moufang–Hopf).

These identities can be written in the form of equations as

$$\sum x_{(1)}x_{(2)(2)(1)} \otimes x_{(2)(1)} \otimes x_{(2)(2)(2)} = \sum x_{(1)(1)(1)}x_{(1)(2)} \otimes x_{(1)(1)(2)} \otimes x_{(2)}$$

and

$$\sum x_{(1)} \otimes x_{(2)(2)(1)} \otimes x_{(2)(1)}x_{(2)(2)(2)} = \sum x_{(1)(1)(1)} \otimes x_{(1)(2)} \otimes x_{(1)(1)(2)}x_{(2)}.$$

They arise in the study of the algebraic seven-dimensional sphere [25].

Surprisingly, $U(\mathbb{M}(\alpha, \beta, \gamma))$ turns out to be rigid in the sense that any deformation which satisfies Moufang-Hopf and co-Moufang–Hopf identities is equivalent to a trivial deformation. Even universal enveloping algebras of finite-dimensional central simple Lie algebras exhibit certain rigidity if considered as Moufang–Hopf algebras since all their deformations are necessarily associative and coassociative; in other words, are quantized enveloping algebras in the usual sense [29,43].

The rigidity characteristic of non-associativity is still poorly understood and needs an adequate cohomological interpretation.

Remark 4 There are a number of deformations of non-associative algebras that carry the adjective "quantum", such as the quantum octonions of [4] and of [5]. They do not seem to fit in the framework of hypothetical quantum loops that we discuss here.

Acknowledgments The authors were supported by the following grants: FAPESP 2012/21938-4 and CONACyT grant 168093-F (J.M.), FAPESP 2012/22537-3 (J.M.P.-I.), CNPq 3305344/2009-9 and FAPESP 2010/50347-9 (I.P.Sh.), Spanish Government project MTM 2010-18370-C04-03 (all three). The first two authors would like to thank the Institute of Mathematics and Statistics of the University of São Paulo for hospitality.

References

1. Akivis, M., Goldberg, V.: Algebraic aspects of web geometry. Commentat. Math. Univ. Carol. **41**, 205–236 (2000)
2. Beck, J.: Triples, Algebras, and Cohomology, Ph.D. thesis, Columbia University, 1967. Repr. Theory Appl. Categ. **2** 1–59 (2003)
3. Benkart, G., Madariaga, S., Pérez Izquierdo, J.M.: Hopf algebras with triality. Trans. Am. Math. Soc. **365**, 1001–1023 (2013)
4. Benkart, G., Pérez-Izquierdo, J.M.: A quantum octonion algebra. Trans. Am. Math. Soc. **352**, 935–968 (2000)
5. Bremner, M.R.: Quantum octonions. Commun. Algebra **27**, 2809–2831 (1999)

6. Bremner, M.R., Hentzel, I.R., Peresi, L.A.: Dimension formulas for the free nonassociative algebra. Commun. Algebra **33**, 4063–4081 (2005)
7. Bremner, M.R., Hentzel, I.R., Peresi, L.A., Usefi, H.: Universal enveloping algebras of the four-dimensional Malcev algebra. Algebras, representations and applications, American Mathematical Society, Providence, RI. Contemp. Math. **483**, 73–89 (2009)
8. Bremner, M.R., Madariaga, S.: Polynomial identities for tangent algebras of monoassociative loops. Commun. Algebra (2013)
9. Bremner, M.R., Usefi, H.: Enveloping algebras of the nilpotent Malcev algebra of dimension five. Algebr. Represent. Theory **13**, 407–425 (2010)
10. Bruck, R.H.: A survey of binary systems. Springer, Berlin (1958)
11. Chibrikov, E.: A right normed basis for free Lie algebras and Lyndon–Shirshov words. J. Algebra **302**, 593–612 (2006)
12. Chibrikov, E.: On free Sabinin algebras. Commun. Algebra **39**, 4014–4035 (2011)
13. Doro, S.: Simple Moufang loops. Math. Proc. Camb. Philos. Soc. **83**, 377–392 (1978)
14. Gerritzen, L., Holtkamp, R.: Hopf co-addition for free magma algebras and the non-associative Hausdorff series. J. Algebra **265**, 264–284 (2003)
15. Glauberman, G.: On loops of odd order II. J. Algebra **8**, 393–414 (1968)
16. Grishkov, A., Zavarnitsine, A.: Groups with triality. J. Algebra Appl. **5**, 441–463 (2006)
17. Hartl, M., Loiseau, B.: Internal object actions in homological categories. arXiv:1003.0096 [math.CT]
18. Hartl, M., Van der Linden, T.: The ternary commutator obstruction for internal crossed modules. Adv. Math. **232**, 571–607 (2013)
19. Hentzel, I., Peresi, L.: Special identities for Bol algebras. Linear Algebra Appl. **436**, 2315–2330 (2012)
20. Hofmann, K.H., Strambach, K.: Topological and analytic loops. In: Chein, O., Pflugfelder, H., Smith, J.D.H. (eds) Quasigroups and loops: Theory and Applications, pp. 205–262. Heldermann, Berlin (1990)
21. Jacobson, N.: Exceptional Lie Algebras. Marcel-Dekker, New York (1971)
22. Jennings, S.A.: The group ring of a class of infinite nilpotent groups. Can. J. Math. **7**, 169–187 (1955)
23. Kerdman, F.S.: Analytic Moufang loops in the large. Algebra Logic **18**, 325–347 (1980) [translation from Algebra Logika 18, 523–555 (1979)]
24. Kikkawa, M.: On local loops in affine manifolds. J. Sci. Hiroshima Univ. Ser. A-I Math. **28** 199–207 (1964)
25. Klim, J., Majid, S.: Hopf quasigroups and the algebraic 7-sphere. J. Algebra **323**, 3067–3110 (2010)
26. Kuzmin, E.: On the relation between Mal'tsev algebras and analytic Mufang groups. Algebra Logika **10**, 3–22 (1971)
27. Loos, O.: Symmetric Spaces I: General Theory. Benjamin, New York (1969)
28. Lemieux, F., Moore, C., Thérien, D.: Subtree-counting loops. Quasigroups Relat. Syst. **8**, 45–65 (2001)
29. Madariaga, S., Perez Izquierdo, J.M.: Non-existence of coassociative quantized universal enveloping algebras of the traceless octonions. Commun. Algebra **40**, 1009–1018 (2012)
30. Malcev, A.I.: Analytic loops (Russian). Mat. Sb. N.S. **36**(78), 569–576 (1955)
31. Mikhailov, R., Passi, I.B.S.: Lower central and dimension series of groups. In: Lecture Notes in Mathematics, vol. 1952. Springer, Berlin (2009)
32. Mikheev, P.O.: Enveloping groups of Moufang loops. Russ. Math. Surv. **48** 195–196 (1993) [translation from Usp. Mat. Nauk 48, 191–192 (1993)]
33. Mikheev, P.O.: On a problem of Chern–Akivis–Shelekhov on hexagonal three-webs. Aequationes Math. **51**, 1–11 (1996)
34. Miheev, P.O., Sabinin, L.V.: Quasigroups and differential geometry. In: Chein, O., Pflugfelder, H., Smith, J.D.H. (eds.) Quasigroups and Loops: Theory and Applications, pp. 357–430. Heldermann, Berlin (1990)
35. Mostovoy, J.: The Notion of Lower Central Series for Loops. In: Sabinin., L., Sbitneva, L., Shestakov, I. (eds.) Non-associative algebra and its applications. pp. 291–298. Chapman & Hall/CRC-Press, Boca Raton (2006)
36. Mostovoy, J.: Nilpotency and dimension filtration for loops. Commun. Algebra **36**, 1565–1579 (2008)
37. Mostovoy, J., Pérez Izquierdo, J.M.: Dimension filtration on loops. Isr. J. Math. **158**, 105–118 (2007)
38. Mostovoy, J., Pérez Izquierdo, J.M.: Ideals in non-associative universal enveloping algebras of Lie triple systems. Forum Math. **22**, 1–20 (2010)
39. Mostovoy, J., Pérez Izquierdo, J.M.: Formal multiplications, bialgebras of distributions and non-associative Lie theory. Transform. Groups **15**, 625–653 (2010)

40. Mostovoy, J., Pérez Izquierdo, J.M., Shestakov, I.: Nilpotent Sabinin algebras. arXiv:1312.2223 [math.RA]
41. Nagy, P.T., Strambach, K.: Loops in group theory and Lie theory. In: De Gruyter Expositions in Mathematics, vol. 35. Walter de Gruyter & Co., Berlin (2002)
42. Pérez Izquierdo, J.M., Shestakov, I.P.: An envelope for Malcev algebras. J. Algebra **272** 379–393 (2004)
43. Pérez Izquierdo, J.M., Shestakov, I.P.: in preparation
44. Passman, D.S.: The algebraic structure of group rings. Wiley, New York (1977)
45. Pérez Izquierdo, J.M.: An envelope for Bol algebras. J. Algebra **284**, 480–493 (2005)
46. Pérez Izquierdo, J.M.: Algebras, hyperalgebras, nonassociative bialgebras and loops. Adv. Math. **208**, 834–876 (2007)
47. Pérez Izquierdo, J.M.: Right ideals in non-associative universal enveloping algebras of Lie triple systems. J. Lie Theory **18**, 375–382 (2008)
48. Reutenauer, C.: Free Lie algebras. Clarendon Press, Oxford (1993)
49. Sabinin, L.V.: Odules as a new approach to a geometry with a connection. Sov. Math., Dokl. **18** 515–518 (1977) [translation from Dokl. Akad. Nauk SSSR 233, 800–803 (1977)]
50. Sabinin, L.V., Mikheev, P.O.: On the infinitesimal theory of local analytic loops. Sov. Math., Dokl. **36** 545–548 (1988) [translation from Dokl. Akad. Nauk SSSR 297, 801–804 (1987)]
51. Sabinin, L.V.: Smooth quasigroups and loops: forty-five years of incredible growth. Commentat. Math. Univ. Carol. **41**, 377–400 (2000)
52. Sabinin, L.V.: Smooth Quasigroups and Loops. In: Mathematics and Its Applications, vol. 492. Kluwer Academic Publishers, Dordrecht (1999)
53. Schafer, R.D.: An introduction to nonassociative algebras. Academic Press, New York-London (1966), Dover Publications Inc, New York (1995)
54. Shelekhov, A.M.: The G-structures associated to a hexagonal 3-web is closed. J. Geom. **35**, 169–176 (1989)
55. Shestakov, I.P.: Every Akivis algebra is linear. Geom. Dedicata **77**, 215–223 (1999)
56. Shestakov, I.P.: Moufang loops and alternative algebras. Proc. Am. Math. Soc. **132**, 313–316 (2004)
57. Shestakov, I.P., Umirbaev, U.U.: Free Akivis algebras, primitive elements, and hyperalgebras. J. Algebra **250**, 533–548 (2002)
58. Smith, J.D.H.: An Introduction to Quasigroups and Their Representations. Chapman & Hall/CRC, Boca Raton (2007)
59. Tvalavadze, M.V., Bremner, M.R.: Enveloping algebras of solvable Malcev algebras of dimension five. Commun. Algebra **39**, 2816–2837 (2011)

3

Guide to nonlinear potential estimates

Tuomo Kuusi · Giuseppe Mingione

Abstract One of the basic achievements in nonlinear potential theory is that the typical linear pointwise estimates via fundamental solutions find a precise analog in the case of nonlinear equations. We give a comprehensive account of this fact and prove new unifying families of potential estimates. We also describe new fine properties of solutions to measure data problems.

Contents

Communicated by Neil Trudinger.

The authors are supported by the ERC grant 207573 "Vectorial Problems" and by the Academy of Finland project "Regularity theory for nonlinear parabolic partial differential equations".

T. Kuusi
Aalto University Institute of Mathematics, P.O. Box 11100, 00076 Aalto, Finland
e-mail: tuomo.kuusi@aalto.fi

G. Mingione (✉)
Dipartimento di Matematica e Informatica, Università di Parma,
Parco Area delle Scienze 53/a, Campus, 43124 Parma, Italy
e-mail: giuseppe.mingione@math.unipr.it

1 A synopsis

The aim of this paper is to give a rather comprehensive introduction to nonlinear potential estimates, i.e., pointwise estimates for solutions to quasilinear, possibly degenerate elliptic equations via linear and nonlinear potentials. The paper contains both new and old results. They fall into two categories. The first consists of those results that have been proved elsewhere, and that are here given in different and/or streamlined version. The second contains new results that are presented for the first time. These fill some of the gaps that were making the current theory still somewhat incomplete.

To ease the reading, the already known results will be stated together with the reference to the corresponding original paper where they have appeared for the first time. The new ones will be presented pointing at the place of this paper where the proof can be found. In general, the first part of the paper is devoted to the presentation of the general setting, with the statements of the main theorems; this goes up to Sect. 8. The remaining parts are instead devoted to the proofs.

We shall start from a presentation of the classical results valid for linear elliptic equations in Sect. 2, and this will serve to give the general guideline to the topics we are going to cover in the nonlinear case. The first potential estimates for nonlinear equations will be introduced in Sect. 3. There the classical pointwise estimates will be presented. In the subsequent Sects. 4 and 5 we shall instead give a class of estimates aimed at unifying the theory. These allow to frame the classical pointwise potential inequalities in a more general setting, allowing for estimates of both size and oscillations of solutions and their derivatives, including fractional derivates. At this stage we shall use "intermediate" Wolff and Riesz potentials and various fractional maximal operators. Especially, we prove Theorem 10, whose role is to unify estimates for u and Du via a family of estimates for suitable fractional operators of u. We shall eventually add a few remarks on the case of equations with coefficients in Sect. 6. All the results up to Sect. 6 are presented for *energy solutions*, that is for solutions belonging to the natural energy space $W^{1,p}$ associated to the problems considered. In Sect. 7 we then turn to the case of general measure data problems, that is when *very weak solutions* come into the play. This will give us the opportunity, in Sect. 8, to present a few interesting consequences of potential estimates. Specifically, we shall prove two theorems about the possibility of describing the fine behaviour of solutions to nonlinear measure data problems and of their gradient via potentials. These theorems are the nonlinear analogs of classical results about fine properties of solutions to linear equations, which are usually derived via linear potential; see Remark 3 below. In Sect. 9 we then gather a series of regularity results that follow as a corollary of the theory presented. Some of these results are well-known and they are now framed in the general and unifying context of nonlinear potential estimates. The remaining sections are devoted to the proofs of the results introduced in the preceding ones, and their titles are self-explaining. In this paper we are not going to deal with parabolic problems, for which we refer to [42–44].

Before starting, we find useful to establish some notation. In what follows we denote by c a general positive constant, possibly varying from line to line; special occurrences will be denoted by $c_1, c_2, \bar{c}_1, \bar{c}_2$ or the like. All these constants will always be *larger or equal than one*; moreover relevant dependencies on parameters will be emphasized using parentheses, i.e., $c_1 \equiv c_1(n, p, \nu, L)$ means that c_1 depends only on n, p, ν, L. We denote by

$$B(x,r) \equiv B_r(x) := \{\tilde{x} \in \mathbb{R}^n \ : \ |\tilde{x} - x| < r\}$$

the open ball with center x and radius $r > 0$. When not important we shall omit denoting the center as follows: $B_r \equiv B(x,r)$. Moreover, with B being a generic ball with radius r, we will denote by σB the ball concentric to B having radius σr, $\sigma > 0$. Unless otherwise stated, different balls in the same context will have the same center. The symbol Ω will denote a bounded open subset of $\mathbb{R}^n$ and we shall always consider the case $n \geq 2$. With $\mathcal{O} \subset \mathbb{R}^n$ being a measurable subset with positive measure, and with $g \colon \mathcal{O} \to \mathbb{R}^n$, $n \geq 1$, being a measurable map, we shall denote by

$$(g)_{\mathcal{O}} \equiv \fint_{\mathcal{O}} g \, d\tilde{x} := \frac{1}{|\mathcal{O}|} \int_{\mathcal{O}} g(\tilde{x}) \, d\tilde{x}$$

its integral average; here $|\mathcal{O}|$ denotes the Lebesgue measure of $\mathcal{O}$. The map g will be typically a gradient. We shall denote

$$|g|(O) := \int_{\mathcal{O}} |g| \, d\tilde{x}$$

therefore adopting a unified notation for both measures and integrals of L^1-maps. In the following, μ will always denote a Borel measure with finite total mass, which is initially defined on a certain open subset $\Omega \subset \mathbb{R}^n$. Since this will not affect the rest, with no loss of generality, all such measures will be considered as defined in the whole $\mathbb{R}^n$ so that $|\mu|(\mathbb{R}^n) < \infty$. We shall denote by $\mathcal{M}_b$ the space of all such measures.

2 From linear to nonlinear

When considering linear elliptic equations, a powerful tool for establishing the qualitative properties of solutions is given by representation formulas via fundamental solutions. Eventually, these lead to consider linear Riesz potentials and singular integrals. Let us recall the situation for the simplest example, which is of course given by the classical Poisson equation

$$-\Delta u = \mu. \tag{1}$$

For the sake of exposition, we initially consider the last equation in the whole $\mathbb{R}^n$, with $n \geq 2$. Here μ, which is eventually taken to be a measure, is again for simplicity assumed to be a smooth and compactly supported function, while u is the unique solution which decays to zero at infinity. The point we are interested in now is that u can be recovered via convolution with the so called fundamental solution

$$G(x, y) \approx \begin{cases} |x-y|^{2-n} & \text{if } n \geq 3 \\ -\log|x-y| & \text{if } n = 2 \end{cases} \tag{2}$$

and this means that the following representation formula holds:

$$u(x) = \int_{\mathbb{R}^n} G(x, y)\, d\mu(y). \tag{3}$$

The formula in the last display allows to shift the study of solutions to the analysis of a related integral operator. It is therefore time for the following:

Definition 1 (*Riesz potentials*) Let $\alpha \in (0, n]$; the linear operator defined by

$$I_\alpha(\mu)(x) := \int_{\mathbb{R}^n} \frac{d\mu(\tilde{x})}{|x-\tilde{x}|^{n-\alpha}}, \tag{4}$$

is called the α-Riesz potential of μ, where μ is a Borel measure defined on $\mathbb{R}^n$.

Now, (3) allows to conclude with the following pointwise potential estimates:

$$|u(x)| \lesssim I_2(|\mu|)(x)| \quad \text{and} \quad |Du(x)| \lesssim I_1(|\mu|)(x). \tag{5}$$

The first inequality actually holds in the case $n \geq 3$; the second one follows differentiating (3). Since the behaviour of Riesz potentials with respect to various relevant function spaces is known (see for instance [1,26]), using (5) allows to infer *size* properties of solutions from those of the relevant potentials, and in a sharp way. For instance, the following regularizing property of Riesz potentials is known for every $q > 1$ and $\alpha > 0$ such that $0 < \alpha q < n$:

$$\|I_\alpha(\mu)\|_{L^{\frac{nq}{n-\alpha q}}(\mathbb{R}^n)} \leq c(n,\alpha,q)\|\mu\|_{L^q(\mathbb{R}^n)}.$$

As an immediate corollary, it follows for instance that $\mu \in L^q$ implies that $Du \in L^{nq/(n-q)}$ whenever $1 < q < n$. Moreover, estimates in further function spaces can be obtained as well, and these yield borderline regularity results. An instance is given by the following Lorentz spaces criterion:

$$\mu \in L(n,1) \Longrightarrow I_1(|\mu|) \in L^\infty \Longrightarrow Du \in L^\infty. \tag{6}$$

We recall that the Lorentz space $L(q,\gamma)(\Omega)$, for $1 \leq q < \infty$ and $\gamma > 0$, consists of those measurable maps μ satisfying the condition

$$\int_0^\infty \left(\lambda^q |\{\tilde{x} \in \Omega : |\mu(\tilde{x})| > \lambda\}|\right)^{\gamma/q} \frac{d\lambda}{\lambda} < \infty \tag{7}$$

so that $\mu \in L(n,1)$ means in particular that

$$\int_0^\infty |\{\tilde{x} \in \Omega : |\mu(\tilde{x})| > \lambda\}|^{1/n}\, d\lambda < \infty. \tag{8}$$

A point, which is less emphasised concerning estimates (5) is that (3) also allows to bound *oscillations* of the solution u in terms of suitable Riesz potentials. This point actually turns out of be of primary interest for us here. To see this, by using the elementary inequality

$$\left||x-\tilde{x}|^{2-n} - |y-\tilde{x}|^{2-n}\right| \leq c(n)\left||x-\tilde{x}|^{2-n-\alpha} + |y-\tilde{x}|^{2-n-\alpha}\right||x-y|^\alpha,$$

which is valid whenever $x, y, \tilde{x} \in \mathbb{R}^n$, we get the new representation formula

$$|u(x)-u(y)| \leq c\left[I_{2-\alpha}(|\mu|)(x) + I_{2-\alpha}(|\mu|)(y)\right]|x-y|^\alpha, \quad 0 \leq \alpha \leq 1. \tag{9}$$

The previous estimate allows to infer, again via potentials, information on the solutions in function spaces measuring smoothness rather than size. For instance, the Hölder continuity criterion

$$I_{2-\alpha}(|\mu|) \in L^\infty \Longrightarrow u \in C^{0,\alpha}$$

is immediate. Again, prescribing that $I_{2-\alpha}(|\mu|) \in L^q$ implies that u belongs to the so called Calderón space C^α_q; see Definition 2 below, and [17,40], for a discussion. In a similar way one gets estimates for the gradient

$$|Du(x) - Du(y)| \le c\left[I_{1-\alpha}(|\mu|)(x) + I_{1-\alpha}(|\mu|)(y)\right]|x-y|^{\alpha}, \quad 0 \le \alpha < 1. \quad (10)$$

As a matter of fact, formulas (9)–(10) can be read as nonlocal, fractional derivatives versions of the classical estimates in (5). This is clarified by the following definition:

Definition 2 (*Calderón spaces*) Let $\alpha \in (0, 1]$, $q \ge 1$, and let $\Omega \subset \mathbb{R}^n$ be an open subset. A measurable function v, finite a.e. in Ω, belongs to the Calderón space $C_q^\alpha(\Omega)$ if and only if there exists a nonnegative function $m \in L^q(\Omega)$ such that

$$|v(x) - v(y)| \le [m(x) + m(y)]|x-y|^{\alpha} \quad (11)$$

holds for almost every couple $(x, y) \in \Omega \times \Omega$.

This is just another way to say that v has "fractional derivatives". The advantage is that nonlocality is reduced to a minimal status: only two points are considered in (11). Moreover, when Ω is a suitably regular domain, Calderón spaces are closely related to the usual fractional Sobolev spaces $W^{\alpha,q}$; see [17]. The function m plays in fact the role of a fractional derivative of v of order α in the L^q-sense. Definition 2 is implicit in the work of DeVore and Sharpley [17], where the authors fix the canonical choice $m = M_\alpha^{\#}(v)$ that is indeed always possible in (11) (see Proposition 1 below). The symbol $M_\alpha^{\#}(v)$ denotes the standard fractional sharp maximal operator introduced in the following (take $\Omega = \mathbb{R}^n$ and $R = \infty$):

Definition 3 (*Fractional sharp maximal operator*) Let $\beta \in [0, 1]$, $x \in \Omega$ and $R \le \operatorname{dist}(x, \partial\Omega)$, and let $f \in L^1(\Omega)$; the function defined by

$$M_{\beta,R}^{\#}(f)(x) := \sup_{0<r\le R} r^{-\beta} \fint_{B(x,r)} |f - (f)_{B(x,r)}|\, d\tilde{x}$$

is called the restricted (centered) sharp fractional maximal function of f.

Notice that for $\beta = 0$ the previous definition gives back the classical sharp maximal operator of Fefferman and Stein (in the restricted version), which is in fact denoted by $M_R^{\#}(f)$. Definition 2 and the interpretation of m as the α-order fractional derivative of v allow to read the inequalities in (9)–(10) as a way to bound fractional derivatives of u via intermediate Riesz potentials. We may express this concept in the following imprecise yet suggestive way:

$$|\partial^\alpha u(x)| \lesssim I_{2-\alpha}(|\mu|)(x), \quad 0 \le \alpha < 2. \quad (12)$$

Needless to say, that last formula has only a symbolic meaning. At this point it is clear that the classical estimates (5) embed in the family of interpolating estimates (12) as particular borderline cases corresponding to the choices $\alpha = 0$ and $\alpha = 1$, respectively. In turn, the estimates in (12) provides a unifying approach to the regularity of solutions to (1). They allow to get results both in spaces aimed at measuring the size of functions and in those measuring smoothness.

One of the aims of this paper is to show that completely similar estimates actually hold for a large class of quasilinear, possibly degenerate equations. A potential theory which is completely analogous to the linear one can be constructed in the nonlinear case too.

3 Basic nonlinear estimates

The main question is now whether and in which sense estimates like (5) and (12) extend to the case of solutions to quasilinear elliptic equations of the type

$$-\operatorname{div} a(Du) = \mu \quad \text{in } \Omega. \tag{13}$$

In the most general case, the right hand side μ is a Borel measure with finite total mass in $\mathbb{R}^n$, i.e. $\mu \in \mathcal{M}_b$. The one in (13) is a nonlinear version of the Poisson equation when the assumptions

$$\nu \operatorname{Id} \leq \partial a(z) \leq L \operatorname{Id} \tag{14}$$

are considered. The numbers $0 < \nu \leq L$ provide bounds for the lowest and highest eigenvalue of the matrix $\partial a(\cdot)$, respectively ($\nu = L = 1$ gives the case of the Laplacean operator). We also want to consider cases where Eq. (13) might be degenerate, thereby examining for instance the so called p-Laplacean operator:

$$-\operatorname{div}(|Du|^{p-2}Du) = \mu. \tag{15}$$

In order to catch the essential properties of the equation in (15), we shall consider general vector fields $a\colon \mathbb{R}^n \to \mathbb{R}^n$. These are assumed to be C^1-regular and to satisfy the following *growth and ellipticity assumptions*:

$$\begin{cases} |a(z)| + |\partial a(z)|(|z|^2+s^2)^{1/2} \leq L(|z|^2+s^2)^{(p-1)/2} \\ \nu(|z|^2+s^2)^{(p-2)/2}|\lambda|^2 \leq \langle \partial a(z)\lambda, \lambda\rangle \\ p \geq 2 \end{cases} \tag{16}$$

whenever $z, \lambda \in \mathbb{R}^n$. The case $p = 2$ gives back (14). We remark that we confine ourselves to the case $p \geq 2$ as we are mainly interested to present the main ideas in the most accessible way; for results in the case $p \leq 2$ we refer to [22,38,40,58]. In the rest of the paper ν, L are fixed parameters whose role is to establish the rate of ellipticity of the vector field $a(\cdot)$. The role of $s \geq 0$ is more peculiar. This parameter serves to distinguish the degenerate ($s = 0$) from the nondegenerate one ($s > 0$). In this last case a model is given, by for instance taking $s = 1$, by the nondegenerate equation p-Laplacean type equation (see [49])

$$-\operatorname{div}((1+|Du|^2)^{(p-2)/2}Du) = \mu.$$

Sometimes we shall also consider equations with measurable coefficients, that is when the vector field $a(\cdot)$ exhibits an explicit dependence on the variable x, i.e.,

$$-\operatorname{div} a(x, Du) = \mu. \tag{17}$$

In this case, we shall use a set of assumptions which is weaker than (16), namely, we shall consider a Carathéodory vector field $a\colon \Omega \times \mathbb{R}^n \to \mathbb{R}^n$ such that

$$\begin{cases} |a(x,z)| \le L(|z|^2 + s^2)^{(p-1)/2} \\ \nu(|z_1|^2 + |z_2|^2 + s^2)^{(p-2)/2}|z_1 - z_2|^2 \le \langle a(x,z_1) - a(x,z_2), z_1 - z_2\rangle \\ p \ge 2 \end{cases} \tag{18}$$

are satisfied whenever $z_1, z_2, x \in \Omega$. These assumptions are, up to adjusting the constants ν, L in a universal way, implied by those in (16). Results for the case when $x \to a(x, \cdot)$ is more regular and satisfy (16) uniformly with respect to $x \in \Omega$ are described in Sect. 6 below.

Our primary emphasis will be on a priori estimates. This means that, according to a scheme which is typical in regularity theory, we shall mainly confine ourselves to present results in form of a priori estimates for more regular solutions and data. We are therefore most of the times considering weak energy solutions to (17), that is functions $u \in W^{1,p}(\Omega)$ satisfying

$$\int_\Omega \langle a(\tilde{x}, Du), D\varphi\rangle \, d\tilde{x} = \int_\Omega \varphi \, d\mu \tag{19}$$

for every choice of $\varphi \in C^\infty(\Omega)$ with compact support in Ω. Here we are still assuming that μ is a Borel measure with finite mass, but we could assume that $\mu \in C^\infty$ as well, stating the same results involving only the total variation of μ, considered as a measure. Results for the original context would then follow via approximation. Such an approach is obviously restrictive when considering general measure data problems. Indeed, observe for instance that due to the monotonicity properties of the operator assumed in (16), considering a solution that belongs to the space $W^{1,p}(\Omega)$ automatically implies that the measure μ belongs to the dual $W^{-1,p'}$, which is certainly not the case for any measure when $p \le n$. As a matter of fact, in general, distributional solutions to measure data problems do not belong to the natural energy space $W^{1,p}(\Omega)$ and for this reasons they are called *very weak solutions*. Treating the case of general measures needs then greater care, already in specifying the notion of solution one is dealing with. We are briefly discussing these aspects in Sects. 7 and 14.4 below, where we shall see how to extend the results presented for energy solutions to the case of general solutions to measure data problems.

Since we are going to deal with local results, we shall use in a standard way the truncated version of the classical Riesz potentials.

Definition 4 (*Truncated Riesz potentials*) Let μ be Borel measure with finite total mass on $\mathbb{R}^n$; the (truncated) Riesz potential is defined by

$$\mathbf{I}^{\mu}_{\beta}(x,R) := \int_0^R \frac{|\mu|(B(x,\varrho))}{\varrho^{n-\beta}} \frac{d\varrho}{\varrho}, \qquad \beta > 0,$$

whenever $x \in \mathbb{R}^n$ and $0 < R \leq \infty$.

The word truncated refers to the inequality $\mathbf{I}^{\mu}_{\beta}(x,R) \leq c(n) I_\beta(|\mu|)(x)$. Now, let us go back to estimates in (5), and let us notice that these cannot hold when $p \neq 2$. Indeed, they clearly do not respect the homogeneity properties of the equation. To see this, we consider a non-null solution to $\operatorname{div}(|Du|^{p-2}Du) = \mu$ with $p > 2$, then, for $\gamma > 0$ we have

$$\tilde{u} := \gamma^{1/(p-1)} u \Longrightarrow \operatorname{div}(|D\tilde{u}|^{p-2} D\tilde{u}) = \gamma\mu := \tilde{\mu}.$$

Assuming now that the first estimate in (5) would hold, this would yield

$$|\tilde{u}(x)| \lesssim I_2(|\tilde{\mu}|)(x),$$

that is, recalling the definitions of $\tilde{u}$, $\tilde{\mu}$ and I_2,

$$|u(x)| \lesssim \gamma^{(p-2)/(p-1)} I_2(\mu)(x).$$

Upon letting $\gamma \to 0$ and using that $p > 2$, we would then conclude with $|u(x)| = 0$ whenever $I_2(\mu)(x)$ is finite. A similar argument—letting $\gamma \to \infty$—also gives that estimate as in (5) cannot hold in the case $p < 2$. In order to overcome this point one is led to consider a new family of potentials. They are obtained by, in some sense, incorporating in Riesz potentials the scaling properties of the p-Laplacean equation.

Definition 5 (*Wolff potentials*) Let μ be Borel measure with finite total mass on $\mathbb{R}^n$; the nonlinear Wolff potential is defined by

$$\mathbf{W}^{\mu}_{\beta,p}(x,R) := \int_0^R \left(\frac{|\mu|(B(x,\varrho))}{\varrho^{n-\beta p}}\right)^{1/(p-1)} \frac{d\varrho}{\varrho}, \quad \beta > 0$$

whenever $x \in \mathbb{R}^n$ and $0 < R \leq \infty$.

Wolff potentials, that despite their name were first considered and studied in [28], reduce to Riesz potentials when $p = 2$, i.e. $\mathbf{I}^{\mu}_{\beta} \equiv \mathbf{W}^{\mu}_{\beta/2,2}$. They play a crucial role in nonlinear potential theory and in the description of the fine properties of solutions to nonlinear equations in divergence form [2,3,28,29,32,59,60].

An important fact about Wolff potentials is that their behaviour can be in several aspects recovered from that of Riesz potentials via so called Havin–Maz'ya potentials $\mathbf{V}_{\beta,p}(|\mu|)(x)$. Indeed, following inequality holds provided $p\beta < n$:

$$\mathbf{W}^{\mu}_{\beta,p}(x,\infty) \lesssim I_\beta\left\{\left[I_\beta(|\mu|)\right]^{1/(p-1)}\right\}(x) =: \mathbf{V}_{\beta,p}(|\mu|)(x).$$

The above estimate allows to derive, in a sharp way, almost all types of local estimates starting by the properties of the Riesz potentials, whose action in several function spaces is in fact known; for this we refer to [28,29].

As first shown in the fundamental works of Kilpeläinen and Malý [37,38] for the case of nonnegative measures, a neat analog of the first estimate in (5) holds using Wolff potentials. Later on, a new and interesting proof has been offered Trudinger and Wang [65,66], and this allows to cover the case of general subelliptic operators. Yet different proofs can be found in [39], and in [21], where an approach covering the case of general signed measures has been developed. The final outcome is summarized in the following:

Theorem 1 [21,37,38,65] *Let $u \in W^{1,p}(\Omega)$ be a weak solution to the equation with measurable coefficients* (17) *under the assumptions* (18). *There exists a constant $c \equiv c(n, p, \nu, L)$ such that the pointwise estimate*

$$|u(x)| \leq c\mathbf{W}^{\mu}_{1,p}(x, R) + c \fint_{B(x,R)} (|u| + Rs)\, d\tilde{x} \tag{20}$$

holds whenever $B(x, R) \subset \Omega$ and the right hand side is finite. Moreover, if

$$\lim_{R\to 0} \mathbf{W}^{\mu}_{1,p}(x, R) = 0 \quad \textit{locally uniformly in } \Omega \textit{ w.r.t. } x \tag{21}$$

then u is continuous in Ω.

A sketchy proof of this theorem is proposed in Sect. 18 below; extensions to more general operators have been given in [50,53]. If $p = 2$, then $\mathbf{W}^{\mu}_{1,p} \equiv \mathbf{I}^{\mu}_{2}$ and we retrieve a local analog of the first estimate in (5). A remarkable point here is that estimate (20) is sharp, and the nonlinear potential $\mathbf{W}^{\mu}_{1,p}$ cannot be replaced by any other smaller potential. This is in fact reported in the following:

Theorem 2 [37,38] *Let $u \in W^{1,p}(\Omega)$ be a nonnegative weak solution to the Eq.* (13) *under the assumptions* (18), *where μ is a positive measure and $s = 0$. There exists a constant $c \equiv c(n, p, \nu, L)$ such that the following pointwise estimate holds whenever $B(x, 2R) \subset \Omega$ and the Wolff potential is finite:*

$$c^{-1}\mathbf{W}^{\mu}_{1,p}(x, R) \leq u(x) \leq c\mathbf{W}^{\mu}_{1,p}(x, 2R) + c \inf_{B(x,R)} u.$$

The possibility of extending pointwise potential estimates to the gradient of solutions has remained an open and discussed issue since the paper [38]. The answer came only recently and here we consider the case of equations of the type (13). The first result in this direction is contained in [57] for the case $p = 2$, that is when assumptions (14) are in force. In [57] it is indeed proved the following analog of the second estimate in (5):

$$|D_\xi u(x)| \leq c\mathbf{I}^{\mu}_{1}(x, R) + c \fint_{B(x,R)} |D_\xi u|\, d\tilde{x}, \qquad \xi \in \{1, \ldots, n\}. \tag{22}$$

For the case $p > 2$ a first result has been given in [21], where the following, apparently natural estimate, has been proved to hold at every Lebesgue point x of Du, under assumptions (16):

$$|Du(x)| \le c\mathbf{W}^{\mu}_{1/p,p}(x,R) + c \fint_{B(x,R)} (|Du| + s)\, d\tilde{x}. \tag{23}$$

The previous estimate seems to put a final word on the problem of a sharp analog of the second estimate in (5) since the orthodoxy of nonlinear potential theory prescribes that Wolff potentials replace Riesz potentials everywhere when $p \neq 2$. Surprisingly enough, in [41] we have shown that this is not the case. Using Wolff potentials is necessary only when estimating solutions, while it is not when passing to their gradients. In fact the following holds:

Theorem 3 [41] *Let $u \in W^{1,p}(\Omega)$ be a solution to the Eq.* (13)*, under the assumptions* (16)*. There exists a constant $c \equiv c(n, p, \nu, L)$ such that the Riesz potential estimate*

$$|Du(x)|^{p-1} \le c\mathbf{I}^{\mu}_{1}(x,R) + c\left(\fint_{B(x,R)} (|Du| + s)\, d\tilde{x}\right)^{p-1} \tag{24}$$

holds whenever $B(x, R) \subset \Omega$ and the right hand side is finite. Moreover, if

$$\lim_{R\to 0} \mathbf{I}^{\mu}_{1}(x,R) = 0 \ \textit{locally uniformly in } \Omega \ \textit{w.r.t. } x \tag{25}$$

then Du is continuous in Ω.

The proof of Theorem 3 will be presented in Sect. 15 below, where it will be obtained as a corollary of more general potential estimates. An extension of the previous result to a class of general operators including the p-Laplacean has been recently given by Baroni in [5].

Remark 1 An obvious ambiguity arises in the statements of Theorems 1–3 when saying that the related estimates hold whenever the right hand side is finite. This might appear obvious. Another point is that estimates in Theorems 1–3 are stated for every x, while both $u(x)$ and $Du(x)$ are only defined if x is Lebesgue point, where the so called precise representative can be defined. Both ambiguities are clarified in Sect. 8. There we prove that, both for u and Du, the set of Lebesgue points coincides with the one for which inequalities (20) and (24) feature a finite right hand side, respectively.

Observe that estimate (24) obviously improves the one in (23) as, if $p \ge 2$ then $1/(p-1) \le 1$, and therefore

$$\begin{aligned}\mathbf{I}^{\mu}_{1}(x,R) &= \int_0^R \frac{|\mu|(B(x_0,\varrho))}{\varrho^{n-1}}\frac{d\varrho}{\varrho} \\ &\lesssim \sum_{i=0}^{\infty} \frac{|\mu|(B(x,R/2^i))}{(R/2^i)^{n-1}}\end{aligned}$$

$$\lesssim \left[\sum_{i=0}^{\infty} \left(\frac{|\mu|(B(x, R/2^i))}{(R/2^i)^{n-1}} \right)^{1/(p-1)} \right]^{p-1}$$

$$\lesssim \left[\int_0^{2R} \left(\frac{|\mu|(B(x_0, \varrho))}{\varrho^{n-1}} \right)^{1/(p-1)} \frac{d\varrho}{\varrho} \right]^{p-1}$$

$$= \left[\mathbf{W}^{\mu}_{1/p,p}(x, 2R) \right]^{p-1} . \tag{26}$$

The implications of Theorem 3 are rather surprising: when switching to the gradient, estimates *linearize* and they reduce to the ones already available for the Poisson equation. The nonlinear, possibly degenerate character of the equations considered plays no role here. In particular, the regularity theory of equations as (13) reduces to that of the Poisson equation up to the C^1-level. Several facts typical of the linear theory can now be reproduced verbatim. For instance, for solutions $u \in W^{1,p}(\mathbb{R}^n)$ to (13) defined in the whole $\mathbb{R}^n$, the classical estimate via classical Riesz potentials

$$|Du(x)|^{p-1} \leq c \int_{\mathbb{R}^n} \frac{d|\mu|(y)}{|x-y|^{n-1}}$$

follows immediately. For a relevant connection between Theorem 3 and fundamental solutions to general nonlinear measure data problems we refer to Sect. 7.1 below. Another very precise analogy with the classical linear theory is a striking nonlinear extension of a fundamental theorem of Stein [63]. This claims that if $v \in W^{1,1}$ is a Sobolev function defined in $\mathbb{R}^n$ with $n \geq 2$, then

$$Dv \in L(n, 1) \implies v \text{ is continuous.} \tag{27}$$

This can be regarded as the limiting case of Sobolev–Morrey embedding theorem that reads $Dv \in L^{n+\varepsilon} \implies v \in C^{0,\varepsilon/(n+\varepsilon)}$ whenever $\varepsilon > 0$. Note indeed that $L^{n+\varepsilon} \subset L(n, 1) \subset L^n$ for every $\varepsilon > 0$ with all the inclusions being strict. Another way to state Stein's theorem concerns the regularity of solutions $u \colon \Omega \to \mathbb{R}^N$ to the Laplacean system, and reads

$$\operatorname{div} Du = \Delta u \in L(n, 1) \implies Du \text{ is continuous.}$$

This follows by (27) and classical Calderón–Zygmund theory. The point is now that if $\mu \in L(n, 1)$, then a computation involving the definition of norm in Lorentz spaces (see [23,41] for discussion) implies that (25) takes place, and therefore we conclude with the following:

Theorem 4 (Nonlinear Stein theorem [41]) *Let $u \in W^{1,p}(\Omega)$ be a solution to the Eq.* (13), *under the assumptions* (16) *and such that $\mu \in L(n, 1)$ locally in Ω. Then Du is continuous in Ω.*

See [12] for a global Lipschitz bound. Without appealing to potentials, but by using different means, the result of the previous theorem also holds for systems.

Theorem 5 (Vectorial nonlinear Stein theorem [45]) *Let $u \in W^{1,p}(\Omega, \mathbb{R}^N)$, $N \geq 1$, be a vector valued solution to the p-Laplacean system*

$$-\Delta_p u = F.$$

Assume that the components of the vector field $F\colon \Omega \to \mathbb{R}^N$ locally belong to the space $L(n, 1)$. Then Du is continuous in Ω.

Finally, let us mention that starting from the techniques developed for the last two theorems, similar results can be proved in the case of fully nonlinear equations [15].

4 Universal potential estimates

The results of this section have a double aim. On the one hand, they show that estimates (9)–(10) have analogs for nonlinear equations. On the other hand, they show that those in (20), (23) and (24) are actually special cases of a more general class of nonlinear potentials estimates. These, in turn, allow to recover in an optimal way all the basic regularity properties of solutions in terms of the regularity of the assigned datum μ. The range of the results implied by such estimates is of course limited by the regularity theory of homogeneous equations as

$$\operatorname{div} a(x, Dw) = 0. \tag{28}$$

Therefore we first recall what is the maximal regularity of solutions to equations as in (28) and in (29) below. In turn, this dramatically changes according to the smoothness assumed on the partial map $x \to a(x, \cdot)$. When measurability is considered, very low degree of regularity is expected and solutions are just Hölder continuous for some exponent. When instead dependence on x becomes more regular then higher regularity follows. In this case, again for sake of simplicity, and since this does not affect the expositions of the main ideas, we shall confine ourselves to the case of equations with no coefficients as

$$\operatorname{div} a(Dw) = 0. \tag{29}$$

The first result we present upgrades estimate (20) to low order fractional derivatives, and allows to give a sharp formulation of the classical De Giorgi's theory via nonlinear Wolff potential estimates. De Giorgi's theory for equations with measurable provides the existence of a *universal Hölder continuity exponent* $\alpha_m \in (0, 1)$, depending only on n, p, ν, L, such that

$$w \in C^{0,\alpha}_{\mathrm{loc}}(\Omega) \qquad \text{for every } \alpha < \alpha_m \tag{30}$$

and

$$|w(x) - w(y)| \leq c \fint_{B_R} (|w| + Rs)\, d\tilde{x} \cdot \left(\frac{|x-y|}{R}\right)^{\alpha}. \tag{31}$$

The previous estimate holds whenever $x, y \in B_{R/2}$ and $B_R \subset \Omega$, for a constant depending only on n, p, ν, L and α. The exponent α_m can be thought as the *maximal Hölder regularity exponent* associated to the vector field $a(\cdot)$. It is universal in the sense that it is independent of $a(\cdot)$ and of the particular solution considered, but just depends only on n, p, ν, L. For more precise information on α_m see Theorem 18 and Remark 6 below. It then holds the following:

Theorem 6 (De Giorgi's theory via potentials [40]) *Let $u \in W^{1,p}(\Omega)$ be a weak solution to the equation with measurable coefficients* (17) *under assumptions* (18). *Let $B_R \subset \Omega$ be such that $x, y \in B_{R/2}$; then*

$$|u(x) - u(y)| \leq c\left[\mathbf{W}^{\mu}_{1-\alpha(p-1)/p,p}(x, R) + \mathbf{W}^{\mu}_{1-\alpha(p-1)/p,p}(y, R)\right]|x-y|^{\alpha} + c \fint_{B_R} (|u| + Rs)\, d\tilde{x} \cdot \left(\frac{|x-y|}{R}\right)^{\alpha} \tag{32}$$

holds provided the right hand side is finite and $0 \leq \alpha < \alpha_m$, where the exponent α_m has been defined in (30)–(31). *Moreover, whenever $\tilde{\alpha} \in [0, \alpha_m)$ is fixed, the dependence of the constant c is uniform for $\alpha \in [0, \tilde{\alpha}]$, in the sense that c depends only on n, p, ν, L and $\tilde{\alpha}$.*

Notice that when $\mu \equiv 0$ estimates (31) and (33) coincide. In general, counterexamples show that in (30) $\alpha_m \to 0$ when $L/\nu \to \infty$. This prevents estimate (33) to hold in general for the full range $\alpha \in [0, 1)$ when $x \to a(x, \cdot)$ is measurable. Things change when the dependence on x is more regular. For instance, when considering Eq. (29), estimate (31) holds for $\alpha = 1$. In this case the validity of (32) extends to the whole interval $\alpha \in [0, 1]$. Moreover, estimates (20) and (23) can be uniformly obtained as limiting endpoint cases of (32) for $\alpha = 0$ and $\alpha = 1$, respectively. In fact, it holds the following:

Theorem 7 (Uniform Wolff potential estimate [40]) *Let $u \in W^{1,p}(\Omega)$ be a weak solution to the Eq.* (13) *under assumptions* (16). *Let $B_R \subset \Omega$ be such that $x, y \in B_{R/2}$; then estimate* (32) *holds for $\alpha \in [0, 1]$, whenever the right hand side is finite. The constant c depends only on n, p, ν, L but is otherwise independent of α.*

Estimate (32) does not catch up, when $\alpha \to 1$, the optimal gradient bound in (24). There is a natural reason for such a lack of endpoint property: it catches the other optimal bound (20) as $\alpha \to 0$, and the two cases involve different potentials. In fact, the change in the nature of the estimates when passing from (20) to (24) requires another theorem, parallel to Theorem 7.

Theorem 8 (Uniform Riesz potential estimate) *Let $u \in W^{1,p}(\Omega)$ be a weak solution to the Eq.* (13) *under assumptions* (16). *Let $B_R \subset \Omega$ be such that $x, y \in B_{R/4}$; then*

$$|u(x) - u(y)| \leq c\left[\mathbf{I}^{\mu}_{p-\alpha(p-1)}(x, R) + \mathbf{I}^{\mu}_{p-\alpha(p-1)}(y, R)\right]^{1/(p-1)}|x-y|^{\alpha} + c \fint_{B_R} (|u| + Rs)\, d\tilde{x} \cdot \left(\frac{|x-y|}{R}\right)^{\alpha} \tag{33}$$

holds provided the right hand side is finite and $0 < \alpha \leq 1$. *Moreover, whenever* $\tilde{\alpha} \in (0, 1]$ *is fixed, the dependence of the constant c is uniform for* $\alpha \in [\tilde{\alpha}, 1]$ *as c depends only* n, p, ν, L *and* $\tilde{\alpha}$.

The previous result is here presented for the first time and for the proof we refer to Sect. 5 below. There Theorem 8 is actually obtained as a corollary of a more general estimate for certain maximal operators of Du. From the discussion made there it will be clear that the points x, y are Lebesgue points of u when the right hand side in (33) is finite; compare with Remark 1. Theorems 6 and 8 provide altogether the optimal analog of estimate (9) and in fact they both unify with Theorem 7 when $p = 2$, when Wolff and Riesz potentials do coincide. We remark that is also possible to quantify the blow-up of the constant c in (33) as $\alpha \to 0$; see Remark 7 below.

We now examine the situation for the gradient, giving an estimate which is again bound to give back (10) when $p = 2$. As already done before Theorem 6, we recall the basic information about the maximal regularity of solutions to homogeneous equations as in (29). These in turn involve the fundamentals of the regularity theory of the p-Laplacean equations and can be summarised by saying that there exists a positive exponent $\alpha_M \in (0, 1)$, depending only on n, p, ν and L, such that whenever $x, y \in B_{R/2}$,

$$Dw \in C^{0,\alpha}_{\mathrm{loc}}(\Omega) \quad \text{for every } \alpha < \alpha_M \tag{34}$$

and

$$|Dw(x) - Dw(y)| \leq c \fint_{B_R} (|Dw| + s)\, d\tilde{x} \cdot \left(\frac{|x-y|}{R}\right)^{\alpha} \tag{35}$$

hold *for any local solution* w to (29). The constant c depends only on n, p, ν, L and α. For estimate (35) and the exponent α_M see Theorem 17 and Remark 5 below. The result in (34) finds its origins in the basic work of Uraltseva [67] while further approaches can be found in [18,24,46,51]. We remark that all these results are again based on the groundbreaking work of De Giorgi [16]. The exponent α_M is in general strictly less than one (see again [67]); while lower bounds on α_M can be obtained by tracking the constant dependence in the various proofs, its precise (optimal) value is not known. An optimal lower bound for α_M is conjectured to be $1/3$, according to the regularity exhibited by the solutions to the so called ∞-Laplacean equation [4]. By taking (34) into account we have the following nonlinear version of estimate (10):

Theorem 9 (Uniform Wolff potential estimate) *Let* $u \in W^{1,p}(\Omega)$ *be a weak solution to the Eq.* (13) *under assumptions* (16). *Let* $B_R \subset \Omega$ *be such that* $x, y \in B_{R/4}$; *then*

$$\begin{aligned} |Du(x) - Du(y)| \leq{}& c \left[\mathbf{W}^{\mu}_{1-\frac{(1+\alpha)(p-1)}{p}, p}(x, R) + \mathbf{W}^{\mu}_{1-\frac{(1+\alpha)(p-1)}{p}, p}(y, R) \right] |x-y|^{\alpha} \\ &+ c \fint_{B_R} |Du - (Du)_{B_R}|\, d\tilde{x} \cdot \left(\frac{|x-y|}{R}\right)^{\alpha} \end{aligned} \tag{36}$$

holds provided the right hand side is finite and

$$0 \leq \alpha < \min\{1/(p-1), \alpha_M\}. \tag{37}$$

The exponent α_M *has been defined in* (34). *Moreover, whenever* $0 \leq \tilde{\alpha} < \min\{1/(p-1), \alpha_M\}$ *is fixed, the dependence of the constant* c *is uniform for* $\alpha \in [0, \tilde{\alpha}]$, *in the sense that* c *depends only* n, p, ν, L *and* $\tilde{\alpha}$.

The proof of this theorem will be presented in Sect. 16 below. Notice that the limitation in (37) makes the statement of Theorem 9 consistent with the definition of Wolff potential given in Definition 5, where it must be $\beta > 0$.

5 Maximal-potential estimates

Let us now explain Theorem 8 proceeding for some while with a few purely heuristic arguments. By considering the function m appearing in (11) as a fractional derivative of order α, we can express the content of Theorem 8 in the following way:

$$|\partial^\alpha u(x)|^{p-1} \lesssim I_{p-\alpha(p-1)}(|\mu|)(x), \quad 0 < \alpha \leq 1 \tag{38}$$

thereby obtaining a nonlinear version of (12). The singularity of the case $\alpha = 0$ essentially stems from the fact that Wolff potentials come then into the play when considering pointwise estimates for u. Let us proceed heuristically and see why Riesz potentials intervene as long as fractional derivatives are considered. We rewrite Eq. (15) as the decoupled system

$$\begin{cases} -\text{div}\, H = \mu \\ H = |Du|^{p-2}Du. \end{cases}$$

This viewpoint tells us that an equation as in (15), which is commonly seen as *a nonlinear equation in the gradient*, can be also seen as linear equation in a *nonlinear vector field of the gradient*, that is H. The analysis of equations as $-\text{div}\, H = \mu$, being $f \to \text{div}\, f$ a differential operator of order one, typically involves estimates of H via operators as $\mathbf{I}_1^\mu$. In turn this explains the appearance of estimates as in (24). The reader will immediately recognize that no similar argument applies to u, as the equation depends directly on Du and not u. This is ultimately the reason for the appearance of Wolff potentials when deriving pointwise for u as (20). Let us now go to intermediate derivatives and let us believe that fractional derivatives $\partial^\alpha u$ can be controlled by in turn controlling quantities as

$$\frac{|u-(u)_{B_R}|}{R^\alpha} \approx R^{1-\alpha}|Du| \approx |\partial^\alpha u| \tag{39}$$

on every possible ball B_R. Using a typical dimension analysis viewpoint, the appearance in the estimates of a multiplicative factor as $R^{-\beta}$ amounts to consider a corresponding derivative of order β in the estimates. Now, let us rewrite Eq. (15) as

$$R^{(\alpha-1)(p-1)}\text{div}\,(|R^{1-\alpha}Du|^{p-2}R^{1-\alpha}Du) = \mu$$

and note that in this scheme of reasoning the operator $f \to R^{(\alpha-1)(p-1)}\text{div}\, f$ has now order $p-\alpha(p-1)$. Its inversion should therefore involve potentials as $\mathbf{I}^\mu_{p-\alpha(p-1)}$. At this point, recalling the identification (39) leads to conclude with an estimate with the dimensional meaning of (38). The point is now to make everything rigorous. A natural way to make the identification in (39) is to use a nonlocal operator as for instance a fractional maximal operator. It is therefore time for the following:

Definition 6 (*Fractional maximal operator*) Let $\beta \in [0, n]$, $x \in \Omega$ and $R < \operatorname{dist}(x, \partial\Omega)$, and let f be an $L^1(\Omega)$-function or a measure with finite mass; the function defined by

$$M_{\beta,R}(f)(x) := \sup_{0<r\leq R} r^{\beta} \frac{|f|(B(x,r))}{|B(x,r)|}$$

is called the restricted (centered) fractional β maximal function of f.

Using fractional maximal operators allows to get Riesz potential estimates which are uniform in the whole range [0, 1], and that provide a rigorous interpretation of previous heuristic arguments. The outcome is summarized in the following theorem, which appears here for the first time. The proof is contained in Sect. 12 below (see Sect. 14.4 for the case of SOLA).

Theorem 10 (Uniform maximal-potential estimates) *Let $u \in W^{1,p}(\Omega)$ be a weak solution to the Eq.* (13) *under assumptions* (16). *Then for every ball $B(x, R) \subset \Omega$ the following estimate:*

$$M_{1-\alpha,R}(Du)(x) \leq c\left[\mathbf{I}^{\mu}_{p-\alpha(p-1)}(x, R)\right]^{1/(p-1)} + cR^{1-\alpha} \fint_{B_R} (|Du| + s)\, d\tilde{x} \tag{40}$$

holds uniformly in $\alpha \in [0, 1]$, with $c \equiv c(n, p, \nu, L)$.

The maximal operators of Definition 3 naturally connect to those presented in Definition 6 via Poincaré inequality

$$M^{\#}_{\alpha,R}(u)(x) \leq cM_{1-\alpha,R}(Du)(x), \quad \alpha \in [0, 1] \tag{41}$$

so that a uniform estimate on $M^{\#}_{\alpha,R}(u)$ follows appealing to Theorem 10, that is

$$M^{\#}_{\alpha,R}(u)(x) \leq c\left[\mathbf{I}^{\mu}_{p-\alpha(p-1)}(x, R)\right]^{1/(p-1)} + cR^{1-\alpha} \fint_{B_R} (|Du| + s)\, d\tilde{x}. \tag{42}$$

We yet remark that in the estimate in the last display the constant c is independent of α, and in fact the estimate holds uniformly for $\alpha \in [0, 1]$.

The turning point to pointwise estimates is now given by the fact that the sharp maximal operator $M^{\#}_{\alpha,R}(u)(x)$ controls the pointwise behaviour of u provided $\alpha > 0$. This fact is expressed in the following Proposition, a first form of which is present in [17]. It in turn relies on some original arguments of Campanato [10].

Proposition 1 *Let $f \in L^1(B_{8R/5})$; for every $\alpha \in (0, 1]$ the inequality*

$$|f(x) - f(y)| \leq \frac{c}{\alpha}\left[M^{\#}_{\alpha,R}(f)(x) + M^{\#}_{\alpha,R}(f)(y)\right]|x - y|^{\alpha} \tag{43}$$

holds whenever $x, y \in B_{2R/5}$, *for a constant* c *depending only on* n. *More precisely,* x *and* y *are Lebesgue points of* f *whenever* $M^{\#}_{\alpha,R}(f)(x)$ *and* $M^{\#}_{\alpha,R}(f)(y)$ *are finite, respectively. Therefore, whenever the right hand side in* (43) *is finite, the values of* f *are defined as follows:*

$$f(x) := \lim_{\varrho\to 0}(f)_{B(x,\varrho)} \quad \textit{and} \quad f(y) := \lim_{\varrho\to 0}(f)_{B(y,\varrho)}.$$

Proposition 1 allows to obtain Theorem 8 as a corollary of Theorem 10 and in particular of (42); see Sect. 12.7 below for the proofs, including the one of Proposition 1. We believe that Theorem 10 helps to understand the peculiar nature of the case $\alpha = 0$ and the occurrence of Wolff potentials instead of Riesz potentials. Indeed, an L^∞-bound on the sharp maximal operator is essentially equivalent to require that u belongs to the space BMO (introduced in [34], see (68) below). On the other hand, a real L^∞-bound on u itself requires an L^∞-bound on $\mathbf{W}^{\mu}_{1,p}$. This situation looks completely natural when considering typical examples of functions lying in BMO$\setminus L^\infty$, as for instance $\log|x|$. The log-scale is indeed the one making a difference between a potentials as $\mathbf{W}^{\mu}_{1,p}$ and $\mathbf{I}^{\mu}_{1}$. Taking for instance

$$\mu(x) = \frac{1}{|x|^p \log^\beta |x|}$$

we see that $\mathbf{I}^{\mu}_{p}(0,1) < \infty$ provided $\beta > 1$ while $\mathbf{W}^{\mu}_{1,p}(0,1) < \infty$ provided $\beta > p-1$. In the next section we shall in fact describe a few additional results showing that also the case $\alpha = 1$ reveals to be special when using maximal operators instead of potentials.

5.1 Non-endpoint maximal estimates

An interesting fact concerning the two cases $\alpha = 0, 1$ of (40) is that when looking for bounds on intermediate derivatives, potentials can be replaced by maximal operators. These give in fact smaller quantities. Maximal operators, as we shall see, allow to replace conditions typically given in Lebesgue or Lorentz spaces, with weaker conditions formulated in terms of Marcinkiewicz spaces. Let us recall the following, elementary inequality which holds for $\gamma \in (0,1)$

$$M_{\beta,\gamma R}(\mu)(x) \le c(n,\gamma,\beta)\mathbf{I}^{\mu}_{\beta}(x,R),$$

see for instance [40, Lemma 4.1]. Then we have the following:

Theorem 11 (Intermediate maximal estimates [40]) *Let* $u \in W^{1,p}(\Omega)$ *be a weak solution to* (13) *under the assumptions* (16). *Let* $B_R \subset \Omega$ *be a ball centred at* x; *then the estimate*

$$\begin{aligned} &M^{\#}_{\alpha,R}(u)(x) + M_{1-\alpha,R}(Du)(x) \\ &\quad \le c\left[M_{p-\alpha(p-1),R}(\mu)(x)\right]^{1/(p-1)} + cR^{-\alpha} \fint_{B_R} |u-(u)_{B_R}|\,d\tilde{x} + cR^{1-\alpha}s \qquad (44) \end{aligned}$$

holds uniformly in $\alpha \in [0, \tilde{\alpha}]$, *whenever* $0 < \tilde{\alpha} < 1$. *The constant* c *depends only on* $n, p, \nu, L, \tilde{\alpha}$.

As for gradient oscillations, we instead have the following result, whose proof is contained in Sect. 13 below.

Theorem 12 (Gradient sharp maximal estimate [40]) *Let* $u \in W^{1,p}(\Omega)$ *be a weak solution to* (13) *under the assumptions* (16). *Let* $B_R \subset \Omega$ *be a ball centered at* x; *then the following estimate:*

$$M^{\#}_{\alpha,R}(Du)(x) \leq c\left[M_{1-\alpha(p-1),R}(\mu)(x)\right]^{1/(p-1)} + cR^{-\alpha} \fint_{B_R} |Du - (Du)_{B_R}|\, d\tilde{x}$$

holds uniformly in $\alpha \in [0, \tilde{\alpha}]$ *whenever*

$$0 \leq \tilde{\alpha} < \min\{1/(p-1), \alpha_M\},$$

for a constant c *depending only on* $n, p, \nu, L, \tilde{\alpha}$. *The exponent* α_M *has been introduced in* (34)–(35).

6 Equations with coefficients

Getting nonlinear potential estimates catching regularity beyond Theorem 6 for equations as in (17), necessitates to assume more regularity on the partial map $x \mapsto a(x, \cdot)$. Keep in mind the discussion before Theorem 6. We therefore define the following averaged vector field:

$$(a)_{x,r}(z) = \fint_{B(x,r)} a(\tilde{x}, z)\, d\tilde{x}, \quad \text{for} \quad z \in \mathbb{R}^n, \tag{45}$$

whenever $B(x, r) \subset \Omega$, and then the averaged (and renormalized) modulus of continuity of $x \mapsto a(x, \cdot)$ as follows:

$$\omega(r) := \left[\sup_{z\in\mathbb{R}^n, B(x,r)\subset\Omega} \fint_{B(x,r)} \left(\frac{|a(\tilde{x}, z) - (a)_{x,r}(z)|}{(|z| + s)^{p-1}}\right)^2 d\tilde{x}\right]^{1/2}. \tag{46}$$

When considering equations of the type

$$-\operatorname{div}(c(x)\tilde{a}(Du)) = \mu \tag{47}$$

the definition in (46) gives something which is comparable to the usual modulus of continuity of the function $c(\cdot)$. We then have

Theorem 13 *Let* $u \in W^{1,p}(\Omega)$ *be a weak solution to* (17), *with assumptions* (16) *(uniformly) verified by the partial map* $z \to a(\cdot, z)$ *for every* $x \in \Omega$. *Then*

- [40] *For every* $\tilde{\alpha} \in (0, 1)$ *there exists a number* $\delta \in (0, 1)$, *depending only on* $n, p, \nu, L, \tilde{\alpha}$, *such that*

$$\limsup_{\varrho \to 0} \omega(\varrho) \leq \delta \tag{48}$$

implies the validity of estimate (32) *whenever* $\alpha \in [0, \alpha_m)$. *In particular, if the limit in* (48) *is zero estimate* (32) *holds for every* $\alpha < 1$

- [6] *If the partial map* $x \to a(x, \cdot)$ *is Dini-continuous, i.e.*

$$\int_0 \omega(\varrho) \frac{d\varrho}{\varrho} < \infty, \tag{49}$$

then estimates (33) *and* (40) *continue to hold, for* $\alpha \in [\tilde{\alpha}, 1]$ *and* $\alpha \in [0, 1]$, *respectively*

- [40] *If the partial map* $x \to a(x, \cdot)$ *is Hölder continuous with exponent* $\tilde{\alpha} < \min\{1/(p-1), \alpha_M\}$ *in the sense that*

$$\sup_r \int_0^r \frac{[\omega(\varrho)]^{2/p}}{\varrho^{\tilde{\alpha}}} \frac{d\varrho}{\varrho} < \infty, \tag{50}$$

then estimate (36) *holds whenever* $\alpha < \tilde{\alpha}$.

In all three cases the constants involved in the estimates additionally depend on the quantities appearing in (48)–(50).

Let us now briefly comment on the assumptions made in the last theorem, remarking that all of them are essentially necessary. Assumption (48) allows to catch the case in which, when referring to Eq. (47), the function $c(\cdot)$ is not even continuous, but of class VMO or BMO with small seminorm. This reconnects the theory presented here to the classical Calderón–Zygmund theory, where VMO-regularity of coefficients is required to establish integrability estimates related to estimate (32); see for instance [9]. Assumption (149) is necessary too. Indeed, estimate (33) implies gradient boundedness when μ is good enough, while Dini continuity of coefficients is known to be necessary to get Lipschitz regularity of solutions; see [33]. The proof of the second statement in the last theorem has the techniques developed for Theorem 10 as starting point, and will appear in the forthcoming paper [6]. Finally, assumption (50) seems to be natural in view of the usual Schauder theory. This prescribes that, in order to have Hölder continuity of the gradient, the Hölder continuity of $x \mapsto a(x, \cdot)$ is necessary.

7 Interlude on measure data problems

When looking at equations as in (17) we have considered distributional solutions lying in $W^{1,p}$. This choice is actually aimed at simplifying the presentation in a way that emphasises the results obtained in the form of a priori estimates. Dealing with $W^{1,p}$-solutions is anyway very much restrictive in view of the fact that, typically, distributional solutions to measure data problems do not enjoy this regularity. A typical instance is the so called *nonlinear Green's function*

$$G_p(x) \approx \begin{cases} \left(|x|^{\frac{p-n}{p-1}} - 1\right) & \text{if } 1 < p \neq n \\ \log\, |x| & \text{if } p = n \end{cases} \tag{51}$$

which solves, in the sense of distributions, the problem

$$\begin{cases} -\Delta_p u = \delta & \text{in } B(0,1) \\ u = 0 & \text{on } \partial B(0,1). \end{cases} \tag{52}$$

In the right hand side of $(52)_1$ there appears the Dirac measure δ charging the origin. A straightforward calculation reveals that $G_p \notin W^{1,p}$ while it holds that

$$|DG_p|^{p-1} \in \mathcal{M}^{n/(n-1)} \Longleftrightarrow |\{|DG_p|^{p-1} > \lambda\}| \lesssim \lambda^{-n/(n-1)}. \tag{53}$$

We recall here that the Marcinkiewicz space $\mathcal{M}^q(\Omega)$, for $1 \leq q < \infty$, is the set of all measurable maps μ satisfying the condition

$$\sup_{\lambda \geq 0} \lambda^q |\{x : |\mu|(x) > \lambda\}| < \infty. \tag{54}$$

The above discussion now motivates the following:

Definition 7 (*Very weak solutions*) A function $u \in W^{1,1}_{\text{loc}}(\Omega)$ is called a very weak solution u to the Eq. (17) in Ω if $a(x, Du) \in L^1_{\text{loc}}(\Omega, \mathbb{R}^n)$ and (19) holds for every $\varphi \in C^\infty_c(\Omega)$.

The terminology "very weak" is used to emphasize the fact that such solutions do not belong, in general, to the natural Sobolev space $W^{1,p}(\Omega)$, i.e. they are not energy solutions. Note that, when considering assumptions (18), in order to guarantee that $a(x, Du) \in L^1(\Omega, \mathbb{R}^n)$ it suffices, for instance, to have $Du \in W^{1,p-1}(\Omega)$. Definition 7, not surprisingly, poses problems. For instance, very weak solutions may exist beside usual energy solutions [61], and are therefore not in general unique. As a matter of fact one of the basic open issues of the theory of measure data problems is to find a function class where to solve in a unique way Dirichlet problems of the type

$$\begin{cases} -\text{div}\, a(x, Du) = \mu & \text{in } \Omega \\ u = 0 & \text{on } \partial\Omega, \end{cases} \tag{55}$$

with μ being in the most general case a (signed) Borel measure with finite total mass. We will not pursue this matter here, rather referring to [14,36] for a more comprehensive discussion. Here we are interested in a class of solutions which we regard to be very natural (see in fact the equivalence results obtained in [36]) and for which all the potential estimates described in this paper continue to hold. This is the class of SOLA (Solutions Obtained by Limits of Approximations), which is introduced in the

Definition 8 (*SOLA*) A function $u \in W^{1,1}_{\text{loc}}(\Omega)$ is a SOLA to the Eq. (17) iff is a very weak solution and there exists a sequence of local energy solutions $\{u_k\} \subset W^{1,p}_{\text{loc}}(\Omega)$ to the equations

$$-\text{div}\, a(x, Du_k) = \mu_k,$$

such that $u_k \to u$ locally in $W^{1,p-1}(\Omega)$.

SOLA are special because they are selected via an approximation procedure using more regular energy solutions, and thereby they inherit a few of their basic properties. For instance: their precise representative is defined out of a null p-capacity set, exactly as $W^{1,p}$-functions; see Theorem 16 below. Let us recall how to build a SOLA to (55), following [8]. One considers solutions $u_k \in W^{1,p}_0(\Omega)$ to regularised problems

$$\begin{cases} -\text{div}\, a(x, Du_k) = \mu_k & \text{in } \Omega \\ u_k = 0 & \text{on } \partial\Omega. \end{cases}$$

The right hand sides data $\mu_k := \mu * \phi_k \in C^\infty$ are canonically obtained by smoothing μ via convolution with a sequence of smooth standard, smooth mollifiers $\{\phi_k\}$. This approach leads to the existence of a very weak solution $u \in W^{1,p-1}_0(\Omega)$ such that $u_k \to u$ in $W^{1,p-1}_0(\Omega)$ (up to a not relabeled subsequence). SOLA are not known to be unique, except in a few special cases (for instance when $\mu \in L^1$). See [8,13,14,36] for a larger discussion. We summarise the basic existence and regularity results available for SOLA in the following:

Theorem 14 [8,55] *Under the assumptions* (18) *with* $p \leq n$, *there exists a SOLA* $u \in W^{1,p-1}_0(\Omega)$ *to* (55). *Moreover, every SOLA* $u \in W^{1,1}_{\text{loc}}(\Omega)$ *to* (17) *is such that*

$$u \in W^{1,q}_{\text{loc}}(\Omega) \quad \textit{for every} \quad q < n(p-1)/(n-1). \tag{56}$$

In the case of equations of the type (13) *under assumptions* (16), *we also have*

$$Du \in W^{(1-\varepsilon)/(p-1),p-1}_{\text{loc}}(\Omega) \quad \textit{for every} \quad \varepsilon \in (0,1). \tag{57}$$

A preliminary integrability result was obtained for a different kind of solutions in the pioneering paper of Lindqvist [48]. The restriction $p \leq n$ is aimed at focusing on the case of SOLA, since when $p > n$ then μ belongs to the dual of $W^{1,p}$ and it is possible to consider standard energy solutions. In display (57) a fractional Sobolev space appears. The meaning is actually that

$$\int_A \int_A \frac{|Du(x) - Du(y)|^{p-1}}{|x-y|^{n+1-\varepsilon}}\, dx\, dy < \infty \qquad \text{for every } \varepsilon \in (0,1)$$

holds whenever $A \Subset \Omega$. We also remark that, in fact, it can be proved that every SOLA u is such that

$$|Du|^{p-1} \in \mathcal{M}^{n/(n-1)}(\Omega). \tag{58}$$

This means that every SOLA exhibits exactly the same integrability of the nonlinear Green's function G_p described in (53) (see [7,20,55]). Both the results in (56) and (57) are sharp, as follows again considering G_p, which can be in fact proved to be the only SOLA to the problem in (52). This last fact follows combining the results in [35,62]. As mentioned above, by mean of approximations arguments, *all the potential estimate in this paper continue to hold for any SOLA* to (55). More details are in Sect. 14.4 below, where we in particular discuss the possibility of characterizing the Lebesgue points of SOLA via linear and nonlinear potentials.

7.1 Potential estimates and fundamental solutions

To check to which extent estimates (20) and (24) replace the usual linear representation formulas in the nonlinear case, the best thing to do is to see how they reproduce the behaviour of the nonlinear fundamental solution G_p. We already know the optimality of estimate (20) from Theorem 2. We therefore concentrate on (24) and show that it actually reverses when considering the nonlinear Green's function G_p. This means that for $x \in B(0,1)\backslash\{0\}$ we have

$$\fint_{B(x,R)} |DG_p|\,d\tilde{x} + \left[\mathbf{I}_1^{\delta}(x,R)\right]^{1/(p-1)} \le c|DG_p(x)|, \tag{59}$$

with $R = 2|x|$, and for a constant c depending only on n and p. Indeed by coarea formula we have

$$\begin{aligned}
\fint_{B(x,R)} |DG_p|\,d\tilde{x} &\le c \fint_{B(0,2R)} |DG_p|\,d\tilde{x} \\
&= \frac{c(n,p)}{R^n} \int_0^{2R} \int_{\partial B_\varrho} |DG_p(y)|\,d\mathcal{H}^{n-1}(y)\,d\varrho \\
&\le \frac{c(n,p)}{R^n} \int_0^{2R} \varrho^{n+(1-n)/(p-1)} \frac{d\varrho}{\varrho} \\
&\le c(n,p)R^{(1-n)/(p-1)} \le c(n,p)|DG_p(x)|.
\end{aligned}$$

On the other hand, as $\delta(B(x,\varrho)) = 0$ for $\varrho \le R/2$ and $\delta(B(x,\varrho)) = 1$ otherwise, we also have

$$\mathbf{I}_1^{\delta}(x,R) = \int_{R/2}^{R} \frac{d\varrho}{\varrho^n} = \frac{2^n-1}{n-1} R^{1-n} \le c(n,p)|DG_p(x)|^{p-1},$$

so that (59) follows combining the last two inequalities.

8 Fine properties of solutions via potentials

We have seen that linear and nonlinear potentials locally control the behaviour of solutions. It is at this point not surprising to discover that potentials also control their so called fine properties. In this respect we present two theorems. The former is concerned with the pointwise behaviour of gradients of SOLA, and in particular with their Lebesgue points. This result employs Riesz potentials. The latter in instead concerned with Lebesgue points of solutions and uses Wolff potentials.

Theorem 15 *Theorem* 3 *continues to hold whenever* $u \in W^{1,p-1}(\Omega)$ *is a SOLA to* (13). *Moreover, the condition*

$$\mathbf{I}_1^{\mu}(x, R) < \infty \tag{60}$$

implies that the following limit exists and therefore defines the precise representative of Du *at the point* x*:*

$$\lim_{\varrho \to 0} (Du)_{B(x,\varrho)} =: Du(x). \tag{61}$$

The proof of this last theorem is included in Sect. 17 below. The analogous statement for solutions is the following:

Theorem 16 *Theorem* 1 *continues to hold whenever* $u \in W^{1,p-1}(\Omega)$ *is a SOLA to* (17). *Moreover, the condition*

$$\mathbf{W}_{1,p}^{\mu}(x, R) < \infty \tag{62}$$

implies that the following limit exists and therefore defines the precise representative of u *at the point* x*:*

$$\lim_{\varrho \to 0} (u)_{B(x,\varrho)} =: u(x). \tag{63}$$

In particular, the set of non-Lebesgue points of u *has* p*-capacity zero.*

We notice that the assertion about the p-capacity follows directly from the fact that set of points where the Wolff potential $\mathbf{W}_{1,p}^{\mu}$ blows-up has zero p-capacity (see for instance [30]).

Remark 2 (Hausdorff dimension of singular sets of SOLA) Theorem 16 allows to define SOLA—which are initially defined only almost everywhere via convergence—outside a singular set (i.e. set of non-Lebesgue points) of Hausdorff dimension not larger than $n - p$, when $p \leq n$. This establishes a connection with another class of solutions to measure data problem. These are called p-superharmonic functions and are defined in the case the measure μ is nonnegative; see [30,38]. For such solutions every point is a Lebesgue point, by construction. The connection is now given by the fact that, in view of [36], every SOLA has a superharmonic representative (that is, they coincide almost everywhere) whenever the measure is nonnegative. In the case of Theorem 15 we instead can conclude that the Hausdorff dimension of the singular set of Du is not larger than $n-1$. The last estimate follows on the other hand also by (57). Indeed, the Hausdorff dimension of the set of non-Lebesgue points of a general $W^{s,\gamma}$-map, with $s\gamma < n$, has Hausdorff dimension not larger that $n - s\gamma$; see [54].

Remark 3 (Analogies with linear potential theory) Theorems 15 and 16 can be also considered as analogs of classical facts in linear potential theory, concerning the pointwise behaviour of solutions to the Poisson equation $-\Delta u = \mu$. As usual, in the linear case they follow from explicit representation formulas and abstract analysis of potentials. Here they are replaced by nonlinear potential estimates. Indeed, define

$$f(x) := (G_\alpha * g)(x) := \int_{\mathbb{R}^n} G_\alpha(x-y)g(y)\,dy,$$

where $g \in L^q(\mathbb{R}^n)$ with $q > 1$. Here G_α denotes the α-Bessel potential, that is defined via

$$G_\alpha := \mathcal{F}^{-1}\left((1+|\xi|^2)^{-\alpha/2}\right),$$

where $\mathcal{F}$ denotes the Fourier transform and $\alpha \in (0, n)$. Then the limit

$$\lim_{\varrho \to 0} (f)_{B(x,\varrho)} =: f(x)$$

exists—and thereby defines the precise representative of f—up to a set of zero (α, q)-capacity. For these facts we refer to [3], and in particular to [3, Proposition 6.1.3]

9 Regularity and corollaries

The estimates presented up to now allow to give a comprehensive and unified picture of the regularity results available for quasilinear equations. We will now briefly describe a few consequences of such estimates. First of all, let us recall that combining Theorem 11 and Proposition 1 yields the following pointwise inequality:

$$\begin{aligned} |u(x)-u(y)| &\le c\left[M_{p-\alpha(p-1),R}(\mu)(x) + M_{p-\alpha(p-1),R}(\mu)(y)\right]^{1/(p-1)}|x-y|^\alpha \\ &\quad + c \fint_{B_R} (|u|+Rs)\,d\tilde{x} \cdot \left(\frac{|x-y|}{R}\right)^\alpha, \end{aligned} \tag{64}$$

that holds whenever $x, y \in B_{R/4}$, $B_R \subset \Omega$, $\alpha \in (0,1)$, where $c \equiv c(n, p, \nu, L, \alpha)$. Notice here that the blow-up of c takes place as soon as α approaches 0 or 1. In a completely similar way, by Theorem 12 and again Proposition 1, we then have that the inequality

$$|Du(x) - Du(y)| \leq c\left[M_{1-\alpha(p-1),R}(\mu)(x) + M_{1-\alpha(p-1),R}(\mu)(y)\right]^{1/(p-1)} |x-y|^{\alpha}$$

$$+ c \fint_{B_R} (|Du| + s)\, d\tilde{x} \cdot \left(\frac{|x-y|}{R}\right)^{\alpha} \tag{65}$$

holds with x, y, r, B_R and c as above, but for $\alpha \in (0, \alpha_M)$. This time the constant c blows-up when $\alpha \to 0$ or $\alpha \to \alpha_M$. We are now interested in seeing how the results presented up to now, including estimates (64)–(65), imply regularity of solutions in various relevant function spaces. Let us briefly recall that the Morrey space $L^{q,\theta}(\Omega)$ for $q \geq 1$ and $0 \leq \theta \leq n$, is defined by requiring that its member $f \in L^q(\Omega)$ satisfies

$$\sup_{B_\varrho \subset \Omega} \varrho^{\theta - n} \int_{B_\varrho} |f|^q \, d\tilde{x} < \infty, \tag{66}$$

while their local version is defined in the usual way. We refer to [1,56] for more on Morrey spaces and their basic properties with related references. We now have the following, comprehensive:

Corollary 1 *Let $u \in W^{1,p}(\Omega)$ be a weak solution—or a SOLA—to the Eq.* (13) *under assumptions* (16). *Then, with all the spaces being meant in their local versions, we have that*

- (C1) *if $\alpha \in (0,1)$ and $q > n(p-1)/[n-p+\alpha(p-1)]$, then*

$$\mu \in L^{\frac{nq}{n(p-1)+[p-\alpha(p-1)]q}} \Longrightarrow u \in C^{\alpha}_{q}$$

(*estimate* (64))

- (C2) *if $p < n$, then*

$$\mu \in \mathcal{M}^{n/p} \Longrightarrow |\mu|(B_\varrho) \lesssim \varrho^{n-p} \Longrightarrow u \in BMO$$

and

$$p = n \Longrightarrow u \in BMO$$

(*Theorem* 11 *with* $\alpha = 0$)

- (C3) $\mathbf{W}^{\mu}_{1,p} \in L^{\infty} \Longrightarrow u \in L^{\infty}$ (*Theorem* 1)
- (C4) *if $p < n$, then $\mu \in L(n/p, 1/(p-1)) \Longrightarrow u \in C^0$* (*Theorem* 1)
- (C5) *if $\alpha \in (0,1)$ and $p \leq n$, then*

$$\mu \in \mathcal{M}^{\frac{n}{p-\alpha(p-1)}} \Longrightarrow |\mu|(B_\varrho) \lesssim \varrho^{n-p+\alpha(p-1)} \Longrightarrow u \in C^{0,\alpha}$$

(*estimate* (64))

- (C6) *if $p \leq n$, then $\mu \in \mathcal{M}_b \Longrightarrow Du \in \mathcal{M}^{n(p-1)/(n-1)}$* (*Theorem* 3)
- (C7) *if $1 < q < n$, then*

$$\mu \in L(q, \gamma) \Longrightarrow Du \in L\left(\frac{nq(p-1)}{n-q}, \gamma(p-1)\right)$$

(*Theorem* 3)

- (C8) *if* $1 < q < \theta$, *then* $|Du|^{p-1} \in L^{\theta q/(\theta-q),\theta}$ (*Theorem* 3)
- (C9) $\mu \in \mathcal{M}^n \Longrightarrow |\mu|(B_\varrho) \lesssim \varrho^{n-1} \Longrightarrow Du \in BMO$ (*Theorem* 12)
- (C10) $\mathbf{I}_1^\mu \in L^\infty \Longrightarrow Du \in L^\infty$ (*Theorem* 8)
- (C11) $\mu \in L(n,1) \Longrightarrow Du \in C^0$ (*Theorem* 3)
- (C12) *if* $0 < \alpha < \min\{1/(p-1), \alpha_M\}$, *then*

$$\mu \in \mathcal{M}^{\frac{n}{1-\alpha(p-1)}} \Longrightarrow |\mu|(B_\varrho) \lesssim \varrho^{n-1+\alpha(p-1)} \Longrightarrow Du \in C^{0,\alpha}$$

(*estimate* (65)).

Proof In order to prove (C1), we begin recalling the following mapping property of the fractional maximal operator:

$$M_{\beta,R} \colon L^t \to L^{nt/(n-\beta t)}, \tag{67}$$

that holds whenever $1 < t < n/\beta$ (see [26,56]). Then we go to estimate (64) and yet recall the definition of Calderón spaces given in Definition 2. This leads to determine

$$\frac{nt}{n-[p-\alpha(p-1)]t} = \frac{q}{p-1} \Longleftrightarrow t = \frac{nq}{n(p-1)+[p-\alpha(p-1)]q}$$

so that $\mu \in L^t$ and (67) with the above choice implies $[M_{p-\alpha(p-1),R}(\mu)]^{1/(p-1)} \in L^q$. At this point (C1) follows using this fact together with estimate (64). Notice that, recalling that we are assuming $p \geq 2$, we have used that

$$t = \frac{nq}{n(p-1)+[p-\alpha(p-1)]q} > 1 \Longleftrightarrow q > \frac{n(p-1)}{n-p+\alpha(p-1)} > 1.$$

We have also use the fact with the previous definitions we in fact have that $t[p-\alpha(p-1)] < n$.

The assertion in (C2) follows immediately by taking $\alpha = 0$ in Theorem 11 when $p = n$. We indeed recall that by the very definition of BMO spaces we have

$$M^{\#}_{0,R}(u) \equiv M^{\#}_R(u) \in L^\infty \Longleftrightarrow u \in \mathrm{BMO}. \tag{68}$$

In the case $2 \leq p < n$, we recall the following Hölder's type inequality valid in Marcinkiewicz spaces $\mathcal{M}^t$, $t > 1$:

$$\|f\|_{L^1(B_\varrho)} \leq \frac{t\varrho^{n(t-1)/t}}{t-1} \|f\|_{\mathcal{M}^t(B_\varrho)} \,. \tag{69}$$

This implies that in our case $|\mu|(B_\varrho) \lesssim \varrho^{n-p}$ holds and, by the definition of fractional maximal operator, we conclude that $M_{p,R}(\mu)$ is locally bounded. At this stage (C2) follows from estimate (44), where we again take $\alpha = 0$.

The statement in (C3) is simply an obvious consequence of Theorem 1.

As for (C4), a direct computation using certain characterisation of Lorentz spaces shows that if $\mu \in L(n/p, 1/(p-1))$ for $p < n$, then (21) holds (see for instance [23] for related computations). At this stage (C4) follows by Theorem 1.

In (C5) the first implication follows again by the Hölder type inequality in (69). On the other hand, we observe that the inequality $|\mu|(B_\varrho) \lesssim \varrho^{n-p+\alpha(p-1)}$ implies that $M_{p-\alpha(p-1),R}(\mu) \in L^\infty$, locally, by the very definition of fractional maximal operator in Definition 6, so that (C5) follows by estimate (64).

(C6) is just an obvious consequence of the following imbedding property of Riesz potentials (see [1]):

$$\mathbf{I}_1 : \mathcal{M}_b \to \mathcal{M}^{n/(n-1)}$$

and of Theorem 3.

Similarly, (C7) follows from Theorem 3 and the classical mapping property

$$\mathbf{I}_1 : L(q, \gamma) \to L(nq/(n-q), \gamma) \qquad \text{for } 1 < q < n \text{ and } \gamma > 0.$$

See also [26,56].

(C8) is essentially a corollary of Theorem 3 and of the mapping property of Riesz potentials

$$\mathbf{I}_1 : L^{q,\theta} \to L^{\theta q/(\theta-q),\theta},$$

originally proved by Adams [1] (see also [56] for a localization).

The first implication in (C9) is a consequence of (69) while the second follows applying Theorem 12 with $\alpha = 0$. Indeed, assuming that $|\mu|(B_\varrho) \lesssim \varrho^{n-1}$ implies that $M_1(\mu)$ is locally bounded, and therefore so is $M^{\#}_{0,R}(Du) \equiv M^{\#}_R(Du)$. By definition of sharp maximal function this implies that the gradient belongs to BMO, locally; see (68).

(C10) follows from Theorem 8 in an obvious way.

(C11) is again a consequence of (25) and of some basic computation involving the definition of Lorentz norm (see [22,41]). Indeed, the condition $\mu \in L(n, 1)$ allows to conclude that (25) holds, again by basic manipulations on Lorentz equivalent norms.

Finally, exactly as for the proof of (C5), (C12) follows again from (69) and estimate (65). □

Some of the points in the previous corollary are well known results when considering for instance the model equation (15). The theory above allows to extend and embed them in a more general context where results follow in a unified way. Specifically, for (C2) see [55,64], for (C6) see [7,8,20], for (C7) see [19,31,56], for (C8) see [56]. We also remark that the previous corollary is just a sample of what is possible to have using the nonlinear potential estimates approach; further spaces, as for instance Lorentz–Morrey or Besov–Morrey spaces are considerable as well (see [55,56] for relevant definitions). More rearrangement invariant function spaces regularity results via potentials are contained in [11].

10 A basic comparison estimate

This section is devoted to the proof of a comparison estimate between a considered solution $u \in W^{1,p}(\Omega)$ to (17) (notice that we are allowing for measurable coefficients here) and the function $v \in u + W^{1,p}_0(B_R)$ defined by solving the Dirichlet problem

$$\begin{cases} \operatorname{div} a(x, Dv) = 0 & \text{in } B_R \\ v = u & \text{on } \partial B_R. \end{cases} \tag{70}$$

Here $B_R \subset \Omega$ denotes a fixed ball with radius R. The monotonicity properties of the vector field $a(\cdot)$ described in (18) can be restated and managed more easily via the use of the auxiliary vector field $V : \mathbb{R}^n \to \mathbb{R}^n$ defined by

$$V(z) := (|z|^2 + s^2)^{(p-2)/4} z. \tag{71}$$

The number s is the one defined in (16) and (18). Indeed, we start observing that the following inequality:

$$\frac{|z_1 - z_2|}{c} \le \frac{|V(z_1) - V(z_2)|}{(|z_1|^2 + |z_2|^2 + s^2)^{(p-2)/4}} \le c|z_1 - z_2| \tag{72}$$

holds, and is valid for all matrixes $z_1, z_2 \in \mathbb{R}^n$ that are not simultaneously null (in the case $s = 0$) and for every $p > 1$. The constant c depends only on n and p. See [27,55] for basic properties and for a discussion on the use of the map $V(\cdot)$ in this context. Since we are assuming $p \ge 2$, in particular we have that

$$|z_1 - z_2|^{p/2} \le c|V(z_1) - V(z_2)|. \tag{73}$$

Combining $(18)_2$ and (72) yields

$$\frac{1}{c}|V(z_1) - V(z_2)|^2 \le \langle a(x, z_2) - a(x, z_1), z_2 - z_1 \rangle \tag{74}$$

again for every choice of $z_1, z_2, x \in \Omega$ and where $c \equiv c(n, p, \nu)$.

The lemmas in this section are already scattered in [21,41,55]. The proofs proposed here are anyway different and shorter. We start with a basic, weighted type energy estimate.

Lemma 1 *Let $u \in W^{1,p}(\Omega)$ be a solution to* (17) *under assumptions* (18), *and let $v \in u + W_0^{1,p}(B_R)$ be as in* (70). *Then*

$$\int_{B_R} \frac{|V(Du) - V(Dv)|^2}{(h + |u - v|)^{\xi}} \, d\tilde{x} \le c \frac{h^{1-\xi}}{\xi - 1} |\mu|(B_R) \tag{75}$$

holds whenever $h > 0$ and $\xi > 1$, where $c \equiv c(n, p, \nu) \ge 1$. The map $V(\cdot)$ has been defined in (71).

Proof The weak formulation

$$\int_{B_R} \langle a(\tilde{x}, Du) - a(\tilde{x}, Dv), D\varphi \rangle \, d\tilde{x} = \int_{B_R} \varphi \, d\mu,$$

is valid whenever $\varphi \in W_0^{1,p}(B) \cap L^{\infty}(B)$, and we use the test functions

$$\eta \equiv \eta_\pm := h^{1-\xi} - (h + (u - v)_\pm)^{1-\xi},$$

which in fact belong to $W_0^{1,p}(B) \cap L^\infty(B)$. We recall the standard notation

$$(u - v)_+ := \max\{u - v, 0\} \quad \text{and} \quad (u - v)_- := \max\{v - u, 0\}.$$

Then, notice that

$$\begin{aligned} |I_\pm| &:= \left| (\xi - 1) \int_{B_R} \frac{\langle a(\tilde{x}, Du) - a(\tilde{x}, Dv), D(u - v)_\pm \rangle}{(h + (u - v)_\pm)^\xi} \, d\tilde{x} \right| \\ &= \left| \int_{B_R} \langle a(\tilde{x}, Du) - a(\tilde{x}, Dv), D\eta_\pm \rangle \, d\tilde{x} \right| \\ &= \left| \int_{B_R} \eta_\pm \, d\mu \right| \le h^{1-\xi} |\mu|(B_R) \end{aligned}$$

and therefore, using (74), we conclude with

$$(\xi - 1) \int_{B_R} \frac{|V(Du) - V(Dv)|^2}{(h + |u - v|)^\xi} \, d\tilde{x} \le c(I_+ - I_-) \le c h^{1-\xi} |\mu|(B_R)$$

so that (75) follows. □

Lemma 1 and Sobolev embedding theorem in turn imply a first comparison estimate.

Lemma 2 *Let $u \in W^{1,p}(\Omega)$ be a solution to* (17) *under assumptions* (18), *and let $v \in u + W_0^{1,p}(B_R)$ be as in* (70). *Then*

$$\fint_{B_R} |Du - Dv|^q \, d\tilde{x} \le c_1 \left[\frac{|\mu|(B_R)}{R^{n-1}} \right]^{q/(p-1)} \tag{76}$$

and therefore

$$\fint_{B_R} |u - v|^q \, d\tilde{x} \le \tilde{c}_1 \left[\frac{|\mu|(B_R)}{R^{n-p}} \right]^{q/(p-1)} \tag{77}$$

hold for constants $c_1, \tilde{c}_1 \equiv c_1, \tilde{c}_1(n, p, \nu, q)$, whenever

$$0 < q < \min\left\{ p, \frac{n(p-1)}{n-1} \right\} =: p_m. \tag{78}$$

Proof We restrict ourselves to the proof of (76), since this obviously implies (77) via Poincaré's inequality. Moreover, we can restrict ourselves to prove (76) assuming $q > 1$, since the result for the remaining values would follow by Hölder's inequality. Now, let us observe that we can always reduce to the situation when there exists $\xi > 1$ such that

$$\xi q/(p-q) = q^*, \tag{79}$$

where q^* is the Sobolev conjugate of q. Indeed, when $p \leq n$ then $p_m = n(p-1)/(n-1)$ and therefore $q < p_m$ implies that $q/(p-q) < q^* = nq/(n-q)$, so that (79) follows for the corresponding choice of ξ. When instead $p > n$ then we first observe that we can prove (76) in the case $n \leq q < p$, since (76) for lower values of q would then follow by Hölder's inequality. Now, since by definition of Sobolev embedding exponent we can take q^* as large as we please, we again find a number $\xi > 1$ for which (79) holds. We then plan to apply Lemma 2 with this choice of ξ and with the following one of h:

$$h = R\left(\fint_{B_R} |Du - Dv|^q \, d\tilde{x}\right)^{1/q}.$$

Notice we can always assume that $h > 0$, otherwise the assertion of Lemma 2 trivializes. By using the definitions in the last two displays, inequalities (73) and (75), and finally Sobolev embedding theorem, we then have

$$\begin{aligned}
&\fint_{B_R} |Du - Dv|^q \, d\tilde{x} \\
&\quad = \fint_{B_R} \frac{|Du - Dv|^q}{(h + |u - v|)^{\xi q/p}} (h + |u - v|)^{\xi q/p} \, d\tilde{x} \\
&\quad \leq c \fint_{B_R} \frac{|V(Du) - V(Dv)|^{2q/p}}{(h + |u - v|)^{\xi q/p}} (h + |u - v|)^{\xi q/p} \, d\tilde{x} \\
&\quad \leq c \left(\fint_{B_R} \frac{|V(Du) - V(Dv)|^2}{(h + |u - v|)^{\xi}} \, d\tilde{x}\right)^{q/p} \left(\fint_{B_R} (h + |u - v|)^{\xi q/(p-q)} \, d\tilde{x}\right)^{(p-q)/p} \\
&\quad \leq c \left(\frac{h^{1-\xi}}{\xi - 1} \frac{|\mu|(B_R)}{R^n}\right)^{q/p} \left[h^{\xi q/p} + \left(\fint_{B_R} |u - v|^{\xi q/(p-q)} \, d\tilde{x}\right)^{(p-q)/p}\right] \\
&\quad \leq c \left(\frac{h^{1-\xi}}{\xi - 1} \frac{|\mu|(B_R)}{R^n}\right)^{q/p} \left[h^{\xi q/p} + R^{\xi q/p} \left(\fint_{B_R} |Du - Dv|^q \, d\tilde{x}\right)^{\xi/p}\right]
\end{aligned}$$

$$\leq ch^{q/p}\left[\frac{|\mu|(B_R)}{R^n}\right]^{q/p}$$

$$= c\left(\fint_{B_R} |Du - Dv|^q \, d\tilde{x}\right)^{1/p} \left[\frac{|\mu|(B_R)}{R^{n-1}}\right]^{q/p}$$

$$\leq \frac{1}{2}\fint_{B_R} |Du - Dv|^q \, d\tilde{x} + c\left[\frac{|\mu|(B_R)}{R^{n-1}}\right]^{q/(p-1)}$$

so that (76) follows. □

Remark 4 By examining the arguments of the previous lemma it is possible to see that the constant c_1 appearing in (76) shows the following natural asymptotic for $q \to p_m$:

$$c_1 = c(n, p, \nu)\left(\frac{1}{p_m - q}\right)^{q/(p-1)}, \tag{80}$$

where c is a constant that remains bounded whenever p varies in a compact subset of $(1, \infty)$; see (79) and the constant appearing in Sobolev–Morrey embedding theorem. The asymptotic in (80) is typical in situations when a borderline estimate in Lebesgue spaces fails, being replaced by an estimate in Marcinkiewicz spaces; recall (58).

11 A sequence of comparison estimates

Given a number $\delta_1 \in (0, 1/4)$ and a ball $B(x, r) \subset \Omega$, we define the sequence of shrinking balls

$$B_j := B(x, r_j), \qquad r_j = \delta_1^j r, \tag{81}$$

whenever $j \geq 0$ is an integer. The related comparison solutions $v_j \in u + W_0^{1,p}(B_j)$ are as in (70), that is

$$\begin{cases} \operatorname{div} a(x, Dv_j) = 0 & \text{in } B_j \\ v_j = u & \text{on } \partial B_j. \end{cases} \tag{82}$$

Then we have

Lemma 3 *Let $u \in W^{1,p}(\Omega)$ be a solution to* (17) *under assumptions* (18), *and let v_{j-1}, v_j be as in* (82) *with $j \geq 1$. Suppose further that*

$$\frac{|\mu|(B_{j-1})}{\lambda^{p-1} r_{j-1}^{n-1}} \leq 1 \tag{83}$$

and that the bounds

$$\frac{\lambda}{A} \leq |Dv_{j-1}| \leq A\lambda \quad \text{in } B_j \tag{84}$$

hold for constants $A \geq 1$ and $\lambda > 0$. Then there exists a constant c_2 depending only on n, p, ν, δ_1, A such that

$$\fint_{B_j} |Du - Dv_j|\, d\tilde{x} \leq c_2 \lambda^{2-p} \left[\frac{|\mu|(B_{j-1})}{r_{j-1}^{n-1}} \right]. \tag{85}$$

Proof We start fixing the following quantities:

$$\gamma := \frac{1}{4(p-1)(n+1)} \quad \text{and} \quad \xi := 1 + 2\gamma, \tag{86}$$

in view of the application of Lemmas 1 and 2, that we shall use with exponents q such that

$$0 < q \leq \xi(p-1) = p - 1 + \frac{1}{2(n+1)}.$$

This in particular fixes the constant c_1 from Lemma 2. We also set

$$\bar{v}_{j-1} := \frac{v_{j-1}}{\lambda}, \qquad \bar{v}_j := \frac{v_j}{\lambda}. \tag{87}$$

In the rest of the proof constants denoted by c will only depend on n, p, ν, δ_1, A and will in general vary from line to line, as usual. We start estimating the term on the left in (85) with the aid of (84) as follows:

$$\begin{aligned} \fint_{B_j} |Du - Dv_j|\, d\tilde{x} &\leq A^{(p-2)(1+\gamma)} \fint_{B_j} |D\bar{v}_{j-1}|^{(p-2)(1+\gamma)} |Du - Dv_j|\, d\tilde{x} \\ &\leq c \fint_{B_j} |D\bar{v}_j - D\bar{v}_{j-1}|^{(p-2)(1+\gamma)} |Du - Dv_j|\, d\tilde{x} \\ &\quad + c \fint_{B_j} |D\bar{v}_j|^{(p-2)(1+\gamma)} |Du - Dv_j|\, d\tilde{x}. \end{aligned} \tag{88}$$

To continue, we estimate the second-last integral appearing in the previous display. For this let us preliminary note that, by using (76) and recalling that $r_j = \delta_1 r_{j-1}$, the following estimate holds indeed for the same range of exponents q working in (76):

$$\begin{aligned} &\fint_{B_j} |D\bar{v}_j - D\bar{v}_{j-1}|^q\, d\tilde{x} \\ &\quad \leq c\lambda^{-q} \fint_{B_j} |Dv_j - Du|^q\, d\tilde{x} + c\lambda^{-q} \frac{|B_{j-1}|}{|B_j|} \fint_{B_{j-1}} |Dv_{j-1} - Du|^q\, d\tilde{x} \end{aligned}$$

$$\le c\lambda^{-q}\left[\frac{|\mu|(B_j)}{r_j^{n-1}}\right]^{q/(p-1)} + c\lambda^{-q}\left[\frac{|\mu|(B_{j-1})}{r_{j-1}^{n-1}}\right]^{q/(p-1)}$$

$$\le c\left[\frac{|\mu|(B_{j-1})}{\lambda^{p-1}r_{j-1}^{n-1}}\right]^{q/(p-1)}. \tag{89}$$

Then, appealing to Hölder's inequality together with (76), gives us

$$\fint_{B_j} |D\bar{v}_j - D\bar{v}_{j-1}|^{(p-2)(1+\gamma)}|Du - Dv_j|\,d\tilde{x}$$

$$\le \left(\fint_{B_j} |D\bar{v}_j - D\bar{v}_{j-1}|^{(p-1)(1+\gamma)}\,d\tilde{x}\right)^{(p-2)/(p-1)}$$

$$\cdot \left(\fint_{B_j} |Du - Dv_j|^{p-1}\,d\tilde{x}\right)^{1/(p-1)}$$

$$\le c\lambda\left[\frac{|\mu|(B_{j-1})}{\lambda^{p-1}r_{j-1}^{n-1}}\right]^{[(1+\gamma)(p-2)+1]/(p-1)}. \tag{90}$$

But now, as

$$\frac{(1+\gamma)(p-2)+1}{p-1} > 1 \tag{91}$$

then (83) implies

$$\fint_{B_j} |D\bar{v}_j - D\bar{v}_{j-1}|^{(p-2)(1+\gamma)}|Du - Dv_j|\,d\tilde{x} \le c\lambda^{2-p}\left[\frac{|\mu|(B_{j-1})}{r_{j-1}^{n-1}}\right]$$

with $c \equiv c(n, p, \nu, \delta_1)$. Combining this last estimate with (88) gives us

$$\fint_{B_j} |Du - Dv_j|\,d\tilde{x}$$

$$\le c\fint_{B_j} |D\bar{v}_j|^{(p-2)(1+\gamma)}|Du - Dv_j|\,d\tilde{x} + c\lambda^{2-p}\left[\frac{|\mu|(B_{j-1})}{r_{j-1}^{n-1}}\right]. \tag{92}$$

It remains to estimate the first term on the right hand side in the above display. Applying Hölder's inequality, together with (72) and (75), and recalling that $\xi = 1 + 2\gamma$, we obtain for any $h > 0$ that

$$
\begin{aligned}
&\fint_{B_j} |D\bar{v}_j|^{(p-2)(1+\gamma)}|Du - Dv_j|\,d\tilde{x} \\
&\leq c \fint_{B_j} \left[\lambda^{2-p}\frac{(|Dv_j| + |Du|)|^{p-2}|Du - Dv_j|}{(h + |u - v_j|)^{\xi}}\right]^{1/2} \\
&\quad \cdot \left[|D\bar{v}_j|^{(p-2)(1+2\gamma)}(h + |u - v_j|)^{\xi}\right]^{1/2} d\tilde{x} \\
&\leq c \fint_{B_j} \left[\lambda^{2-p}\frac{|V(Du) - V(Dv_j)|^2}{(h + |u - v_j|)^{\xi}}\right]^{1/2} \left[|D\bar{v}_j|^{(p-2)\xi}(h + |u - v_j|)^{\xi}\right]^{1/2} d\tilde{x} \\
&\leq c \left(\lambda^{2-p} \fint_{B_j} \frac{|V(Du) - V(Dv_j)|^2}{(h + |u - v_j|)^{\xi}}\,d\tilde{x}\right)^{1/2} \\
&\quad \cdot \left(\fint_{B_j} |D\bar{v}_j|^{(p-2)\xi}(h + |u - v_j|)^{\xi}\,d\tilde{x}\right)^{1/2} \\
&\leq c\sqrt{\lambda^{2-p}}\left[h^{1-\xi}\frac{|\mu|(B_j)}{r_j^n}\right]^{1/2} \left(\fint_{B_j} |D\bar{v}_j|^{(p-2)\xi}(h + |u - v_j|)^{\xi}\,d\tilde{x}\right)^{1/2}.
\end{aligned}
\tag{93}
$$

Now we choose

$$
h := \left(\fint_{B_j} |D\bar{v}_j|^{(p-2)\xi}|u - v_j|^{\xi}\,d\tilde{x}\right)^{1/\xi} + \delta \tag{94}
$$

for some small positive $\delta \in (0, 1)$ to get

$$
\begin{aligned}
&\left(\fint_{B_j} |D\bar{v}_j|^{(p-2)\xi}(h+|u - v_j|)^{\xi}\,d\tilde{x}\right)^{1/2} \\
&\leq 2h^{\xi/2}\left(\fint_{B_j} |D\bar{v}_j|^{(p-2)\xi} d\tilde{x}\right)^{1/2} + 2h^{\xi/2}.
\end{aligned}
$$

Notice that the presence in (94) of the parameter δ is aimed at guaranteeing that h is positive; we shall eventually let $\delta \to 0$ at the end of the proof. In the above display, we use (89) and (84) to gain

$$\fint_{B_j} |D\bar{v}_j|^{(p-2)\xi} d\tilde{x} \leq c \fint_{B_j} |D\bar{v}_{j-1} - D\bar{v}_j|^{(p-2)\xi} d\tilde{x} + c \fint_{B_j} |D\bar{v}_{j-1}|^{(p-2)\xi} d\tilde{x}$$
$$\leq c \left[\frac{|\mu|(B_{j-1})}{\lambda^{p-1} r_{j-1}^{n-1}} \right]^{(p-2)\xi/(p-1)} + cA^{(p-2)\xi} \leq c,$$

so that we conclude with

$$\left(\fint_{B_j} |D\bar{v}_j|^{(p-2)\xi} (h + |u - v_j|)^{\xi} \, d\tilde{x} \right)^{1/2} \leq ch^{\xi/2},$$

where the constant c ultimately depends only on n, p, ν, δ_1, A. Plugging the last estimate in (93) now yields

$$\fint_{B_j} |D\bar{v}_j|^{(p-2)(1+\gamma)} |Du - Dv_j| \, d\tilde{x} \leq c \sqrt{\frac{h}{r_j}} \left[\frac{|\mu|(B_j)}{\lambda^{p-2} r_j^{n-1}} \right]^{1/2}$$

while applying Young's inequality and recalling again that $r_j = \delta_1 r_{j-1}$ gives

$$\fint_{B_j} |D\bar{v}_j|^{(p-2)(1+\gamma)} |Du - Dv_j| \, d\tilde{x} \leq \frac{c\lambda^{2-p}}{\varepsilon} \left[\frac{|\mu|(B_{j-1})}{r_{j-1}^{n-1}} \right] + \frac{\varepsilon h}{r_{j-1}} \tag{95}$$

whenever $\varepsilon \in (0, 1)$, where $c \equiv c(n, p, \nu, \delta_1, A)$ is in particular independent of ε. It finally remains to estimate h, which has been defined in (94). We have

$$h \leq c \left(\fint_{B_j} |D\bar{v}_{j-1} - D\bar{v}_j|^{(p-2)\xi} |u - v_j|^{\xi} \, d\tilde{x} \right)^{1/\xi}$$
$$+ c \left(\fint_{B_j} |D\bar{v}_{j-1}|^{(p-2)\xi} |u - v_j|^{\xi} \, d\tilde{x} \right)^{1/\xi} + \delta =: I_1 + I_2 + \delta. \tag{96}$$

Using (76) and (89) to estimate I_1 leads to

$$
\begin{aligned}
I_1 &\leq c\left(\fint_{B_j} |D\bar{v}_{j-1} - D\bar{v}_j|^{\xi(p-1)}\, d\tilde{x}\right)^{(p-2)/[\xi(p-1)]} \\
&\quad \cdot \left(\fint_{B_j} |u - v_j|^{\xi(p-1)}\, d\tilde{x}\right)^{1/[\xi(p-1)]} \\
&\leq c r_j \left[\frac{|\mu|(B_{j-1})}{\lambda^{p-1} r_{j-1}^{n-1}}\right]^{(p-2)/(p-1)} \left(\fint_{B_j} |Du - Dv_j|^{\xi(p-1)}\, d\tilde{x}\right)^{1/[\xi(p-1)]} \\
&\leq c r_{j-1} \lambda^{2-p} \left[\frac{|\mu|(B_{j-1})}{r_{j-1}^{n-1}}\right]. \qquad (97)
\end{aligned}
$$

In turn we bound I_2 appealing to (76) and using (84) repeatedly as follows:

$$
\begin{aligned}
I_2 &\leq c A^{p-2} \left(\fint_{B_j} |u - v_j|^{\xi}\, d\tilde{x}\right)^{1/\xi} \\
&\leq c \left(\fint_{B_j} |u - v_j|^{n/(n-1)}\, d\tilde{x}\right)^{(n-1)/n} \\
&\leq c r_j \fint_{B_j} |Du - Dv_j|\, d\tilde{x} \\
&\leq c A^{(p-2)\xi} r_{j-1} \fint_{B_j} |D\bar{v}_{j-1}|^{(p-2)\xi} |Du - Dv_j|\, d\tilde{x} \\
&\leq c r_{j-1} \fint_{B_j} |D\bar{v}_j|^{(p-2)\xi} |Du - Dv_j|\, d\tilde{x} \\
&\quad + c r_{j-1} \fint_{B_j} |D\bar{v}_{j-1} - D\bar{v}_j|^{(p-2)\xi} |Du - Dv_j|\, d\tilde{x}.
\end{aligned}
$$

Similarly to (90) we find

$$
\fint_{B_j} |D\bar{v}_j - D\bar{v}_{j-1}|^{(p-2)\xi} |Du - Dv_j|\, d\tilde{x} \leq c\lambda^{2-p} \left[\frac{|\mu|(B_{j-1})}{r_{j-1}^{n-1}}\right]
$$

so that

$$I_2 \leq c r_{j-1} \fint_{B_j} |D\bar{v}_j|^{(p-2)\xi} |Du - Dv_j|\, d\tilde{x} + c r_{j-1} \lambda^{2-p} \left[\frac{|\mu|(B_{j-1})}{r_{j-1}^{n-1}} \right].$$

Combining this last estimate with (97) and eventually with (96) yields

$$\frac{h}{r_{j-1}} \leq c_* \fint_{B_j} |D\bar{v}_j|^{(p-2)\xi} |Du - Dv_j|\, d\tilde{x} + c_* \lambda^{2-p} \left[\frac{|\mu|(B_{j-1})}{r_{j-1}^{n-1}} \right] + \frac{\delta}{r_{j-1}},$$

where c_* depends only on $n, p, \nu, L, A, \delta_1$. Plugging the inequality in the above display in (95), choosing $\varepsilon = 1/(2c_*)$ and reabsorbing terms leads to

$$\fint_{B_j} |D\bar{v}_j|^{(p-2)\xi} |Du - Dv_j|\, d\tilde{x} \leq c \lambda^{2-p} \left[\frac{|\mu|(B_{j-1})}{r_{j-1}^{n-1}} \right] + \frac{c\delta}{r_{j-1}}.$$

Now (85) follows using this last inequality in combination with the one in display (92), and finally letting $\delta \to 0$. □

12 Proof of Theorems 8 and 10

The proof of Theorem 10 falls in six steps, going through Sects. 12.1–12.6 respectively. It requires the use of several different tools, and ultimately relies on a delicate iteration technique. Finally, in Sect. 12.7, we shall briefly show how to get Theorem 8 from Theorem 10.

In the following, given a ball $B \subset \Omega$, we define the excess of a vector field $f \in L^1(B, \mathbb{R}^k)$ over B as

$$E(f, B) := \fint_B |f - (f)_B|\, d\tilde{x}.$$

This functional, roughly speaking, provides an integral measure of the oscillations of the vector field f in the ball B. An elementary property of the excess is given by the following inequality:

$$\fint_B |f - (f)_B|\, d\tilde{x} \leq 2 \fint_B |f - \gamma|\, d\tilde{x} \quad \text{for every } \gamma \in \mathbb{R}^n. \tag{98}$$

We remark that here and in the following we shall use the comparison estimates presented in Sects. 10 and 11 that in fact applies in particular to the case of the solutions considered in Theorem 10.

12.1 A density property of a-harmonic functions

In this section we shall consider a solution $v \in W^{1,p}_{\mathrm{loc}}(B) \cap W^{1,1}(B)$ to the equation

$$\operatorname{div} a(Dv) = 0 \qquad \text{in } B, \tag{99}$$

where $B \subset \mathbb{R}^n$ is a given ball. We here restate in a suitable form the basic regularity properties a priori estimates for solutions to equation in (99); proofs can be for instance found in [18,21,51,55]. The first is the classical gradient $L^\infty - L^1$ bound

$$\sup_{\gamma B} |Dv| + s \leq \frac{c_l}{(1-\gamma)^n} \fint_B (|Dv| + s)\, d\tilde{x}, \tag{100}$$

which is valid whenever $\gamma \in (0,1)$ and for constant c_l depending only on n, p, ν, L. Next, we restate in suitable way the gradient Hölder continuity estimate included in display (35); more precisely, we select $\beta \equiv \beta(n, p, \nu, L) \in (0, \alpha_M)$ such that

$$|Dv(x_1) - Dv(x_2)| \leq c_h \sigma^\beta \fint_B (|Dv| + s)\, d\tilde{x} \tag{101}$$

holds whenever $\sigma B \subset B$ is a ball concentric to B with $\sigma \in (0, 1/2)$, such that $x_1, x_2 \in \sigma B$, where again $c_h \geq 1$ depends only on n, p, ν, L. Based on the previous regularity estimates, we can prove the following result, that roughly tells that something that happens in average, happens actually everywhere, and in a precisely quantifiable way.

Proposition 2 (Density improvement) *Let $v \in W^{1,p}_{\mathrm{loc}}(B) \cap W^{1,1}(B)$ be a solution to* (99) *such that*

$$\frac{\lambda}{\Gamma} \leq \fint_{\sigma^k B} |Dv|\, d\tilde{x} \quad \text{and} \quad \sup_{B/2} (|Dv| + s) \leq \Gamma\lambda \tag{102}$$

hold for some integer $k \geq 1$ and for numbers $\Gamma \geq 1$, $\lambda > 0$ and $\sigma \in (0, 1/4)$ such that

$$0 < \sigma \leq \left(\frac{1}{4 c_h \Gamma^2}\right)^{1/\beta}, \tag{103}$$

where β and c_h are the constants appearing in (101). *Then*

$$\frac{\lambda}{4\Gamma} \leq |Dv| \qquad \text{holds in } \sigma B.$$

Proof The first inequality in (102) implies that there exists a point $x_0 \in \sigma^k B$ such that

$$|Dv(x_0)| > \frac{\lambda}{2\Gamma}.$$

On the other hand, as (101) holds, the second inequality in (102) gives

$$|Dv(x) - Dv(x_0)| \leq c_h \Gamma \lambda \sigma^\beta$$

whenever $x \in \sigma B$, the last two inequalities and (103) then give

$$|Dv(x)| \geq |Dv(x_0)| - |Dv(x) - Dv(x_0)| \geq \frac{\lambda}{2\Gamma} - \frac{\lambda}{4\Gamma} = \frac{\lambda}{4\Gamma}$$

for all $x \in \sigma B$. □

Finally, we recall two decay estimates that hold for the excess functional of the gradient and of the solutions. They are again related to the Hölder continuity of the gradient of solutions to (99) and of the solutions themselves, in the case we are considering homogeneous equations with measurable coefficients of the type

$$\operatorname{div} a(x, Dv) = 0 \quad \text{in } B. \tag{104}$$

We now connect two basic regularity results for solutions to (99) and (104), respectively. They are bound to provide

Theorem 17 [19,21,47] *Let $v \in W^{1,p}_{\mathrm{loc}}(B) \cap W^{1,1}(B)$ be a weak solution to* (99) *under the assumptions* (16). *There exist constants $\beta \in (0, 1]$ and $c_d \geq 1$, both depending only on n, p, ν, L, such that the estimate*

$$\fint_{\sigma B} |Dv - (Dv)_{\sigma B}|\, d\tilde{x} \leq c_d \sigma^\beta \fint_B |Dv - (Dv)_B|\, d\tilde{x} \tag{105}$$

holds whenever $\sigma B \subset B$ are concentric balls.

Remark 5 The exponent β appearing in (105) can be chosen arbitrarily close to the exponent α_M appearing in (34). As a matter of fact, estimate (35) is in fact obtained as a corollary of the one in (105) via the standard Campanato's integral characterisation of Hölder continuous functions [10]. The exponent α_M considered in Theorem 9 is indeed nothing but the sup of the numbers β considerable in Theorem 17. Explicit, though not optimal, estimates on these numbers via the structure parameters n, p, ν, L, are retrievable tracking the dependence of the constants in the proofs in [19,21,47].

Theorem 18 *Let $v \in W^{1,p}_{\mathrm{loc}}(B) \cap W^{1,1}(B)$ be a weak solution to* (104) *under the assumptions* (18), *and where the vector field $a\colon \Omega \times \mathbb{R}^n \to \mathbb{R}^n$ has measurable dependence on x. There exist constants $\tilde{\beta} \in (0, 1]$ and $\tilde{c}_d \geq 1$, both depending only on n, p, ν, L, such that the estimate*

$$\fint_{\sigma B} |v - (v)_{\sigma B}|\, d\tilde{x} \le \tilde{c}_d \sigma^{\tilde{\beta}} \fint_{B} |v - (v)_B|\, d\tilde{x} + \tilde{c}_d \sigma^{\tilde{\beta}} \tilde{r} s \tag{106}$$

holds whenever $\sigma B \subset B$ *are concentric balls, where* $\tilde{r}$ *is the radius of* B.

Proof The proof combines a few well-known facts in the regularity theory of the equation considered. Let us recall the following Caccioppoli type inequality, that states there exists a constant $c \equiv c(n, p, \nu, L)$ such that

$$\fint_{B/2} (|Dv| + s)\, d\tilde{x} \le \frac{c}{|B|^{1/n}} \fint_{B} |v - (v)_B|\, d\tilde{x} + cs \tag{107}$$

holds; see [40, Proposition 4.1]. On the one hand, the following gradient decay estimate holds whenever $\sigma \in (0, 1/2]$ (see for instance [55, Lemma 3.3]):

$$\fint_{\sigma B} (|Dv| + s)\, d\tilde{x} \le c\sigma^{-1+\tilde{\beta}} \fint_{B/2} (|Dv| + s)\, d\tilde{x} \tag{108}$$

where $\tilde{\beta} < \alpha_m$ and $c \equiv c(n, p, \nu, L, \tilde{\beta})$. Both the inequalities in the last displays are typical in the literature (see for instance [25, Chapter 7]) when using L^p-norms instead of the L^1-ones. Combining them then yields

$$\begin{aligned} \fint_{\sigma B} |v - (v)_{\sigma B}|\, d\tilde{x} &\le c\sigma |B|^{1/n} \fint_{\sigma B} |Dv|\, d\tilde{x} \\ &\le c\sigma^{\tilde{\beta}} |B|^{1/n} \fint_{B/2} (|Dv| + s)\, d\tilde{x} \\ &\le c\sigma^{\tilde{\beta}} \fint_{B} |v - (v)_B|\, d\tilde{x} + c\sigma^{\tilde{\beta}} |B|^{1/n} s \end{aligned}$$

that proves (107) in the case $\sigma \in (0, 1/2]$. On the other hand, the remaining case $\sigma \in (1/2, 1]$ is trivial by using (98). □

Remark 6 Exactly as in Remark 5, the exponent $\tilde{\beta}$ appearing in Theorem 18 can be actually taken as close to α_m—appearing in (30)–(31)—as we please. As a matter of fact estimates (31) and (106) are equivalent by noticing that if v solves then so does $v - k$, whenever k is a real number. Again, the exponent α_m used in Theorem 6 is actually defined as the sup of the numbers β for which (106) works. Estimates for such numbers are available in terms on the structure parameters n, p, ν, L.

12.2 Setting of the constants

In the following all the balls considered will be centered at the point $x \in \Omega$, and we start form a ball $B_R \equiv B(x, R) \subset \Omega$ as in the statement of the theorem. We introduce λ_M as

$$\lambda_M := HR^{1-\alpha} \fint_{B_R} (|Du| + s)\, d\tilde{x} + H \left[\mathbf{I}^{\mu}_{p-\alpha(p-1)}(x, R) \right]^{1/(p-1)}, \tag{109}$$

and fix the constant $H \geq 1$ in a few lines (see (114) below), in a way that makes it depending only on n, p, ν, L; we will then prove that

$$M_{1-\alpha,R}(Du)(x) \leq c\lambda_M \tag{110}$$

holds uniformly in $\alpha \in [0, 1]$, for yet another constant $c \equiv c(n, p, \nu, L)$. Clearly, we may assume without loss of generality that $\lambda_M > 0$, otherwise there is nothing to prove. Let us now fix a few constants that will be relevant in the following; the constants c_l, c_h, c_d have been defined in (100), (101) and Theorem 17, respectively, while c_1 has been introduced in Lemma 2 (that here will be used with $q = 1$); all these constants depend only on n, p, ν, L. Again, $\beta \equiv \beta(n, p, \nu, L) \in (0, 1)$ is the exponent appearing in (101) and (105). Then we fix

$$\delta_1 := \left(\frac{1}{10^{8n} c_l^2 c_d c_h} \right)^{1/\beta}. \tag{111}$$

It follows that also δ_1 depends only on n, p, ν, L. With such a choice of δ_1 we consider the balls in (81), that is

$$B_j := B(x, r_j), \qquad r_j = \delta_1^j r, \qquad r := R/2,$$

and the related comparison solutions $v_j \in u + W_0^{1,p}(B_j)$ in (82), that is

$$\begin{cases} \operatorname{div} a(Dv_j) = 0 & \text{in } B_j \\ v_j = u & \text{on } \partial B_j. \end{cases} \tag{112}$$

Again, with δ_1 now being fixed we determine the constant c_2 from Lemma 3 as follows:

$$c_2 := c_2(n, p, \nu, \delta_1, A) \quad \text{with} \quad A := 10^{4n} c_l. \tag{113}$$

The following inclusions hold for every $j \geq 0$, and the will be used throughout the proof: $\delta_1 B_j = B_{j+1} \subset B_j \subset (1/4)B_j \subset B_j$. We finally set

$$H := 10^8 c_1 c_2 c_d \delta_1^{-5n} \tag{114}$$

and again notice that H only depends on n, p, ν, L. Now, let us set

$$\kappa := \min\left\{1, \delta_1^{n-p+\alpha(p-1)}\right\} \quad \text{and} \quad \tilde{r}_j := \frac{r_j + r_{j+1}}{2}$$

for $j \geq 0$, and observe that

$$
\begin{aligned}
\mathbf{I}^{\mu}_{p-\alpha(p-1)}(x, R) &= \int_0^{2r} \frac{|\mu|(B_\varrho)}{\varrho^{n-p+\alpha(p-1)}} \frac{d\varrho}{\varrho} \\
&= \sum_{j=0}^{\infty} \int_{\tilde{r}_j}^{r_j} \frac{|\mu|(B_\varrho)}{\varrho^{n-p+\alpha(p-1)}} \frac{d\varrho}{\varrho} + \int_{8r/7}^{2r} \frac{|\mu|(B_\varrho)}{\varrho^{n-p+\alpha(p-1)}} \frac{d\varrho}{\varrho} \\
&\quad + \sum_{j=0}^{\infty} \int_{r_{j+1}}^{\tilde{r}_j} \frac{|\mu|(B_\varrho)}{\varrho^{n-p+\alpha(p-1)}} \frac{d\varrho}{\varrho} + \int_{r}^{8r/7} \frac{|\mu|(B_\varrho)}{\varrho^{n-p+\alpha(p-1)}} \frac{d\varrho}{\varrho} \\
&\geq \sum_{j=0}^{\infty} \int_{\tilde{r}_j}^{r_j} \frac{|\mu|(B_\varrho)}{\varrho^{n-p+\alpha(p-1)}} \frac{d\varrho}{\varrho} + \int_{8r/7}^{2r} \frac{|\mu|(B_\varrho)}{\varrho^{n-p+\alpha(p-1)}} \frac{d\varrho}{\varrho} \\
&\geq \kappa \sum_{j=0}^{\infty} \frac{|\mu|(\bar{B}_{j+1})}{r_{j+1}^{n-p+\alpha(p-1)}} \int_{\tilde{r}_j}^{r_j} \frac{d\varrho}{\varrho} + \frac{|\mu|(\bar{B}_0)}{2^{n-1} r^{n-p+\alpha(p-1)}} \int_{8r/7}^{2r} \frac{d\varrho}{\varrho} \\
&\geq \delta_1^{n-1} \log\left(\frac{2}{1+\delta_1}\right) \sum_{j=0}^{\infty} \frac{|\mu|(\bar{B}_{j+1})}{r_{j+1}^{n-p+\alpha(p-1)}} + \frac{\log(7/4)}{2^{n-1}} \left[\frac{|\mu|(\bar{B}_0)}{r_0^{n-p+\alpha(p-1)}}\right] \\
&\geq \delta_1^{n} \sum_{j=0}^{\infty} \frac{|\mu|(\bar{B}_j)}{r_j^{n-p+\alpha(p-1)}}.
\end{aligned} \tag{115}
$$

Therefore, by (109) and the choice in (114) it follows that

$$
10^8 c_1 c_2 c_d \delta_1^{-4n} \left[\sum_{j=0}^{\infty} \frac{|\mu|(\bar{B}_j)}{r_j^{n-p+\alpha(p-1)}}\right]^{1/(p-1)} \leq \lambda_M. \tag{116}
$$

Of course we are using constants like 10^8 to emphasize the fact that in certain places of the proof what it matters is to take large/small quantities. In the rest of the proof we shall denote, for every $j \geq 0$,

$$
E_j := E(Du, B_j) \quad \text{and} \quad a_j := |(Du)_{B_j}| \tag{117}
$$

and

$$
\lambda_j := r_j^{\alpha-1} \lambda_M. \tag{118}
$$

12.3 Iterating quantities

The proof is based on the fact that certain quantities, related to the fractional maximal operator we are considering, stay bounded. Specifically, we shall prove that the following inequality holds:

$$\sup_{j\geq 1} r_j^{1-\alpha} \fint_{B_j} (|Du| + s)\, d\tilde{x} \leq \lambda_M. \tag{119}$$

Let us briefly recall that the last inequality implies the one in display (110) with $c = (2/\delta_1)^n$. Indeed, for $0 < \varrho \leq r_1$ we find $i \geq 1$ such that $r_{i+1} = \delta_1 r_i < \varrho \leq r_i$, therefore

$$\varrho^{1-\alpha} \fint_{B_\varrho} (|Du| + s)\, d\tilde{x} \leq r_i^{1-\alpha} \left(\frac{r_i}{\varrho}\right)^n \fint_{B_i} (|Du| + s)\, d\tilde{x} \leq \frac{\lambda_M}{\delta_1^n}.$$

On the other hand, if $r_1 < \varrho \leq R$, recalling that $r = R/2$, then we similarly have

$$\varrho^{1-\alpha} \fint_{B_\varrho} (|Du| + s)\, d\tilde{x} \leq R^{1-\alpha} \left(\frac{R}{\varrho}\right)^n \fint_{B_R} (|Du| + s)\, d\tilde{x} \leq \frac{2^n \lambda_M}{\delta_1^n}$$

where this time we have used directly (109). All in all (110) follows with $c = (2/\delta_1)^n$. Let us now define

$$C_j := r_j^{1-\alpha} \fint_{B_j} |Du|\, d\tilde{x} + r_{j-1}^{1-\alpha} \fint_{B_{j-1}} |Du|\, d\tilde{x} + \delta_1^{-n} r_j^{1-\alpha} \fint_{B_j} |Du - (Du)_{B_j}|\, d\tilde{x}$$

whenever $j \geq 1$. Notice that, since $r_{j-1}^{1-\alpha} \leq R^{1-\alpha}$ for every $j \geq 1$, then the definitions in (109) and (114) imply that

$$C_1 \leq 4\delta_1^{-3n} R^{1-\alpha} \fint_{B_R} (|Du| + s)\, d\tilde{x} \leq \frac{\lambda_M}{100}. \tag{120}$$

Notice that the last inequality implies, in particular, that

$$R^{1-\alpha} s \leq \frac{\lambda_M}{100}. \tag{121}$$

The rest of the proof is now devoted to establish (119). This will be done using induction on certain *interaction chains* defined in the next step.

12.4 Iteration chains

We consider the set $\mathcal{L}$ defined by

$$\mathcal{L} := \left\{ j \in \mathbb{N} \, : \, C_j \leq \frac{\lambda_M}{100} \right\}, \tag{122}$$

and, accordingly, for $k \geq 1$, we then define the set

$$\mathcal{C}_i^k = \{ j \in \mathbb{N} : i \leq j \leq i+k, \ i \in \mathcal{L}, \ i+k+1 \in \mathcal{L}, \ \ j \notin \mathcal{L} \text{ if } j > i \} \tag{123}$$

and call it *maximal iteration chain* of length k, starting at i. In other words, we have $\mathcal{C}_i^k = \{i, \dots, i+k\}$ and each element of $\mathcal{C}_i^k$ but i lies outside of $\mathcal{L}$; $\mathcal{C}_i^k$ is maximal in the sense that there cannot be another set of the same type properly containing it. Obviously, such sets do not exist when $\mathcal{L} = \mathbb{N}$. In the same way we define

$$\mathcal{C}_i^\infty = \{ j \in \mathbb{N} : i \leq j < \infty, \ i \in \mathcal{L}, \ \ j \notin \mathcal{L} \text{ if } j > i \} \tag{124}$$

as *the infinite maximal chain* starting at i. Notice that, in every case, the smallest element of such a chain always belongs to $\mathcal{L}$, being then the only one of the chain to have such a property. Now, observe that if $\mathcal{L} = \mathbb{N} \backslash \{0\}$, then we are finished as in this case we have $C_j \leq \lambda_M/100$ for every $j \geq 1$ and therefore

$$r_j^{1-\alpha} \fint_{B_j} (|Du| + s) \, d\tilde{x} \leq C_j + R^{1-\alpha} s \leq \frac{\lambda_M}{100} + \frac{\lambda_M}{100} \leq \lambda_M \tag{125}$$

where we also used (121). Let us then assume that $\mathcal{L} \neq \mathbb{N} \backslash \{0\}$. Since anyway $1 \in \mathcal{L}$ by (120), then there must be at least one iteration chain. Let us therefore consider one of these, $\mathcal{C}_i^k$, $k \leq \infty$. We shall use finite induction to show that

$$\sup_{i \leq j \leq i+k} r_j^{1-\alpha} \fint_{B_j} (|Du| + s) \, d\tilde{x} \leq \lambda_M. \tag{126}$$

This, due to the fact that $\mathcal{C}_i^k$ is arbitrary, will finally prove (119) since if a certain index j does not belong to $\mathcal{L}$, then it must belong to some chain. On the other hand, if j does not belong to any chain then it belongs to $\mathcal{L}$ and the corresponding inequality in (119) is automatically verified by the definition of C_j and (125). We are now reduced to show the validity of (126) for any iteration chain $\mathcal{C}_i^k$.

12.5 Density and decay properties along iteration chains

The idea is now that along iteration chains, a condition of the type $C_j \geq \lambda_M/100$ guarantees that the equation becomes nondegenerate and can be in a sense linearized. The consequence is that the excess functional of the gradient decays in a way that resembles the one of solutions to the Poisson equation. To implement this we need to get a few density estimates. Let us preliminary prove a *cheap decay estimate*, that actually holds also outside the context of the present proof, but whose assumptions are obviously satisfied here by (111).

Lemma 4 (Cheap decay estimate) *Let the maps $v_j \in u + W_0^{1,p}(B_j)$ be defined in* (112) *and the balls $\{B_j\}$ be defined in* (81). *Assume that the number δ_1 satisfies*

$$\delta_1 \leq \left(\frac{1}{10^7 c_d}\right)^{1/\beta} . \tag{127}$$

Then the following estimate holds for every $j \geq 0$ and for a constant depending only on n, p, ν, L but not on $\alpha \in [0, 1]$:

$$r_{j+1}^{1-\alpha} E_{j+1} \leq \frac{1}{10^6} r_j^{1-\alpha} E_j + 8c_1 c_d \delta_1^{-n} \left[\frac{|\mu|(B_j)}{r_j^{n-p+\alpha(p-1)}}\right]^{1/(p-1)} . \tag{128}$$

Here the numbers $\{E_j\}$ have been defined in (117), *and the constants c_1, c_d have been introduced in* (76) *and* (105), *respectively.*

Proof By using, in order, triangle inequality, Lemma 2 and Theorem 17 with $v \equiv v_j$, and also using (98) repeatedly, we have

$$\begin{aligned}
r_{j+1}^{1-\alpha} E_{j+1} &\leq 2r_{j+1}^{1-\alpha} \fint_{B_{j+1}} |Du - (Dv_j)_{B_{j+1}}|\, d\tilde{x} \\
&\leq 2r_{j+1}^{1-\alpha} \fint_{B_{j+1}} |Dv_j - (Dv_j)_{B_{j+1}}|\, d\tilde{x} \\
&\quad + 2\delta_1^{-n} r_{j+1}^{1-\alpha} \fint_{B_j} |Dv_j - Du|\, d\tilde{x} \\
&\leq 2c_d \delta_1^{\beta} r_j^{1-\alpha} \fint_{B_j} |Dv_j - (Dv_j)_{B_j}|\, d\tilde{x} \\
&\quad + 2c_1 \delta_1^{-n} r_{j+1}^{1-\alpha} \left[\frac{|\mu|(B_j)}{r_j^{n-1}}\right]^{1/(p-1)} \\
&\leq 4c_d \delta_1^{\beta} r_j^{1-\alpha} \fint_{B_j} |Du - (Du)_{B_j}|\, d\tilde{x} \\
&\quad + 8c_1 c_d \delta_1^{-n} r_j^{1-\alpha} \left[\frac{|\mu|(B_j)}{r_j^{n-1}}\right]^{1/(p-1)} \\
&= 4c_d \delta_1^{\beta} r_j^{1-\alpha} E_j + 8c_1 c_d \delta_1^{-n} r_j^{1-\alpha} \left[\frac{|\mu|(B_j)}{r_j^{n-1}}\right]^{1/(p-1)}
\end{aligned}$$

so that (4) follows by (127) and noticing that

$$r_j^{1-\alpha}\left[\frac{|\mu|(B_j)}{r_j^{n-1}}\right]^{1/(p-1)} = \left[\frac{|\mu|(B_j)}{r_j^{n-p+\alpha(p-1)}}\right]^{1/(p-1)}. \tag{129}$$

□

The following lemma is an easy consequence of the previous one:

Lemma 5 *Let $j \geq 0$ be an integer; if*

$$\delta_1^{-n} r_j^{1-\alpha} E_j \leq \frac{\lambda_M}{100}$$

where $\lambda_M > 0$ is defined in (109), *then it also holds that*

$$\delta_1^{-n} r_{j+1}^{1-\alpha} E_{j+1} \leq \frac{\lambda_M}{10^5}. \tag{130}$$

Proof Using (128) we have

$$\delta_1^{-n} r_{j+1}^{1-\alpha} E_{j+1} \leq \frac{\delta_1^{-n}}{10^6} r_j^{1-\alpha} E_j + 8c_1 c_d \delta_1^{-2n}\left[\frac{|\mu|(B_j)}{r_j^{n-p+\alpha(p-1)}}\right]^{1/(p-1)} \leq \frac{\lambda_M}{10^5}$$

where in the last estimate we have employed also (116). □

The cheap decay estimate for the excess functional of Lemma 4 is not sufficient to get gradient estimates via linear Riesz potentials. Indeed, it only implies Wolff potential estimates as in (23), as shown in [21]. Anyway, by using it together with the density properties of Proposition 2 and Lemma 3, we deduce another, better decay estimate, which is this time implying linear potentials estimates.

Lemma 6 (Linearized decay estimate) *Let $j \geq 1$ be an integer; if*

$$r_{j-1}^{1-\alpha} \fint_{B_{j-1}} (|Du| + s)\, d\tilde{x} \leq \lambda_M \tag{131}$$

holds and if it happens that

$$\delta_1^{-n} r_j^{1-\alpha} E_j \leq \frac{\lambda_M}{100} \quad \textit{and} \quad C_{j+1} \geq \frac{\lambda_M}{100} \tag{132}$$

then we have

$$r_{j+1}^{1-\alpha} E_{j+1} \leq \frac{1}{4} r_j^{1-\alpha} E_j + 8c_2 c_d \delta_1^{-n} \lambda_M^{2-p}\left[\frac{|\mu|(B_{j-1})}{r_{j-1}^{n-p+\alpha(p-1)}}\right], \tag{133}$$

where c_2 depends only on n, p, ν, L and has been defined in (113).

Proof Using triangle inequality, the second inequality in (132) and Lemma 2 we estimate

$$\begin{aligned}
&\delta_1^{-n} r_{j+1}^{1-\alpha} E_{j+1} + \sum_{k=j,j+1} r_k^{1-\alpha} \fint_{B_k} |Dv_{j-1}| \, d\tilde{x} \\
&\geq C_{j+1} - \sum_{k=j,j+1} r_k^{1-\alpha} \fint_{B_k} |Du - Dv_{j-1}| \, d\tilde{x} \\
&\geq \frac{\lambda_M}{100} - \sum_{k=j,j+1} r_k^{1-\alpha} \fint_{B_k} |Du - Dv_{j-1}| \, d\tilde{x} \\
&\geq \frac{\lambda_M}{100} - 2\delta_1^{-2n} r_{j-1}^{1-\alpha} \fint_{B_{j-1}} |Du - Dv_{j-1}| \, d\tilde{x} \\
&\geq \frac{\lambda_M}{100} - 2\delta_1^{-2n} c_1 r_{j-1}^{1-\alpha} \left[\frac{|\mu|(B_{j-1})}{r_{j-1}^{n-1}} \right]^{1/(p-1)} \\
&\geq \frac{\lambda_M}{100} - 2\delta_1^{-2n} c_1 \left[\frac{|\mu|(B_{j-1})}{r_{j-1}^{n-p+\alpha(p-1)}} \right]^{1/(p-1)} \\
&\geq \frac{\lambda_M}{200}
\end{aligned}$$

where we have used (129) and (116) to perform the last estimations; we have also used that $r_k^{1-\alpha} \leq r_{j-1}^{1-\alpha}$. The previous inequality and (130) then give

$$\frac{\lambda_M}{200} \leq \sum_{k=j,j+1} r_k^{1-\alpha} \fint_{B_k} |Dv_{j-1}| \, d\tilde{x} + \frac{\lambda_M}{10^5}$$

so that we conclude with

$$\max_{k=j,j+1} r_k^{1-\alpha} \fint_{B_k} |Dv_{j-1}| \, d\tilde{x} \geq \frac{\lambda_M}{10^3}$$

and in any case, since $r_k^{1-\alpha} \leq r_{j-1}^{1-\alpha}$, we have that

$$\max_{k=j,j+1} \fint_{B_k} |Dv_{j-1}| \, d\tilde{x} \geq \frac{r_{j-1}^{\alpha-1} \lambda_M}{10^3} = \frac{\lambda_{j-1}}{10^3} \geq \frac{\lambda_{j-1}}{10^3 c_l} \tag{134}$$

holds, where the numbers λ_j have been defined in (118). On the other hand, observe that using Lemma 2 and (131) we have

$$
\begin{aligned}
r_{j-1}^{1-\alpha} \fint_{B_{j-1}} (|Dv_{j-1}|+s)\,d\tilde{x} &\le r_{j-1}^{1-\alpha} \fint_{B_{j-1}} (|Du|+s)\,d\tilde{x} \\
&\quad + c_1 r_{j-1}^{1-\alpha}\left[\frac{|\mu|(B_{j-1})}{r_{j-1}^{n-1}}\right]^{1/(p-1)} \\
&\le \lambda_M + c_1\left[\frac{|\mu|(B_{j-1})}{r_{j-1}^{n-p+\alpha(p-1)}}\right]^{1/(p-1)}
\end{aligned}
$$

and taking again (116) into account we have

$$
\fint_{B_{j-1}} (|Dv_{j-1}|+s)\,d\tilde{x} \le 2r_{j-1}^{\alpha-1}\lambda_M .
$$

Applying estimate (100) then yields

$$
\big(|Dv_{j-1}|+s\big) \le 2^{n+1}c_l r_{j-1}^{\alpha-1}\lambda_M = 2^{n+1}c_l\lambda_{j-1} \quad \text{in } (1/2)B_{j-1}. \tag{135}
$$

We are now in position to apply the density improvement results from Proposition 2; indeed conditions in (102) are satisfied with the choices $v \equiv v_{j-1}$, $B \equiv B_{j-1}$, $\sigma = \delta_1$, $k = 1$ or $k = 2$ (by (134)), $\lambda \equiv \lambda_{j-1}$ and $\Gamma = 10^{3n}c_l$; observe that this is possible by the choice of δ_1 made in (111). Notice that (103) is satisfied due to the choice of δ_1 in (111). Proposition 2 now yields

$$
\frac{\lambda_{j-1}}{10^{4n}c_l} \le |Dv_{j-1}| \quad \text{in } B_j = \delta_1 B_{j-1}.
$$

The previous inequality and (135) summarise in the following:

$$
\frac{\lambda_{j-1}}{A} \le |Dv_{j-1}| \le A\lambda_{j-1} \ \text{ in } B_j , \qquad A = 10^{4n}c_l.
$$

On the other hand, notice that again by (116) we have

$$
\frac{|\mu|(B_{j-1})}{\lambda_{j-1}^{p-1}r_{j-1}^{n-1}} = \frac{|\mu|(B_{j-1})}{\lambda_M^{p-1}r_{j-1}^{n-p+\alpha(p-1)}} \le 1.
$$

We are thus in position to apply Lemma 3 thereby obtaining

$$
\begin{aligned}
r_j^{1-\alpha}\fint_{B_j} |Du - Dv_j|\,d\tilde{x} &\le c_2 r_{j-1}^{1-\alpha}\lambda_{j-1}^{2-p}\left[\frac{|\mu|(B_{j-1})}{r_{j-1}^{n-1}}\right] \\
&= c_2\lambda_M^{2-p}\left[\frac{|\mu|(B_{j-1})}{r_{j-1}^{n-p+\alpha(p-1)}}\right],
\end{aligned}
$$

where the constant c_2 has been fixed in (113). With the last comparison estimate we can conclude as in Lemma 4, that is using Theorem 17 and the inequality in the last display as follows:

$$\begin{aligned} r_{j+1}^{1-\alpha}E_{j+1} &\le 2r_{j+1}^{1-\alpha}\fint_{B_{j+1}} |Du-(Dv_j)_{B_{j+1}}|\,d\tilde{x} \\ &\le 2r_{j+1}^{1-\alpha}\fint_{B_{j+1}} |Dv_j-(Dv_j)_{B_{j+1}}|\,d\tilde{x} + 2\delta_1^{-n}r_j^{1-\alpha}\fint_{B_j}|Du-Dv_j|\,d\tilde{x} \\ &\le 2c_d\delta_1^{\beta}r_j^{1-\alpha}\fint_{B_j}|Dv_j-(Dv_j)_{B_j}|\,d\tilde{x} + 2c_2\delta_1^{-n}\lambda_M^{2-p}\left[\frac{|\mu|(B_{j-1})}{r_{j-1}^{n-p+\alpha(p-1)}}\right] \\ &\le 4c_d\delta_1^{\beta}r_j^{1-\alpha}E_j + 8c_2c_d\delta_1^{-n}\lambda_M^{2-p}\left[\frac{|\mu|(B_{j-1})}{r_{j-1}^{n-p+\alpha(p-1)}}\right] \end{aligned} \tag{136}$$

and the proof of (133) is complete recalling that $4c_d\delta_1^{\beta} \le 1/4$ by (111). □

12.6 Iteration and conclusion

Here we finally prove that (126) holds for any iteration chain C_i^k, thereby concluding the proof; we remark that by construction any chain C_i^k is such that $i \ge 1$ since the numbers C_j are defined only for $j \ge 1$. We shall first prove by induction that

$$\begin{cases} r_j^{1-\alpha}\fint_{B_j}(|Du|+s)\,d\tilde{x} \le \lambda_M \\ \delta_1^{-n}r_j^{1-\alpha}E_j \le \frac{\lambda_M}{100} \\ r_{j+1}^{1-\alpha}E_{j+1} \le \frac{1}{4}r_j^{1-\alpha}E_j + 8c_2c_d\delta_1^{-n}\lambda_M^{2-p}\left[\frac{|\mu|(B_{j-1})}{r_{j-1}^{n-p+\alpha(p-1)}}\right] \end{cases} \tag{137}$$

holds for every $j \in \{i, \ldots, i+k-1\}$. Notice that by definition of iteration chains we always have $k \ge 1$. Moreover, notice also that we are going to use finite induction when $k < \infty$, that is when the length of the iteration chain is finite; when $k = \infty$ we are simply going to prove (137) for every $j \ge i$. We start by considering the case $j = i$, which is the induction basis. This follows by the very definitions of C_i^k and C_i as far as $(137)_2$ is concerned. As for $(137)_1$, notice that using (121) we have

$$r_i^{1-\alpha}\fint_{B_i}(|Du|+s)\,d\tilde{x} \le C_i + R^{1-\alpha}s \le \lambda_M.$$

Exactly in the same way we also have

$$r_{i-1}^{1-\alpha} \fint_{B_{i-1}} (|Du| + s)\, d\tilde{x} \le C_i + R^{1-\alpha} s \le \lambda_M .$$

This last inequality allows to get $(137)_3$. Indeed, we need Lemma 6 with $j = i$; this applies by the inequality in the last display and since by definition of iteration chain $\mathcal{C}_i^k$ we have $C_{i+1} \ge \lambda_M/100$; note that here we are also applying $(137)_2$. Next, we proceed verifying the induction step; notice that this case occurs only when $k \ge 2$ while when $k = 1$ the first step already concludes the proof of (137) since $\{i\} = \{i, \ldots, i+k-1\}$. We assume that (137) holds for all indexes $j \in \{i, \ldots, h\}$ with $h < i+k-1$, and then we prove it for the index $h+1$. By Lemma 5 and $(137)_2$ (for $j = h$), we immediately have

$$\delta_1^{-n} r_{h+1}^{1-\alpha} E_{h+1} \le \frac{\lambda_M}{100}. \tag{138}$$

Next, we are going to prove that

$$r_{h+1}^{1-\alpha} \fint_{B_{h+1}} (|Du| + s)\, d\tilde{x} \le \lambda_M . \tag{139}$$

To this aim we start observing that, whenever $0 \le k_1 < k_2$ are integers, we have

$$\begin{aligned} |(Du)_{B_{k_2}} - (Du)_{B_{k_1}}| &= \left| \sum_{j=k_1}^{k_2-1} (Du)_{B_{j+1}} - (Du)_{B_j} \right| \\ &\le \sum_{j=k_1}^{k_2-1} \left|(Du)_{B_{j+1}} - (Du)_{B_j}\right| \\ &\le \sum_{j=k_1}^{k_2-1} \fint_{B_{j+1}} |Du - (Du)_{B_j}|\, dy \\ &\le \sum_{j=k_1}^{k_2-1} \frac{|B_j|}{|B_{j+1}|} E_j \\ &= \delta_1^{-n} \sum_{j=k_1}^{k_2-1} E_j , \end{aligned} \tag{140}$$

so that by letting $k_1 = i$ and $k_2 = h+1$ we obtain

$$a_{h+1} - a_i \le \delta_1^{-n} \sum_{j=i}^{h} E_j . \tag{141}$$

By adding up inequalities $(137)_3$ for $j \in \{i, \ldots, h\}$ and yet adding $r_i^{1-\alpha} E_i$ to both sides we get

$$\sum_{j=i}^{h+1} r_j^{1-\alpha} E_j \leq r_i^{1-\alpha} E_i + \frac{1}{4} \sum_{j=i}^{h} r_j^{1-\alpha} E_j + 8 c_2 c_d \delta_1^{-n} \lambda_M^{2-p} \sum_{j=0}^{\infty} \frac{|\mu|(B_j)}{r_j^{n-p+\alpha(p-1)}},$$

so that reabsorbing terms yields

$$\sum_{j=i}^{h+1} r_j^{1-\alpha} E_j \leq 2 r_i^{1-\alpha} E_i + 16 c_2 c_d \delta_1^{-n} \lambda_M^{2-p} \sum_{j=0}^{\infty} \frac{|\mu|(B_j)}{r_j^{n-p+\alpha(p-1)}}.$$

Using then $(137)_2$ for $j = i$ and (116) allows to conclude with

$$\sum_{j=i}^{h+1} r_j^{1-\alpha} E_j \leq \frac{\lambda_M \delta_1^n}{20}.$$

The last inequality together (141) gives (recall that $r_{h+1}^{1-\alpha} \leq r_j^{1-\alpha}$ for $j \leq h$)

$$\begin{aligned} r_{h+1}^{1-\alpha} a_{h+1} &\leq r_i^{1-\alpha} a_i + \delta_1^{-n} \sum_{j=i}^{h} r_j^{1-\alpha} E_j \\ &\leq r_i^{1-\alpha} \fint_{B_i} |Du| \, d\tilde{x} + \delta_1^{-n} \sum_{j=i}^{h} r_j^{1-\alpha} E_j \\ &\leq \frac{\lambda_M}{100} + \frac{\lambda_M}{20} \leq \frac{\lambda_M}{10}. \end{aligned}$$

Merging the last inequality with (121) and (138), and again using Jensen's inequality, we estimate

$$\begin{aligned} r_{h+1}^{1-\alpha} \fint_{B_{h+1}} (|Du| + s) \, d\tilde{x} &= r_{h+1}^{1-\alpha} \fint_{B_{h+1}} (|Du| - |(Du)_{B_{h+1}}|) \, d\tilde{x} \\ &\quad + r_{h+1}^{1-\alpha} a_{h+1} + r_{h+1}^{1-\alpha} s \\ &\leq r_{h+1}^{1-\alpha} E_{h+1} + r_{h+1}^{1-\alpha} a_{h+1} + R^{1-\alpha} s \\ &\leq \lambda_M, \end{aligned} \tag{142}$$

so that (139) follows. To complete the induction step we shall finally prove $(137)_3$ with $j = h + 1$, that is

$$r_{h+2}^{1-\alpha} E_{h+2} \leq \frac{1}{4} r_{h+1}^{1-\alpha} E_{h+1} + 8 c_2 c_d \delta_1^{-n} \lambda_M^{2-p} \left[\frac{|\mu|(B_h)}{r_h^{n-p+\alpha(p-1)}} \right]. \tag{143}$$

For this we want to apply Lemma 6 in the case $j = h+1$; let us verify that assumptions are satisfied. Notice that since here we are assuming that $h \leq i + k - 2$ then $C_{h+2} \geq \lambda_M/100$; moreover, (138) holds and by induction assumption we have

$$r_{j-1}^{1-\alpha} \fint_{B_{j-1}} (|Du| + s)\, d\tilde{x} = r_h^{1-\alpha} \fint_{B_h} (|Du| + s)\, d\tilde{x} \leq \lambda_M.$$

At this point Lemma 6 applies and (143) follows. All in all, using induction (in particular, finite induction when $k < \infty$) this proves

$$\sup_{i \leq j \leq i+k-1} r_j^{1-\alpha} \fint_{B_j} (|Du| + s)\, d\tilde{x} \leq \lambda_M. \tag{144}$$

In particular, this implies (126) when $k = \infty$, otherwise we still have to prove that

$$r_{i+k}^{1-\alpha} \fint_{B_{i+k}} (|Du| + s)\, d\tilde{x} \leq \lambda_M. \tag{145}$$

This has been on the other hand implicitly proved before. Indeed, since now (137) holds in the full range $j \in \{i, \ldots, i+k-1\}$ we repeat the computation from (141) to (142) with $h = i + k - 1$, that indeed gives (145). This, together with (144), finally gives (126). The proof is complete.

12.7 Proposition 1 and Theorem 8

Proof of Proposition 1 Let us define $\varrho_i = r/2^i$, for every integer $i \geq 0$ and $r = 101|x - y|/100$. Then observe that

$$\begin{aligned} |(f)_{B(x,\varrho_i)} - (f)_{B(x,\varrho_{i+1})}| &\leq \fint_{B(x,\varrho_{i+1})} |f - (f)_{B(x,\varrho_i)}|\, d\tilde{x} \\ &\leq 2^n \fint_{B(x,\varrho_i)} |f - (f)_{B(x,\varrho_i)}|\, d\tilde{x} \\ &\leq c(n) M^{\#}_{\alpha,\varrho_i}(f)(x)\varrho_i^{\alpha} \leq c(n) M^{\#}_{\alpha,R}(f)(x)\varrho_i^{\alpha} \end{aligned} \tag{146}$$

and, upon summation, we get

$$\begin{aligned} \sum_{i \geq 0} |(f)_{B(x,\varrho_i)} - (f)_{B(x,\varrho_{i+1})}| &\leq c(n) M^{\#}_{\alpha,R}(f)(x) \sum_{i \geq 0} \varrho_i^{\alpha} \\ &\leq \frac{c}{\alpha} M^{\#}_{\alpha,R}(f)(x)|x - y|^{\alpha}. \end{aligned} \tag{147}$$

We have used the elementary inequality

$$\sum_{i\geq 0}\frac{1}{2^{i\alpha}}\leq\frac{c}{\alpha}.$$

Notice that this shows that $\{(f)_{B(x,\varrho_i)}\}$ is a Cauchy sequence and therefore the following limit exists:

$$\lim_{i\to\infty}(f)_{B(x,\varrho_i)}. \tag{148}$$

Let us now show that the limit

$$\lim_{\varrho\to 0}(f)_{B(x,\varrho)} \tag{149}$$

exists too, and therefore defines the precise representative of f and x, which is indeed denoted by $f(x)$. To this aim, take $0<\varrho\leq\varrho_0$. There exists an integer i such that $\varrho_{i+1}<\varrho\leq\varrho_i$; therefore, as for (146), we have

$$|(f)_{B(x,\varrho_i)}-(f)_{B(x,\varrho)}|\leq c(n)M^{\#}_{\alpha,R}(f)(x)\varrho^{\alpha}.$$

The last inequality, together with the existence of the limit in (148), proves the existence of the limit in (149). Changing in the previous argument x by y and choosing this time $r=|x-y|/100$ gives

$$\sum_{i\geq 0}|(f)_{B(y,\varrho_i)}-(f)_{B(y,\varrho_{i+1})}|\leq\frac{c}{\alpha}M^{\#}_{\alpha,R}(f)(y)|x-y|^{\alpha} \tag{150}$$

and in the same way we can prove that the precise representative of f can be defined at y. Notice now that telescoping summation and (147) gives

$$|f(x)-(f)_{B(x,\varrho_0)}|\leq\sum_{i\geq 0}|(f)_{B(x,\varrho_i)}-(f)_{B(x,\varrho_{i+1})}|\leq\frac{c}{\alpha}M^{\#}_{\alpha,R}(f)(x)|x-y|^{\alpha}$$

with a similar estimate for balls centred at y following by (150). Therefore, using triangle inequality, and recalling the choices of r made for the points x and y, we conclude with

$$\begin{aligned}|f(x)-f(y)|\leq{}&|(f)_{B(y,101|x-y|/100)}-(f)_{B(y,|x-y|/100)}|\\&+\frac{c}{\alpha}\left[M^{\#}_{\alpha,R}(f)(x)+M^{\#}_{\alpha,R}(f)(y)\right]|x-y|^{\alpha}.\end{aligned}$$

To estimate the first term in the right hand side of the previous inequality we proceed as follows:

$$\begin{aligned}
&|(f)_{B(x,101|x-y|/100)} - (f)_{B(y,|x-y|/100)}| \\
&\quad \le \fint_{B(y,|x-y|/100)} |f - (f)_{B(x,101|x-y|/100)}|\, d\tilde{x} \\
&\quad \le c \fint_{B(x,101|x-y|/100)} |f - (f)_{B(x,101|x-y|/100)}|\, d\tilde{x} \\
&\quad \le c M^{\#}_{\alpha,R}(f)(x)|x-y|^{\alpha}.
\end{aligned}$$

Collecting the inequalities in the last two displays yields (43). Notice that in the last display we have used that $B(y, |x-y|/100) \subset B(x, 101|x-y|/100)$. We have also used that

$$\frac{101|x-y|}{100} \le \frac{404R}{500} \le R \quad \text{and} \quad B(x, 101|x-y|/100) \subset B_{8R/5}.$$

Proof of Theorem 8 Applying Proposition 1 in the ball $B_{R/4}$ and then (42), we get that the estimate

$$\begin{aligned}
|u(x) - u(y)| \le \frac{c}{\alpha}\left[\mathbf{I}^{\mu}_{p-\alpha(p-1)}(x, R) + \mathbf{I}^{\mu}_{p-\alpha(p-1)}(y, R)\right]^{1/(p-1)} |x-y|^{\alpha} \\
+\frac{cR}{\alpha} \fint_{B(x,5R/8)} (|Du| + s)\, d\tilde{x} \cdot \left(\frac{|x-y|}{R}\right)^{\alpha} \\
+\frac{cR}{\alpha} \fint_{B(y,5R/8)} (|Du| + s)\, d\tilde{x} \cdot \left(\frac{|x-y|}{R}\right)^{\alpha} \qquad (151)
\end{aligned}$$

holds for every choice $x, y \in B_{R/4}$. The constant c appearing in the previous estimate just depends on n, but is independent of $\alpha \in [0, 1]$. In order to estimate the last two integrals, we recall the following Caccioppoli type inequality:

$$\fint_{B} |Du|\, d\tilde{x} \le \frac{c}{r} \fint_{\gamma B} |u - (u)_{\gamma B}|\, d\tilde{x} + c\left[\frac{|\mu|(\gamma B)}{(\gamma r)^{n-1}}\right]^{1/(p-1)} + cs \qquad (152)$$

that holds whenever $\gamma B \subset \Omega$ is a ball with radius γr, and $\gamma > 1$. The constant c depends only on n, p, ν, L and γ. See [40, Proposition 4.1] for a proof. We apply (152) to $B(x, 5R/8)$ and $B(y, 5R/8)$ with the choice $\gamma = 7/6$ and we use it in (151). Observe that this is possible since

$$x, y \in B_{R/4} \Longrightarrow \gamma B(x, 5R/8) \cup \gamma B(y, 5R/8) \subset B_R.$$

Then we notice that the following elementary estimate holds whenever $\sigma \in (0, 1)$:

$$
\begin{aligned}
&R\left[\frac{|\mu|(B(x,\sigma R))}{(\sigma R)^{n-1}}\right]^{1/(p-1)}\left(\frac{|x-y|}{R}\right)^{\alpha} \\
&\leq -\frac{1}{\min\{\sigma^{n-1}, \sigma^{(p-1)(1-\alpha)}\}\log\sigma}\left[\int_{\sigma R}^{R}\frac{|\mu|(B(x,\varrho))}{\varrho^{n-p+\alpha(p-1)}}\frac{d\varrho}{\varrho}\right]^{1/(p-1)}|x-y|^{\alpha} \\
&\leq c(\sigma)\left[\mathbf{I}^{\mu}_{p-\alpha(p-1)}(x,R)\right]^{1/(p-1)}|x-y|^{\alpha}
\end{aligned}
$$

(with a similar one centred at y), and, using it with $\sigma = 35/48$, we finally conclude with (33). □

Remark 7 There is a variant of estimate (33), which is now uniform in the whole range $\alpha \in [0, 1]$, and it is the following:

$$
\begin{aligned}
|u(x)-u(y)| \leq{}& \frac{c}{\alpha}\left[\mathbf{I}^{\mu}_{p-\alpha(p-1)}(x,R)+\mathbf{I}^{\mu}_{p-\alpha(p-1)}(y,R)\right]^{1/(p-1)}|x-y|^{\alpha} \\
&+\frac{c}{\alpha}\fint_{B_R}(|u|+Rs)\,d\tilde{x}\cdot\left(\frac{|x-y|}{R}\right)^{\alpha}
\end{aligned}
\tag{153}
$$

The constant c appearing in the above estimate is this time independent of α, and only exhibits a dependence on n, p, ν, L. Estimate (153) follows exactly as estimate (33), but tracking the dependence on the constants on α starting from (151). Estimate (153) describes in a precise way the blow-up rate of the constant c in (33) when $\alpha \to 0$. More precisely, the coefficient $1/\alpha$ appears to be related to the loss of L^∞-estimates towards BMO-ones already discussed after Proposition 1. Moreover, it is the same blow-up rate appearing when fractional estimates get lost as the fractional differentiability parameter disappears. See for instance [52].

13 Proof of Theorem 12

This theorem has been proved in [40] in a slightly different form and with a different proof; Theorem 12 is essential for the proof of Theorem 9. We consider the sequence of shrinking balls introduced in (81) with the corresponding solutions $v_j \in u + W^{1,p}_0(B_i)$ defined in (112); the number $\delta_1 \in (0, 1/4)$ is this time chosen as

$$
\delta_1 := \left(\frac{\varepsilon}{4c_d}\right)^{1/(\beta-\tilde{\alpha})}, \qquad \beta := \frac{\alpha_M+\tilde{\alpha}}{2}, \tag{154}
$$

while we take $r := R$. Here we take $\varepsilon \in (0, 1/2)$ to be a fixed number. Finally, we recall the definition of the excess quantities $\{E_j\}$ in (117). Before going on let us record an identity of later use, that is

$$r_j^{-\alpha}\left[\frac{|\mu|(B_j)}{r_j^{n-1}}\right]^{1/(p-1)} = \left[\frac{|\mu|(B_j)}{r_j^{n-1+\alpha(p-1)}}\right]^{1/(p-1)}. \tag{155}$$

We are now going to use estimate (105) as we have done for Lemma 4. In particular, in light of Remark 5, with $\tilde{\alpha} < \alpha_M$, we can choose in (105) a number β such that $\tilde{\alpha} < \beta < \alpha_M$ and we take the one in (154). Therefore, recalling that $r_{j+1}^{-\alpha} = \delta_1^{-\alpha} r_j^{-\alpha}$, we have

$$\begin{aligned}
r_{j+1}^{-\alpha} E_{j+1} &\leq 2r_{j+1}^{-\alpha} \fint_{B_{j+1}} |Dv_j - (Dv_j)_{B_{j+1}}|\, d\tilde{x} \\
&\quad + 2\delta_1^{-n} r_{j+1}^{-\alpha} \fint_{B_j} |Dv_j - Du|\, d\tilde{x} \\
&\leq 2c_d \delta_1^{\beta-\alpha} r_j^{-\alpha} \fint_{B_j} |Dv_j - (Dv_j)_{B_j}|\, d\tilde{x} \\
&\quad + 2c\delta_1^{-n-\alpha} r_j^{-\alpha}\left[\frac{|\mu|(B_j)}{r_j^{n-1}}\right]^{1/(p-1)} \\
&\leq 4c_d \delta_1^{\beta-\tilde{\alpha}} r_j^{-\alpha} E_j + 8c\delta_1^{-n-\tilde{\alpha}}\left[\frac{|\mu|(B_j)}{r_j^{n-1+\alpha(p-1)}}\right]^{1/(p-1)}.
\end{aligned}$$

By recalling (154) we conclude with

$$r_{j+1}^{-\alpha} E_{j+1} \leq \varepsilon r_j^{-\alpha} E_j + c\left[\frac{|\mu|(B_j)}{r_j^{n-1+\alpha(p-1)}}\right]^{1/(p-1)} \tag{156}$$

where $c \equiv c(n, p, \nu, L, \tilde{\alpha})$ as a consequence of the fact that the choice in (154) determines a number δ_1 which again depends on n, p, ν, L, $\tilde{\alpha}$ and ε via the dependence of the quantities β and c_d. Observe that the resulting estimate is independent of α as long as this belongs to $[0, \tilde{\alpha}]$. From (156) it follows that

$$\sup_{j\geq 1} r_j^{-\alpha} E_j \leq \varepsilon \sup_{j\geq 1} r_j^{-\alpha} E_j + \varepsilon R^{-\alpha} E_0 + c(\varepsilon) \sup_{j\geq 0}\left[\frac{|\mu|(B_j)}{r_j^{n-1+\alpha(p-1)}}\right]^{1/(p-1)}. \tag{157}$$

Taking for instance $\varepsilon = 1/4$, reabsorbing terms and yet adding $R^{-\alpha}E_0$ to both sides of the resulting inequality, we conclude with

$$\sup_{j\geq 0} r_j^{-\alpha} E_j \leq c R^{-\alpha} \fint_{B_R} |Du - (Du)_{B_R}|\, d\tilde{x} + c\left[M_{1-\alpha(p-1),R}(\mu)(x)\right]^{1/(p-1)} =: cS. \tag{158}$$

This last estimate implies the statement of the theorem, that is

$$M^{\#}_{\alpha,R}(Du)(x) \leq cS\,, \tag{159}$$

where $c \equiv c(n, p, \nu, L, \tilde{\alpha})$ and S in fact denotes the quantity appearing in the right hand side of (158). Indeed, let us consider a positive radius $\varrho \leq R$, and determine $j \geq 0$ such that $r_{j+1} < \varrho \leq r_j$; then using also (98) we have

$$\begin{aligned}
\varrho^{-\alpha} \fint_{B_\varrho} |Du - (Du)_{B_\varrho}|\, d\tilde{x} &\leq 2\varrho^{-\alpha} \fint_{B_\varrho} |Du - (Du)_{B_j}|\, d\tilde{x} \\
&\leq 2\left(\frac{\varrho}{r_j}\right)^{-n-\alpha} r_j^{-\alpha} \fint_{B_j} |Du - (Du)_{B_j}|\, d\tilde{x} \\
&\leq 2\delta_1^{-n-\tilde{\alpha}} r_j^{-\alpha} \fint_{B_j} |Du - (Du)_{B_j}|\, d\tilde{x} \leq cS
\end{aligned}$$

and (159) follows keeping in mind the dependence of δ_1 on $n, p, \nu, L, \tilde{\alpha}$. The proof is complete.

13.1 A VMO-type result

A modification to the proof of Theorem 12 allows to get a corollary that is interesting in itself and that will be useful in order to obtain subsequent results.

Theorem 19 (VMO gradient regularity) *Let $u \in W^{1,p}(\Omega)$ be a weak solution to the Eq.* (13) *under the assumptions* (16)*. Assume that*

$$S_1(x) := \sup_{B(x,\varrho)\subset\Omega} \fint_{B(x,\varrho)} |Du|\, d\tilde{x} < \infty \tag{160}$$

and

$$\lim_{R\to 0} \frac{|\mu|(B(x,R))}{R^{n-1}} = 0 \tag{161}$$

hold for a certain $x \in \Omega$, then

$$\lim_{R\to 0} \fint_{B(x,R)} |Du - (Du)_{B(x,R)}|\, d\tilde{x} = 0. \tag{162}$$

Moreover, if Du is locally bounded and if the limit in (161) *is uniform with respect to x then Du is (locally) VMO in Ω.*

Proof In order to prove (162) we show that, for every choice of $\tilde{\varepsilon}$, there exists $\tilde{R} > 0$, depending on $n, p, \nu, L, \tilde{\varepsilon}, \mu$ and the point x, such that

$$\varrho \le \tilde{R} \Longrightarrow E(Du, B(x,\varrho)) < \tilde{\varepsilon}. \tag{163}$$

We revisit the proof of Theorem 12, where we consider the case $\alpha = \tilde{\alpha} = 0$. Assume that $B(x,R) \subset \Omega$. Estimate (157) then gives

$$\sup_{j\ge 1} E_j \le 2\varepsilon E(Du, B(x,R)) + c(\varepsilon)\left[M_{1,R}(\mu)(x)\right]^{1/(p-1)},$$

which is valid as soon as δ_1 and ε are linked as in (154), and $\varepsilon \in (0, 1/2)$. In particular, we also have

$$\sup_{j\ge 1} E_j \le 4\varepsilon S_1 + c(\varepsilon)\left[M_{1,R}(\mu)(x)\right]^{1/(p-1)}, \tag{164}$$

where we have used the abbreviation $S_1 \equiv S_1(x)$. Now, let us notice how the previous argument implies that

$$\sup_{\varrho\le\delta_1 R} E(Du, B(x,\varrho)) \le 4\varepsilon S_1 + c(\varepsilon)\left[M_{1,R}(\mu)(x)\right]^{1/(p-1)} \tag{165}$$

holds for every $\varrho \le \delta_1 R = r_1$ and δ_1 is again chosen as in (154). Indeed, consider a number $\varrho \le \delta_1 R$ with $\tilde{\alpha} = 0$; then there exists an integer $k \ge 1$ such that $r_{k+1} < \varrho \le r_k$. This means that $\varrho = \delta_1^k R'$ for some $R' \in (\delta_1 R, R]$. Now, in the proof of Theorem 12 replace R by R' and we gain a new chain of cylinders B_j for which (164) obviously holds; indeed, notice that $M_{1,R'}(\mu)(x) \le M_{1,R}(\mu)(x)$ since $R' \le R$. With this new definition we have $E(Du, B_\varrho) = E_k$ and therefore (165) is nothing but (164) with this new choice of the radii $\{r_j\} \equiv \{\delta_1^j R'\}$. All in all we have proved that (165) holds with δ_1 depending only on n, p, ν, L and ε. Let us recall that estimate (165) holds with any choice of the initial radius R, which is a free parameter that is going to be chosen in the next lines; notice indeed that all the constants determined up top now are independent of the starting radius R. Now, first choose $\varepsilon \equiv \varepsilon(n, p, \nu, L, \tilde{\varepsilon})$ small enough in order to have

$$4\varepsilon S_1 \le \frac{\tilde{\varepsilon}}{2}. \tag{166}$$

This determines the constant $c(\varepsilon)$ as a function of $n, p, \nu, L, \tilde{\varepsilon}$; notice that this is possible by the first assumption in (161), while using the second we can determine a radius R_1, depending essentially on $n, p, \nu, L, \tilde{\varepsilon}$ and the point x (via the rate of convergence in (161)), such that

$$R \leq R_1 \Longrightarrow c(\varepsilon)\left[M_{1,R}(\mu)(x)\right]^{1/(p-1)} \leq \frac{\tilde{\varepsilon}}{2}. \tag{167}$$

Combining the content of the last two displays with the one of (165) (where we now take with $R \equiv R_1$), and recalling the definition of the excess functional $E(Du, B_R)$, yields (163) with the choice $\tilde{R} = \delta_1 R_1$. As for the local VMO-regularity of Du we recall that this means that, for every $\tilde{\varepsilon}$, the choice of $\tilde{R}$ in (163) can be done independently of the point x as long as $B_{\tilde{R}}$ varies in a fixed compact subset $\Omega' \Subset \Omega$. We can now check this by showing that all the choices of the constants above can be done independently of the point x, but depending only on the fixed subset Ω'. Indeed (166) can be replaced by

$$4\varepsilon S_1 \leq 4\varepsilon \|Du\|_{L^\infty(\Omega')} \leq \frac{\tilde{\varepsilon}}{2}. \tag{168}$$

The choice of R_1 in (167) can be made uniform as well by (161), which is now uniform with respect to x, and the proof of the VMO-regularity of Du is complete, too.

14 Proof of Theorem 15

We first prove the theorem in the case we have an energy solution $u \in W^{1,p}(\Omega)$; this will be done through Steps 14.1–14.3. Then, in the final Step 14.4, we describe the approximation argument to prove the theorem for any SOLA $u \in W_0^{1,p-1}(\Omega)$ to problem (55). We indeed prefer a separate treatment of the two cases since in this second part we shall also present the general arguments to prove that the other potential estimates presented in this paper for energy solutions actually hold for SOLA too.

14.1 Proof in the case of energy solutions: beginning

We fix a ball $B_R \equiv B(x, R) \subset \Omega$ and then consider in the rest of the proof balls which are concentric to it. We notice that by assumption (60) the quantity

$$\lambda_M := c\left[\mathbf{I}_1^{\mu}(x, R)\right]^{1/(p-1)} + c \fint_{B_R} (|Du| + s)\, d\tilde{x} \tag{169}$$

is finite, where $c \equiv c(n, p, \nu, L)$ is the constant appearing in (40). We shall then prove that, for every $\varepsilon > 0$, there exists a radius $r_\varepsilon \leq R$, in general depending only on n, p, ν, L and the point x, such that

$$|(Du)_{B(x,\varrho)} - (Du)_{B(x,\tau)}| \leq \lambda_M \varepsilon \quad \text{holds whenever } 0 < \tau < \varrho \leq r_\varepsilon. \tag{170}$$

This will finally prove (61). Notice that with the above definitions, by (40) we have

$$\fint_{B(x,\varrho)} (|Du| + s)\, d\tilde{x} \leq \lambda_M \qquad \text{for every } \varrho \leq R \tag{171}$$

while, and this time by Theorem 19, we have that

$$\lim_{\varrho \to 0} E(Du, B(x, \varrho)) = 0. \tag{172}$$

Notice that, in order to verify the assumptions in (161) we have implicitly used the fact that $\mathbf{I}_1^{\mu}(x, R) < \infty$ implies that

$$\lim_{\varrho \to 0} \mathbf{I}_1^{\mu}(x, \varrho) = 0 \tag{173}$$

and this, thanks to an argument similar to the one on (115), also implies

$$\lim_{\varrho \to 0} \frac{|\mu|(B(x, \varrho))}{\varrho^{n-1}} = 0 \tag{174}$$

which is in fact the second part of the main assumption in Theorem 161.

14.2 Setting of the constants and a sequence

We revisit the arguments of the proof of Theorem 10, while in the following the number ε is fixed as the one introduced in (170). As usual the constants c_l, c_h, c_d have been defined in (100), (101) and (105), respectively; c_1 is from Lemma 2 (where we take $q = 1$) and $\beta \equiv \beta(n, p, \nu, L) \in (0, 1)$ appears in (101) and (105). We this time fix $\delta_1 \equiv \delta_1(n, p, \nu, L, \varepsilon)$ as

$$\delta_1 := \left(\frac{\varepsilon^2}{10^{8n} c_l^2 c_d c_h} \right)^{1/\beta}. \tag{175}$$

With such a choice of δ_1 we determine the constant c_2 from Lemma 3 as follows:

$$c_2 := c_2(n, p, \nu, L, \delta_1, A) \qquad \text{with} \qquad A := \frac{10^{4n} c_l}{\varepsilon}. \tag{176}$$

Notice that all the constants appearing up to now show a global dependence on the constants n, p, ν, L and ε. Now we can take a radius $R_l \equiv R_l(n, p, \nu, L, \varepsilon, x) \leq R$ such that the following two conditions are satisfied:

$$\left[\int_0^{2R_l} \frac{|\mu|(B(x, \varrho))}{\varrho^{n-1}} \frac{d\varrho}{\varrho} \right]^{1/(p-1)} \leq \frac{\delta_1^{5n} \lambda_M \varepsilon}{10^8 c_1 c_2 c_d} \tag{177}$$

and

$$\sup_{0<\varrho\le R_l} E(Du, B(x,\varrho)) \le \frac{\delta_1^{5n}\lambda_M\varepsilon}{10^8}. \tag{178}$$

Notice that we can satisfy (177) by (173), while the choice in (178) is possible due to (172). Finally, for integers $j \ge 0$ we consider the balls as in (81) where this time

$$B_j := B(x, r_j), \quad r_j = \delta_1^j r, \quad r := R_l$$

and the related comparison solutions $v_j \in u + W_0^{1,p}(B_j)$ in (112). Accordingly, the numbers $\{E_i\}$ are again defined as in (117). Notice also that (177) and a computation totally similar to the one in (115) (where we take $\alpha = 1$) give

$$\left[\sum_{j=0}^{\infty} \frac{|\mu|(\bar{B}_j)}{r_j^{n-1}}\right]^{1/(p-1)} \le \frac{\delta_1^{4n}\lambda_M\varepsilon}{10^8 c_1 c_2 c_d}. \tag{179}$$

In Step 14.3 below, we will prove that

$$|(Du)_{B_h} - (Du)_{B_k}| \le \frac{\lambda_M\varepsilon}{12} \quad \text{holds whenever } 2 \le k \le h. \tag{180}$$

Let us assume for a moment that (180) and let us finish the proof of (170) with the choice

$$r_\varepsilon := \delta_1^2 R_l. \tag{181}$$

Indeed, let us fix $0 < \tau < \varrho \le r_\varepsilon$. This means that there exist two integers k and h, such that $2 \le k \le h$,

$$\delta_1^{k+1} R_l < \varrho \le \delta_1^k R_l \quad \text{and} \quad \delta_1^{h+1} R_l < \tau \le \delta_1^h R_l \tag{182}$$

hold. Applying (178), and taking (182) into account, we get

$$\begin{aligned}
|(Du)_{B(x,\varrho)} - (Du)_{B_{k+1}}| &\le \fint_{B_{k+1}} |Du - (Du)_{B(x,\varrho)}|\, d\tilde{x} \\
&\le \frac{|B(x,\varrho)|}{|B_{k+1}|} \fint_{B(x,\varrho)} |Du - (Du)_{B(x,\varrho)}|\, d\tilde{x} \\
&\le \delta_1^{-n} E(Du, B(x,\varrho)) \le \frac{\lambda_M\varepsilon}{10}.
\end{aligned}$$

Similarly, we also gain

$$|(Du)_{B(x,\tau)} - (Du)_{B_{h+1}}| \leq \frac{\lambda_M \varepsilon}{10}.$$

Using the inequalities in the last two displays together with (180) and triangle inequality establishes (170). It remains to prove (180) and this will be done in the next step.

14.3 Non-degenerate iteration chains and proof of (180)

We set, for $j \geq 1$

$$C_j := \fint_{B_j} |Du| \, d\tilde{x}$$

and then, as in (122), we define

$$\mathcal{L} := \left\{ j \in \mathbb{N} \, : \, C_j \leq \frac{\lambda_M \varepsilon}{100} \right\}. \tag{183}$$

Accordingly, we also define the chains $\mathcal{C}_i^k$ and $\mathcal{C}_i^\infty$ as in (123) and (124), respectively. The difference here is the appearance of the parameter ε in the new definition of the set $\mathcal{L}$ in (183); this allows to interpret the sets $\mathcal{C}_i^k$ as maximal chains along which the equation becomes nondegenerate. In this case it will be useful to consider the following number:

$$j_m := \min \mathcal{L}. \tag{184}$$

Needless to say, in case $\mathcal{L}$ is empty we have $j_m = \infty$ and this is actually a favourable case since the problem never becomes degenerate. The following lemma is similar to Lemma 6:

Lemma 7 *Let $j \geq 1$ be an integer such that*

$$C_{j+1} \geq \frac{\lambda_M \varepsilon}{100}$$

holds; then we have

$$E_{j+1} \leq \frac{\varepsilon}{4} E_j + 8 c_2 c_d \delta_1^{-n} \lambda_M^{2-p} \left[\frac{|\mu|(B_{j-1})}{r_{j-1}^{n-1}} \right], \tag{185}$$

where c_2 depends only on n, p, ν, L, ε and has been defined in (176).

Proof As for Lemma 6 we estimate

$$\begin{aligned}
\fint_{B_{j+1}} |Dv_{j-1}|\,d\tilde{x} &\geq C_{j+1} - \fint_{B_{j+1}} |Du - Dv_{j-1}|\,d\tilde{x} \\
&\geq \frac{\lambda_M \varepsilon}{100} - \fint_{B_{j+1}} |Du - Dv_{j-1}|\,d\tilde{x} \\
&\geq \frac{\lambda_M \varepsilon}{100} - \delta_1^{-2n} \fint_{B_{j-1}} |Du - Dv_{j-1}|\,d\tilde{x} \\
&\geq \frac{\lambda_M \varepsilon}{100} - \delta_1^{-2n} c_1 \left[\frac{|\mu|(B_{j-1})}{r_{j-1}^{n-1}}\right]^{1/(p-1)} \\
&\geq \frac{\lambda_M \varepsilon}{200} \qquad (186)
\end{aligned}$$

where we used (76) and (179). On the other hand, using triangle inequality, (171), (179) and yet Lemma 2, we have

$$\fint_{B_{j-1}} (|Dv_{j-1}| + s)\,d\tilde{x} \leq \fint_{B_{j-1}} (|Du| + s)\,d\tilde{x} + c_1 \left[\frac{|\mu|(B_{j-1})}{r_{j-1}^{n-1}}\right]^{1/(p-1)} \leq 2\lambda_M .$$

Applying estimate (100) with $v \equiv v_{j-1}$ then yields

$$\big(|Dv_{j-1}| + s\big) \leq 2^{n+1} c_l \lambda_M \quad \text{in } B_{j-1}/2. \qquad (187)$$

In view of (186)–(187) we are in position to apply Proposition 2 with the choices $v \equiv v_{j-1}$, $B \equiv B_{j-1}$, $\sigma = \delta_1$, $k = 1$, $\lambda \equiv \lambda_M$ and $\Gamma = 10^{3n} c_l/\varepsilon$; observe that this is again possible by the choice of δ_1 made in (175). This yields

$$\frac{\lambda_M \varepsilon}{10^{4n} c_l} \leq |Dv_{j-1}| \quad \text{in } B_j$$

The inequality in the above display and (187) summarise in the following line:

$$\frac{\lambda}{A} \leq |Dv_{j-1}| \leq A\lambda \ \text{ in } B_j, \qquad A = \frac{10^{4n} c_l}{\varepsilon}$$

and therefore we are in position to apply Lemma 3 thereby obtaining

$$\fint_{B_j} |Du - Dv_j|\,d\tilde{x} \leq c_2 \lambda_M^{2-p} \left[\frac{|\mu|(B_{j-1})}{r_{j-1}^{n-1}}\right],$$

where the constant c_2 has been fixed in (176). With the last comparison estimate we can conclude with (185) as in (136). □

We now proceed with the proof of (180), obviously assuming $k < h$. We then analyse three different cases.

Case 1: $k < h \leq j_m$, where we recall that the number j_m has been defined in (184). By the very definition of j_m it follows that $C_{j+1} \geq \lambda_M \varepsilon/100$ holds for every $j \in \{k-1, \ldots, h-2\}$ and therefore we apply Lemma 7 to get that

$$E_{j+1} \leq \frac{1}{4} E_j + 8 c_2 c_d \delta_1^{-n} \lambda_M^{2-p} \left[\frac{|\mu|(B_{j-1})}{r_{j-1}^{n-1}} \right] \tag{188}$$

holds, again for every $j \in \{k-1, \ldots, h-2\}$. Summing up yields

$$\sum_{j=k}^{h-1} E_j \leq \frac{1}{4} \sum_{j=k-1}^{h-2} E_j + 8 c_2 c_d \delta_1^{-n} \lambda_M^{2-p} \sum_{j=0}^{\infty} \frac{|\mu|(B_j)}{r_j^{n-1}}$$

so that

$$\sum_{j=k}^{h-1} E_j \leq \frac{1}{3} E_{k-1} + 16 c_2 c_d \delta_1^{-n} \lambda_M^{2-p} \sum_{j=0}^{\infty} \frac{|\mu|(B_j)}{r_j^{n-1}}.$$

Finally, recalling (178) and (179), we conclude with

$$\sum_{j=k}^{h-1} E_j \leq \frac{\delta_1^{2n} \lambda_M \varepsilon}{50}.$$

On the other hand, taking $k_1 = k$ and $k_2 = h$ in (140) and using the inequality in the last display we can estimate

$$|(Du)_{B_h} - (Du)_{B_k}| \leq \delta_1^{-n} \sum_{j=k}^{h-1} E_j \leq \lambda_M \varepsilon \tag{189}$$

and (180) follows.

Case 2: $j_m \leq k < h$. Here (180) follows as a consequence of the inequalities

$$|(Du)_{B_h}| \leq \frac{\lambda_M \varepsilon}{25} \quad \text{and} \quad |(Du)_{B_k}| \leq \frac{\lambda_M \varepsilon}{25}. \tag{190}$$

If $h \in \mathcal{L}$, the first inequality in (190) follows immediately from the definition of $\mathcal{L}$; we can therefore assume $h \notin \mathcal{L}$. Then, as $h > j_m$, it is possible to consider a nondegenerate iteration chain $\mathcal{C}_{i_h}^{m_h}$ with $m_h > 0$, such that $h \in \mathcal{C}_{i_h}^{m_h}$; notice that $h > i_h$ as $h \notin \mathcal{L} \ni i_h$, since by definition the first index in a chain does belong to $\mathcal{L}$. We are again in position to apply Lemma 7 for $j \in \{i_h, \ldots, i_h + m_h - 1\}$, so that (188) holds for the corresponding indexes. Summing up, proceeding as after (188), and yet summing E_{i_h} to both sides of the resulting inequality, we arrive at

$$\sum_{j=i_h}^{i_h+m_h} E_j \le 2E_{i_h} + 8c_2 c_d \delta_1^{-n} \lambda_M^{2-p} \sum_{j=0}^{\infty} \frac{|\mu|(B_j)}{r_j^{n-1}} \le \frac{\delta_1^{2n} \lambda_M \varepsilon}{50}.$$

We have again used (178) and (179). Proceeding as in (140) (taking $k_1 = i_h$ and $k_2 = h$), and using the definition of $\mathcal{C}_{i_h}^{m_h}$ to estimate

$$|(Du)_{B_{i_h}}| \le C_{i_h} \le \frac{\lambda_M \varepsilon}{100}$$

we have

$$\begin{aligned} |(Du)_{B_h}| &\le |(Du)_{B_h} - (Du)_{B_{i_h}}| + |(Du)_{B_{i_h}}| \\ &\le \delta_1^{-n} \sum_{i=i_h}^{h-1} E_i + \frac{\lambda_M \varepsilon}{100} \\ &\le \delta_1^{-n} \sum_{i=i_h}^{i_h+m_h} E_i + \frac{\lambda_M \varepsilon}{100} \le \frac{\lambda_M \varepsilon}{25}, \end{aligned}$$

that is the first inequality in (190). The proof of the second inequality in (190) is completely similar; we just observe that we can assume that $k > j_m$ otherwise the inequality itself is trivial as $k \in \mathcal{L}$.

Case 3: $k < j_m < h$. This can be actually treated by a combination of the first two cases. It is indeed sufficient to prove that the inequalities in display (190) still hold. The first inequality in (190) follows exactly as in Case 2. As for the second estimate in (190), let us remark that, as $j_m \in \mathcal{L}$, we have that

$$|(Du)_{B_{j_m}}| \le C_{j_m} \le \frac{\lambda_M \varepsilon}{50}.$$

On the other hand, we can repeat the argument of Case 1, with h replaced by j_m, iterating from h to j_m and thereby obtaining

$$|(Du)_{B_{j_m}} - (Du)_{B_k}| \le \frac{\lambda_M \varepsilon}{50}$$

as in (189). Finally, the inequalities in the last two displays and yet triangle inequality give

$$|(Du)_{B_k}| \le |(Du)_{B_k} - (Du)_{B_{j_m}}| + |(Du)_{B_{j_m}}| \le \frac{\lambda_M \varepsilon}{50} + \frac{\lambda_M \varepsilon}{100} \le \frac{\lambda_M \varepsilon}{25}$$

and therefore the second inequality in (190) follows. The proof of Theorem 15 is therefore complete in the case of an energy solution $u \in W^{1,p}(\Omega)$.

14.4 Very weak solutions

In this section we are going to prove Theorem 15 for SOLA. We shall also obtain the extensions of Theorems 10 and 19 to SOLA, since these are indeed needed as preliminary results. The method we propose here is of course based on an approximation argument, but, rather than passing to the limits in the final estimates, we will pass to the limits "in the proofs". Strictly speaking, we shall pass to the limits in some estimates from the proofs. The same methods developed for energy solutions will then work for SOLA. We start with a preliminary result. Let us consider a ball $B(x, R) \subset \Omega$, and a sequence of shrinking balls $\{B_j\}$, concentric to $B(x, R)$:

$$\cdots B_{j+1} \subset B_j \subset B_{j-1} \cdots B_0 \subset B(x, R). \tag{191}$$

Then we have the following:

Lemma 8 *Let u be a SOLA to* (17) *under the assumptions* (18), *and let $\{B_j\}$ be the sequence of balls considered in* (191). *There exists a sequence of functions $\{v_j\}$ such that for every $j \geq 0$ it holds*

- $v_j \in W^{1,p}_{\mathrm{loc}}(B_j) \cap u + W^{1,p-1}_0(B_j)$
- v_j *is a local energy solution in B_j i.e.* $\mathrm{div}\, a(x, Dv_j) = 0$ *in* B_j
- *There exist constants $c_1, \bar{c}_1 \equiv c_1, \bar{c}_1(n, p, \nu)$ such that*

$$\fint_{B_j} |Du - Dv_j|^q \, d\tilde{x} \leq c_1 \left[\frac{|\mu|(\bar{B}_j)}{r_j^{n-1}} \right]^{q/(p-1)} \tag{192}$$

and

$$\fint_{B_j} |u - v_j|^q \, d\tilde{x} \leq \bar{c}_1 \left[\frac{|\mu|(\bar{B}_j)}{r_j^{n-p}} \right]^{q/(p-1)} \tag{193}$$

hold whenever q is as in (78)
- *There exists a constant $c \equiv c(n, p, \nu)$ such that*

$$\int_{B_j} \frac{|V(Du) - V(Dv_j)|^2}{(h + |u - v_j|)^{\xi}} \, d\tilde{x} \leq c \frac{h^{1-\xi}}{\xi - 1} |\mu|(\bar{B}_j) \tag{194}$$

holds whenever $h > 0$ and $\xi > 1$, where $c \equiv c(n, p, \nu) \geq 1$. The function $V(\cdot)$ has been defined in (71).

Proof The proof goes via approximation and we confine ourselves to explain the arguments for the inequality in display (192), those for the other ones being completely analogous. We preliminary recall that by the results in [8,55]—see also Theorem 14—any SOLA u to (55) belongs to $W^{1,q}_{\mathrm{loc}}(\Omega)$ for the numbers q in the range described in

(76). With u being now a fixed SOLA to (17), we consider the associated sequence $\{u_k\} \subset W^{1,p}_{\mathrm{loc}}(\Omega)$ from Definition 8 and note that using a standard diagonal we may assume, up to passing to not relabelled subsequences, that $Du_k \to Du$ a.e.. With B_j being fixed we now build the function v_j as required in the statement of the lemma. We start defining $w_k \in u_k + W^{1,p}_0(B_j)$ as the unique solution to the Dirichlet problem

$$\begin{cases} \operatorname{div} a(x, Dw_k) = 0 & \text{in } B_j \\ w_k = u_k & \text{on } \partial B_j. \end{cases} \tag{195}$$

Applying Lemma 2 in this context gives

$$\fint_{B_j} |Du_k - Dw_k|^q \, d\tilde{x} \le c_1 \left[\frac{|\mu_k|(\bar{B}_j)}{r_j^{n-1}} \right]^{q/(p-1)} . \tag{196}$$

The next step is now letting $k \to \infty$ in the previous inequality, and for this we need to recall a few regularity properties of the functions w_k. An important point is that the way the functions w_k have been defined gives that they belong to the Sobolev space $W^{1,p}(B_j)$ and this allows to apply the regularity theory available for such solutions. By (196) and the very definition of the functions u_k (they are converging to u in $W^{1,p-1}_0(\Omega)$) it follows that the sequence $\{w_k\}$ is bounded in $W^{1,p-1}(B_j)$; in turn this fact and the bound in (100) allows to get a uniform bound of the type $\|Dw_k\|_{L^\infty(\gamma B_j)} \le c(\gamma)$. In the same way we deduce, by using (101), that the sequence of maps $\{Dw_k\}$ is equicontinuous in γB_j. We can therefore invoke Ascoli–Arzelá's theorem that, together with a standard diagonal argument (recall that here $\gamma < 1$ is arbitrary), gives that there exists a function $w \in W^{1,\infty}_{\mathrm{loc}}(B_j) \cap u + W^{p-1}_0(B_j)$ such that, up to a not relabelled subsequence, we have $w_k \to w$ and $Dw_k \to Dw$ locally uniformly in B_j. Now, the locally uniform convergence of the gradients Dw_k allows to pass to the limits in the equations (195) and to conclude that $\operatorname{div} a(x, Dw) = 0$. Fatou's lemma now gives

$$\begin{aligned} \fint_{B_j} |Du - Dw|^q \, d\tilde{x} &\le \liminf_{k\to\infty} \fint_{B_j} |Du_k - Dw_k|^q \, d\tilde{x} \\ &\le \limsup_{k\to\infty} c_1 \left[\frac{|\mu_k|(B_j)}{r_j^{n-1}} \right]^{q/(p-1)} \le c_1 \left[\frac{|\mu|(\bar{B}_j)}{r_j^{n-1}} \right]^{q/(p-1)} , \end{aligned}$$

and in the last estimate we have used a standard property of the weak*-convergence of measures. A similar reasoning can be worked out with respect to (193) and (194). All in all we have proved the lemma with the choice $v_j \equiv w$. □

With the previous lemma at our disposal the proofs of Theorems 10, 19 and 15 for SOLA proceed verbatim as in the case of standard energy solutions, with the sequence v_j replacing everywhere the one in (112) whenever this is used. Indeed all the remaining facts used about the functions v_j follow by Lemma 8. In particular, all the facts reported in Sects. 11 and 12.1 are still valid since the functions v_j belong to $W^{1,p}_{\mathrm{loc}}(B_j) \cap W^{1,p-1}(B_j)$ and therefore standard regularity theory applies to them. The only difference is that the quantities $|\mu|(\bar{B}_j)$ must replace everywhere the analogous $|\mu|(B_j)$. This is clearly not a problem since the basic inequalities the quantities $|\mu|(\bar{B}_j)$ have to satisfy are implied by (116) and (179), which are already involving $|\mu|(\bar{B}_j)$. Needless to say, whenever we are considering vanishing limits as in (161) the same thing hold when considering closed balls.

15 Theorem 3 as a corollary

The assertion of Theorem 3 can be now obtained as a corollary of the results presented up to now, and, by the approach in the previous section, directly for SOLA. Indeed, estimate (24) follows directly by estimate (40) and the fact that the finiteness of $\mathbf{I}_1^{\mu}(x, R)$ implies that x is a Lebesgue point of Du. This is ensured by Theorem 15, see also Remark 1. It remains to prove that (25) implies the continuity of Du. For this we shall essentially revisit the proof of Theorem 15 given in the previous section and we shall see that the Cauchy sequence information in (170) holds locally uniformly in x. This means that the radius r_ε can be chosen independently of the point x as long as this varies in a fixed open subset $\Omega' \Subset \Omega$. The continuity of Du then follows by showing that it can be obtained as the locally uniform limit of the net of continuous maps defined by

$$x \to (Du)_{B(x,\varrho)} \qquad \text{for} \quad 0 < \varrho \le d := \operatorname{dist}(\Omega', \partial\Omega)\,, \tag{197}$$

which are obviously continuous. Note that we already know that the above maps converge pointwise to Du (that is, they converges to its precise representative at every point) as established in Theorem 15. Now, a standard covering argument and the assumption (25) imply that Du is locally bounded in Ω. Therefore, selecting another open subset Ω'' such that $\Omega' \Subset \Omega'' \Subset \Omega$, we set

$$\lambda_M := c\|\mathbf{I}_1^{\mu}(\cdot, R)\|_{L^\infty(\Omega'')}^{1/(p-1)} + c\||Du| + s\|_{L^\infty(\Omega'')} \tag{198}$$

for some $R > 0$, where c is the constant appearing in (40). This number is going to replace the analogous one picked in (169), making it independent of x. Then, we observe that assuming (25) allows to satisfy condition (174), uniformly in x (with an argument for instance based on computations as in (115)) and this implies that (172) holds again locally uniformly in x via Theorem 19. This, and the assumption (25), imply that we can replace (177) and (178) by the stronger conditions

$$\sup_{x\in\Omega'} \left[\int_0^{2R_l} \frac{|\mu|(B(x,\varrho))}{\varrho^{n-1}} \frac{d\varrho}{\varrho} \right]^{1/(p-1)} \le \frac{\delta_1^{5n}\lambda_M \varepsilon}{10^8 c_1 c_2 c_d}$$

and

$$\sup_{x\in\Omega'}\ \sup_{0<\varrho\leq R_l}\ E(Du, B(x,\varrho)) \leq \frac{\delta_1^{5n}\lambda_M\varepsilon}{10^8}$$

respectively, where now in addition we take $R_l < \mathrm{dist}(\Omega', \partial\Omega'')$ and λ_M is defined as in (198). With the inequalities of the last two displays being now in force it is easy to check that all the inequalities after (179) become uniform with respect to the point x considered. In particular, the numbers δ_1 and R_l that determine r_ε via (181), that is the radius for which (170) holds, are independent of x. This means that the maps defined in (197) converge uniformly to Du and the continuity of Du therefore follows.

16 Proof of Theorem 9

Theorem 9 has been again proved in a slightly different form in [40]. The proof of Theorem 9 goes essentially in two stages: we first obtain a pointwise estimate for a quantity like $|Du - G|$ (for any vector $G \in \mathbb{R}^n$). Dimensionally speaking, this quantity represents an oscillation of the gradient when G is suitably chosen, i.e. when for instance is taken to be an average of the gradient. Then we conclude applying the sharp maximal inequality of Theorem 12 that in fact allows to estimate, in an integral way, oscillations of the gradient. Notice that in the rest of the proof we shall assume that

$$\mathbf{W}^{\mu}_{1-\frac{(1+\alpha)(p-1)}{p},p}(x,R) + \mathbf{W}^{\mu}_{1-\frac{(1+\alpha)(p-1)}{p},p}(y,R) < \infty \tag{199}$$

holds, otherwise there is nothing to prove. We notice that this condition allows to conclude that both x and y are Lebesgue points of the gradient Du; this fact is indeed implied by Theorem 15, as the Wolff potentials appearing in (199) control the Riesz potential $\mathbf{I}^{\mu}_1$ at the points x, y via a computation similar to the one in (26). We will anyway give a shorter proof of this fact to make the proof of Theorem 9 self-contained; see Lemma 9 below.

16.1 General setting

We readopt the general setting of Theorem 10. Therefore we start by $x, y \in B_{R/4}$ and we select $r < R/2$ to be determined later. Then we follow the scheme described in Sect. 11. In particular, we define the shrinking balls $\{B_i\}$ as in (81); they are all centred at x, while eventually all the arguments will be reapplied when all the balls will be centred at y. The functions v_j are defined in (82) while as usual we denote

$$E_j := \fint_{B_j} |Du - (Du)_{B_j}|\, dx\,, \qquad j \geq 0.$$

The number δ_1 determining the shrinking rate of the balls B_j is instead defined as

$$\delta_1 := \left(\frac{1}{10^7 c_d}\right)^{1/\beta}, \tag{200}$$

where the constants c_d and β are from Theorem 17. This in turn yields a dependence of δ_1 on n, p, ν, L only.

16.2 Estimates on the excess

By the choice of δ_1 we can apply Lemma 4 with $\alpha = 1$; this gives

$$E_{j+1} \leq \frac{1}{10^6} E_j + 8c_1 c_d \delta_1^{-n} \left[\frac{|\mu|(B_j)}{r_j^{n-1}}\right]^{1/(p-1)} \tag{201}$$

for every $j \geq 0$. By summing inequalities (201) for $j \in \{k, \dots, h\}$, where $1 \leq k \leq h$ are two arbitrary nonnegative integers, we gain

$$\sum_{j=k+1}^{h+1} E_j \leq \frac{1}{10^6} \sum_{j=k}^{h} E_j + c \sum_{j=k}^{\infty} \left[\frac{|\mu|(B_j)}{r_j^{n-1}}\right]^{1/(p-1)}.$$

so that

$$\sum_{j=k+1}^{h+1} E_j \leq E_k + c \sum_{j=k}^{\infty} \left[\frac{|\mu|(B_j)}{r_j^{n-1}}\right]^{1/(p-1)}$$

and, ultimately

$$\sum_{j=k}^{h+1} E_j \leq 2E_k + c \sum_{j=k}^{\infty} \left[\frac{|\mu|(B_j)}{r_j^{n-1}}\right]^{1/(p-1)},$$

where $c \equiv c(n, p, \nu, L)$. By noticing that

$$\left[\frac{|\mu|(B_j)}{r_j^{n-1}}\right]^{1/(p-1)} = r_j^{\alpha} \left[\frac{|\mu|(B_j)}{r_j^{n-1+\alpha(p-1)}}\right]^{1/(p-1)} \leq r^{\alpha} \left[\frac{|\mu|(B_j)}{r_j^{n-1+\alpha(p-1)}}\right]^{1/(p-1)}$$

we conclude with

$$\sum_{j=k}^{h+1} E_j \leq 2E_k + cr^{\alpha} \sum_{j=k}^{\infty} \left[\frac{|\mu|(B_j)}{r_j^{n-1+\alpha(p-1)}}\right]^{1/(p-1)}. \tag{202}$$

Next, we estimate the last sum in terms of Wolff potentials as follows:

$$
\begin{aligned}
\mathbf{W}^{\mu}_{1-\frac{(1+\alpha)(p-1)}{p},p}(x,2r) &= \int_0^{2r} \left[\frac{|\mu|(B(x,\varrho))}{\varrho^{n-1+\alpha(p-1)}}\right]^{1/(p-1)} \frac{d\varrho}{\varrho} \\
&= \sum_{j=0}^{\infty} \int_{r_{j+1}}^{r_j} \left[\frac{|\mu|(B(x,\varrho))}{\varrho^{n-1+\alpha(p-1)}}\right]^{1/(p-1)} \frac{d\varrho}{\varrho} \\
&\quad + \int_r^{2r} \left[\frac{|\mu|(B(x,\varrho))}{\varrho^{n-1+\alpha(p-1)}}\right]^{1/(p-1)} \frac{d\varrho}{\varrho} \\
&\geq \sum_{j=0}^{\infty} \left[\frac{|\mu|(B_{j+1})}{r_j^{n-1+\alpha(p-1)}}\right]^{1/(p-1)} \int_{r_{j+1}}^{r_j} \frac{d\varrho}{\varrho} \\
&\quad + \left[\frac{|\mu|(B_0)}{(2r)^{n-1+\alpha(p-1)}}\right]^{1/(p-1)} \int_r^{2r} \frac{d\varrho}{\varrho} \\
&\geq \delta_1^{(n-1)/(p-1)+\alpha} \log\left(\frac{1}{\delta_1}\right) \sum_{j=0}^{\infty} \left[\frac{|\mu|(B_{j+1})}{r_{j+1}^{n-1+\alpha(p-1)}}\right]^{1/(p-1)} \\
&\quad + \frac{\log 2}{2^{(n-1)/(p-1)+\alpha}} \left[\frac{|\mu|(B_0)}{r_0^{n-1+\alpha(p-1)}}\right]^{1/(p-1)} \\
&\geq \delta_1^n \sum_{j=0}^{\infty} \left[\frac{|\mu|(B_j)}{r_j^{n-1+\alpha(p-1)}}\right]^{1/(p-1)}.
\end{aligned}
\tag{203}
$$

Now, by taking $k = 0$ and $h = i$ in (202) we have

$$
\sum_{j=0}^{i+1} E_j \leq 2E_0 + cr^{\alpha} \sum_{j=0}^{\infty} \left[\frac{|\mu|(B_j)}{r_j^{n-1+\alpha(p-1)}}\right]^{1/(p-1)}
$$

and merging the content of the last two displays we conclude with

$$
\sum_{j=0}^{i+1} E_j \leq 2E_0 + cr^{\alpha} \mathbf{W}^{\mu}_{1-\frac{(1+\alpha)(p-1)}{p},p}(x,R) \tag{204}
$$

where we have used the fact that $2r \leq R$ and the constant c depends only on n, p, ν, L thanks the choice of δ_1 made in (200). In particular, again taking $\alpha = 0$ and eventually letting $i \to \infty$ in the previous inequality gives

$$\sum_{j=0}^{\infty} E_j \leq 2E_0 + c\mathbf{W}^{\mu}_{1/p,p}(x, R). \tag{205}$$

16.3 Pointwise estimate and Lebesgue point criterion

Let us prove

Lemma 9 (Wolff potentials control Lebesgue points) *The condition*

$$\mathbf{W}^{\mu}_{1/p,p}(x, R) < \infty \tag{206}$$

implies that the following limit exists:

$$\lim_{i\to\infty} (Du)_{B_i} = Du(x).$$

Proof Observe that exactly as in (140) we have, whenever $h > k$

$$|(Du)_{B_h} - (Du)_{B_k}| \leq \delta_1^{-n} \sum_{j=k}^{h-1} E_j. \tag{207}$$

Using (202) with $\alpha = 0$ we gain

$$|(Du)_{B_h} - (Du)_{B_k}| \leq cE_k + c\sum_{j=k}^{\infty} \left[\frac{|\mu|(B_j)}{r_j^{n-1}}\right]^{1/(p-1)}. \tag{208}$$

Now, notice that (203) with $\alpha = 0$ gives

$$\sum_{j=0}^{\infty} \left[\frac{|\mu|(B_j)}{r_j^{n-1}}\right]^{1/(p-1)} \leq c\mathbf{W}^{\mu}_{1/p,p}(x, 2r) < \infty \tag{209}$$

and this, together with (205) gives

$$\lim_{k\to\infty} E_k + \sum_{j=k}^{\infty} \left[\frac{|\mu|(B_j)}{r_j^{n-1}}\right]^{1/(p-1)} = 0.$$

In turn this last fact and (208) imply that $\{(Du)_{B_i}\}$ is a Cauchy sequence and the assertion is proved. Notice that a slight variant of this argument (see for instance

Sect. 14.2) leads to establish that x is a Lebesgue point of Du in the usual sense. See also Sect. 17 below. □

As (199) implies the finiteness condition in (206) the previous lemma implies in particular that

$$\lim_{i\to\infty} |(Du)_{B_i} - G| = |Du(x) - G| \tag{210}$$

holds whenever $G \in \mathbb{R}^n$. Let us use (140) for the choice $k_1 = 0$ and $k_2 = i \geq 1$ to get

$$|(Du)_{B_i} - (Du)_{B_0}| \leq \delta_1^{-n} \sum_{j=0}^{i-1} E_j \equiv c \sum_{j=0}^{i-1} E_j$$

with $c \equiv c(n, p, \nu, L)$. Further use of triangle inequality then gives

$$|(Du)_{B_i} - G| \leq |(Du)_{B_0} - G| + c \sum_{j=0}^{i-1} E_j$$

so that, using (204) we obtain

$$|(Du)_{B_i} - G| \leq |(Du)_{B_0} - G| + cE_0 + cr^\alpha \mathbf{W}^\mu_{1-\frac{(1+\alpha)(p-1)}{p},p}(x, R),$$

where the constant c depends only on n, p, ν, L. Letting $i \to \infty$ and recalling (210) we conclude with

$$\begin{aligned} |Du(x) - G| &\leq c \fint_{B(x,r)} |Du - (Du)_{B(x,r)}| + |Du - G|)\, d\tilde{x} \\ &\quad + cr^\alpha \mathbf{W}^\mu_{1-\frac{(1+\alpha)(p-1)}{p},p}(x, R). \end{aligned}$$

16.4 Use of the maximal estimate and conclusion

Writing inequality (211) for y instead of x and summing up gives

$$\begin{aligned} |Du(x) - Du(y)| &\leq c \fint_{B(x,r)} (|Du - (Du)_{B(x,r)}| + |Du - G|)\, d\tilde{x} \\ &\quad + c \fint_{B(y,r)} (|Du - (Du)_{B(y,r)}| + |Du - G|)\, d\tilde{x} \\ &\quad + c\left[\mathbf{W}^\mu_{1-\frac{(1+\alpha)(p-1)}{p},p}(x, R) + \mathbf{W}^\mu_{1-\frac{(1+\alpha)(p-1)}{p},p}(y, R)\right] r^\alpha. \end{aligned} \tag{211}$$

We recall that we can still choose G and $r \leq R/2$; we therefore take $G := (Du)_{B(x,3r)}$ and $r := |x-y|/2$, so that $B(y,r) \subset B(x,3r)$ and therefore, using also (98) we have

$$
\begin{aligned}
&\fint_{B(x,r)} (|Du - (Du)_{B(x,r)}| + |Du - G|)\, d\tilde{x} \\
&\qquad + \fint_{B(y,r)} (|Du - (Du)_{B(y,r)}| + |Du - G|)\, d\tilde{x} \\
&\qquad \leq c(n) \fint_{B(x,3r)} |Du - (Du)_{B(x,3r)}|\, d\tilde{x}. \qquad (212)
\end{aligned}
$$

Now, notice that as we are assuming that the initial ball B_R is such that $x, y \in B_{R/4}$, then we have necessarily $|x-y| \leq R/2$ so that now $r \leq R/4$ and $B(x,3r) \subset B(x, 3R/4) \subset B_R$. Therefore we use the definition of sharp maximal operator and Theorem 12 to estimate

$$
\begin{aligned}
&\fint_{B(x,3r)} |Du - (Du)_{B(x,3r)}|\, d\tilde{x} \leq c r^{\alpha} M^{\#}_{\alpha,3r}(Du)(x) \\
&\qquad \leq c r^{\alpha} \left[M_{1-\alpha(p-1),3R/4}(\mu)(x)\right]^{1/(p-1)} + c\left(\frac{r}{R}\right)^{\alpha} \fint_{B_{3R/4}} |Du - (Du)_{B_{3R/4}}|\, d\tilde{x}.
\end{aligned}
$$

It is possible to estimate the fractional maximal operators by Wolff potentials via completely elementary means; for instance, the following inequality is proved in [40, Lemma 4.1] and holds whenever $\gamma \in (0,1)$:

$$
\left[M_{\beta,\gamma R}(\mu)(x)\right]^{1/(p-1)} \leq \frac{\max\{\gamma^{(\beta-n)/(p-1)}, 1\}}{(-\log\gamma)|B(0,1)|^{1/(p-1)}} \mathbf{W}^{\mu}_{\beta/p,p}(x,R). \qquad (213)
$$

The content of the last two displays and (98) yields

$$
\begin{aligned}
&\fint_{B(x,3r)} |Du - (Du)_{B(x,3r)}|\, d\tilde{x} \\
&\qquad \leq c r^{\alpha} \mathbf{W}^{\mu}_{1-\frac{(1+\alpha)(p-1)}{p},p}(x,R) + c\left(\frac{r}{R}\right)^{\alpha} \fint_{B_R} |Du - (Du)_{B_R}|\, d\tilde{x}.
\end{aligned}
$$

Connecting the last inequality to (212) and eventually to (211) completes the proof of Theorem 9.

17 Proof of Theorem 16

The proof goes in two steps, reported in Sects. 17.1 and 17.2, respectively. We here give the whole proof of Theorem 16 in the case of an energy solution $u \in W^{1,p}(\Omega)$;

the case of a general SOLA can be treated adapting the arguments valid for the energy case along the lines described in Sect. 14.4 for Theorem 15. In the rest of the proof all the balls will be centred at the point x, starting from an initial ball $B(x, R) \subset \Omega$ with $R \leq 1$, which is not restrictive of course.

17.1 Smallness of the excess

Here we prove

$$\lim_{\varrho \to 0} E(u, B(x, \varrho)) = \lim_{\varrho \to 0} \fint_{B(x,\varrho)} |u - (u)_{B(x,\varrho)}| \, d\tilde{x} = 0 \tag{214}$$

that is, we prove that for every $\tilde{\varepsilon} > 0$ there exists $\tilde{R} > 0$, basically depending on $\tilde{\varepsilon}$ and on the point x, such that

$$\varrho \leq \tilde{R} \Longrightarrow E(u, B(x, \varrho)) < \tilde{\varepsilon}. \tag{215}$$

To start with, we notice that by using (44) with $\alpha = 0$, together with (213) with $\beta = p$ to establish the finiteness of the relevant maximal operator, we get

$$\sup_{0<\varrho \leq R} \fint_{B(x,\varrho)} |u - (u)_{B(x,\varrho)}| \, d\tilde{x} < \infty. \tag{216}$$

Thanks to this we are able to determine $\varepsilon > 0$ small enough in order to have

$$\varepsilon \sup_{0<\varrho \leq R} \fint_{B(x,\varrho)} |u - (u)_{B(x,\varrho)}| \, d\tilde{x} + 2\varepsilon R s \leq \frac{\tilde{\varepsilon}}{2}. \tag{217}$$

With ε been determined, we define δ_1 as

$$\delta_1 := \left(\frac{\varepsilon}{10^8 \tilde{c}_d} \right)^{1/\tilde{\beta}}, \tag{218}$$

where the constants $\tilde{c}_d$ and $\tilde{\beta}$ are from Theorem 18. In turn, having δ_1 being defined, we choose the radius $r \leq R/2$ such that the smallness condition

$$16 \tilde{c}_1 \tilde{c}_d \delta_1^{-3n} \mathbf{W}^{\mu}_{1,p}(x, 2r) \leq \frac{\tilde{\varepsilon}}{2} \tag{219}$$

is satisfied. The constant $\tilde{c}_1$ has been introduced in (77). Notice that this is possible since the assumption in (62) and the definition of Wolff potential imply that

$$\lim_{\varrho \to 0} \mathbf{W}^{\mu}_{1,p}(x, \varrho) = 0. \tag{220}$$

Determining r allows to consider the sequence of shrinking balls $\{B_j\}$ as defined in (81); the functions $\{v_j\}$ are as in (112). Accordingly, we define

$$\tilde{E}_j := \fint_{B_j} |u - (u)_{B_j}| \, d\tilde{x}.$$

Finally, an argument similar to the ones in (203) yields

$$\delta_1^n \sum_{j=0}^{\infty} \left[\frac{|\mu|(B_j)}{r_j^{n-p}} \right]^{1/(p-1)} \leq \mathbf{W}^{\mu}_{1,p}(x, 2r). \tag{221}$$

This is indeed an inequality that actually works whenever $\delta_1 \in (0, 1/4)$ is a free parameter and the balls $\{B_j\}$ are defined as $B_j := B(x, \delta_1^j r)$.

Now, the first thing we do is looking at Lemma 4 (when $\alpha = 0$) and make computations similar to the one involved there, but using Theorem 18 instead of Theorem 17, and using (77) instead of (76). Specifically, for every $j \geq 0$ we have

$$\begin{aligned}
\tilde{E}_{j+1} &\leq 2 \fint_{B_{j+1}} |v_j - (v_j)_{B_{j+1}}| \, d\tilde{x} + 2\delta_1^{-n} \fint_{B_j} |v_j - u| \, d\tilde{x} \\
&\leq 2\tilde{c}_d \delta_1^{\tilde{\beta}} \fint_{B_j} |v_j - (v_j)_{B_j}| \, d\tilde{x} + 2\tilde{c}_1 \delta_1^{-n} \left[\frac{|\mu|(B_j)}{r_j^{n-p}} \right]^{1/(p-1)} + 2\tilde{c}_d \delta_1^{\tilde{\beta}} r_j s \\
&\leq 4\tilde{c}_d \delta_1^{\tilde{\beta}} \fint_{B_j} |u - (u)_{B_j}| \, d\tilde{x} + 8\tilde{c}_1 \tilde{c}_d \delta_1^{-n} \left[\frac{|\mu|(B_j)}{r_j^{n-p}} \right]^{1/(p-1)} + 2\tilde{c}_d \delta_1^{\tilde{\beta}} r_j s.
\end{aligned}$$

Recalling (218) we conclude that the following estimate holds for every $j \geq 0$:

$$\tilde{E}_{j+1} \leq \frac{\varepsilon}{10^6} \tilde{E}_j + 8\tilde{c}_1 \tilde{c}_d \delta_1^{-n} \left[\frac{|\mu|(B_j)}{r_j^{n-p}} \right]^{1/(p-1)} + \tilde{\varepsilon} r_j s. \tag{222}$$

Summing up the preceding inequalities and proceeding as in Sect. 16.2 yields

$$\sum_{j=k}^{h} \tilde{E}_j \leq \varepsilon \tilde{E}_{k-1} + \varepsilon r s \sum_{j=0}^{\infty} \delta_1^j + 16\tilde{c}_1 \tilde{c}_d \delta_1^{-n} \sum_{j=0}^{\infty} \left[\frac{|\mu|(B_j)}{r_j^{n-p}} \right]^{1/(p-1)}, \tag{223}$$

which is valid whenever $1 \leq k \leq h$ are integers. We then use (217) to estimate the first two terms in the right hand side of the previous inequality and (219)–(221) to estimate the remaining one; therefore we conclude with

$$\sum_{j=1}^{\infty} \tilde{E}_j \leq \tilde{\varepsilon}. \tag{224}$$

This last inequality in particular implies that

$$\fint_{B(x,\varrho)} |u - (u)_{B(x,\varrho)}|\, d\tilde{x} \leq \tilde{\varepsilon} \qquad \text{holds whenever } \varrho = r_j \text{ and } j \geq 1. \tag{225}$$

On the other hand, take now $\varrho' \leq r_1 = \delta_1 r$ and determine $k \geq 1$ such that $r_{k+1} < \varrho \leq r_k$. This means that $\varrho' = \delta_1^k r'$ for $r' \in (\delta_1 r, r)$. Now, in the proof of (225) we replace r by r' and we get a new chain of cylinders $B_j' = B(x, \delta^j r')$ to which we apply the same reasoning as above; then (225) follows for $\varrho = r_j'$ and every $j \geq 1$. In particular it follows that $E(u, B(x, \varrho')) \leq \tilde{\varepsilon}$ and this means now that now (215) holds for $\tilde{R} = \delta_1 r$ where r has been determined via the choice in (219).

17.2 Proof that x is a Lebesgue point of u

We are now ready for the proof of (63), therefore showing that for every $\varepsilon > 0$, there exists a radius r_ε, depending on ε, such that

$$|(u)_{B(x,\varrho)} - (u)_{B(x,\tau)}| \leq \varepsilon \qquad \text{holds whenever } 0 < \tau < \varrho \leq r_\varepsilon. \tag{226}$$

We now define

$$\delta_1 := \left(\frac{1}{10^8 \tilde{c}_d}\right)^{1/\tilde{\beta}} \tag{227}$$

where $\tilde{c}_d$ appears in (106), and then we take a radius $\tilde{R}_l$ such that

$$\sup_{0<\varrho\leq\tilde{R}_l} E(u, B(x, \varrho)) + 16\tilde{c}_1\tilde{c}_d\delta_1^{-2n}\mathbf{W}^{\mu}_{1,p}(x, 2\tilde{R}_l) + 2s\tilde{R}_l \leq \frac{\delta_1^n \varepsilon}{3}. \tag{228}$$

We notice that this is possible thanks to (214) and (220). With δ_1 and $\tilde{R}_l$ having been defined, we then use the usual shrinking chain of balls $\{B_j\}$ defined in (81) with $r = \tilde{R}_l$, so that, in particular we have $r_j = \delta_1^j \tilde{R}_l$; again, the functions $\{v_j\}$ are as in (112). Similarly to (207), whenever $h > k \geq 1$ are integers it holds that

$$|(u)_{B_h} - (u)_{B_k}| \leq \delta_1^{-n} \sum_{j=k}^{h-1} \tilde{E}_j. \tag{229}$$

We now observe that (223) still works in this setting with $\varepsilon = 1$, according to the choice in (227). This is

$$\sum_{j=k}^{h} \tilde{E}_j \le \tilde{E}_{k-1} + rs \sum_{j=0}^{\infty} \delta_1^j + 16\tilde{c}_1\tilde{c}_d\delta_1^{-2n}\mathbf{W}_{1,p}^{\mu}(x, 2r),$$

and we also used (221), which is again an inequality that whenever $\delta_1 \in (0, 1/4)$ is a fixed parameter. Combining the inequalities in the last two displays together with (228), we obtain that

$$|(u)_{B_h} - (u)_{B_k}| \le \frac{\varepsilon}{3} \quad \text{holds for every choice of } h > k \ge 1. \tag{230}$$

We can now prove (226) with the choice $r_\varepsilon := \delta_1 \tilde{R}_l$. Indeed, let us fix $0 < \tau < \varrho \le r_\varepsilon$. This means that there exist two integers k and h, such that $1 \le k \le h$,

$$\delta_1^{k+1}\tilde{R}_l < \varrho \le \delta_1^k \tilde{R}_l \quad \text{and} \quad \delta_1^{h+1}\tilde{R}_l < \tau \le \delta_1^h \tilde{R}_l$$

hold. Applying (228) yields

$$\begin{aligned} |(u)_{B(x,\varrho)} - (u)_{B_{k+1}}| &\le \fint_{B_{k+1}} |u - (u)_{B(x,\varrho)}| \, d\tilde{x} \\ &\le \frac{|B(x,\varrho)|}{|B_{k+1}|} \fint_{B_{k+1}} |u - (u)_{B(x,\varrho)}| \, d\tilde{x} \\ &\le \delta_1^{-n} E(Du, B(x,\varrho)) \le \frac{\varepsilon}{3}, \end{aligned}$$

and, similarly,

$$|(u)_{B(x,\tau)} - (u)_{B_{h+1}}| \le \frac{\varepsilon}{3}.$$

The last two inequalities and (230) establish (226) and the proof of Theorem 16 is complete.

18 Back to the roots: Theorem 1 (sketch)

The arguments developed for Theorem 16 allow to get a quick proof of Theorem 1, that we sketch in the following lines. We repeat the construction of Sects. 17.1–17.2 with δ_1 as in (227), but starting from any ball $B(x, R) \subset \Omega$, with $r = R/2$; the sequence $\{B_j\}$ is once again accordingly defined as in (81). In (229) we take $k = 1$ and let $h \to \infty$; this yields

$$|u(x) - (u)_{B_0}| \le \delta_1^{-n} \sum_{j=1}^{\infty} \tilde{E}_j.$$

Using this last inequality together with (221) and (223) (here it is $\varepsilon = 1$) yields

$$\begin{aligned} |u(x)| &\le |(u)_{B_0}| + crs + \tilde{E}_0 + c\mathbf{W}^{\mu}_{1,p}(x, R) \\ &\le c\mathbf{W}^{\mu}_{1,p}(x, R) + c \fint_{B(x,R)} (|u| + Rs)\, d\tilde{x} \end{aligned}$$

for a constant $c \equiv c(n, p, \nu, L)$, that is (20). The proof of the continuity of u under assumption (21) follows instead making the arguments for the proof of Theorem 16 uniform with respect to x, in the same way we have done in Sect. 15 with respect to the proof of Theorem 15.

Acknowledgments We acknowledge the hospitality of the Universidad Autonoma de Madrid in April and May 2013, where part of this paper was written. In particular, we thank Matteo Bonforte and "Maestro" Juan Luis Vázquez for their kindness. The paper was finalised while we were both guests of the Institut Mittag-Leffler, joining the program "Evolutionary problems" in the fall 2013. Last but not least, we thank Paolo Baroni for precious remarks on a preliminary version of the paper.

References

1. Adams, D.R.: A note on Riesz potentials. Duke Math. J. **42**, 765–778 (1975)
2. Adams, D.R., Meyers, N.G.: Thinnes and Wiener criteria for nonlinear potentials. Indiana Univ. Math. J. **22**, 169–197 (1972)
3. Adams, D.R., Hedberg, L.I.: Function spaces and potential theory. Grundlehren der Mathematischen Wissenschaften, vol. 314. Springer, Berlin (1996)
4. Aronsson, G., Crandall, M.G., Juutinen, P.: A tour of the theory of absolutely minimizing functions. Bull. Am. Math. Soc. (N.S.) **41**, 439–505 (2004)
5. Baroni, P.: Linear potential estimates for a class of elliptic equations (2013, preprint)
6. Baroni, P., Di Castro, A., Palatucci, G.: (in preparation)
7. Bénilan, P., Boccardo, L., Gallouët, T., Gariepy, R., Pierre, M., Vázquez, J.L.: An L^1-theory of existence and uniqueness of solutions of nonlinear elliptic equations. Ann. Scuola Norm. Sup. Pisa Cl. Sci. (IV) **22**, 241–273 (1995)
8. Boccardo, L., Gallouët, T.: Nonlinear elliptic and parabolic equations involving measure data. J. Funct. Anal. **87**, 149–169 (1989)
9. Byun, S.S., Wang, L., Zhou, S.: Nonlinear elliptic equations with BMO coefficients in Reifenberg domains. J. Funct. Anal. **250**, 167–196 (2007)
10. Campanato, S.: Proprietà di hölderianità di alcune classi di funzioni. Ann. Scuola Norm. Sup. Pisa (III) **17**, 175–188 (1963)
11. Cianchi, A.: Nonlinear potentials, local solutions to elliptic equations and rearrangements. Ann. Scu. Norm. Sup. Cl. Sci. (V) **10**, 335–361 (2011)
12. Cianchi, A., Maz'ya, V.: Global Lipschitz regularity for a class of quasilinear elliptic equations. Commun. PDE **36**, 100–133 (2011)
13. Dall'Aglio, A.: Approximated solutions of equations with L^1 data. Application to the H-convergence of quasi-linear parabolic equations. Ann. Mat. Pura Appl. (IV) **170**, 207–240 (1996)
14. Dal Maso, G., Murat, F., Orsina, L., Prignet, A.: Renormalized solutions of elliptic equations with general measure data. Ann. Scuola Norm. Sup. Pisa Cl. Sci. (IV) **28**, 741–808 (1999)
15. Daskalopoulos, T., Kuusi, T., Mingione, G.: Borderline estimates for fully nonlinear elliptic equations. Commun. PDE (to appear)

16. De Giorgi, E.: Sulla differenziabilità e l'analiticità delle estremali degli integrali multipli regolari. Mem. Accad. Sci. Torino Cl. Sci. Fis. Mat. Nat. (III) **125**(3), 25–43 (1957)
17. DeVore, R.A., Sharpley, R.C.: Maximal functions measuring smoothness. Mem. Am. Math. Soc. **47**, no. 293 (1984)
18. DiBenedetto, E.: $C^{1+\alpha}$ local regularity of weak solutions of degenerate elliptic equations. Nonlinear Anal. TMA **7**, 827–850 (1983)
19. DiBenedetto, E., Manfredi, J.J.: On the higher integrability of the gradient of weak solutions of certain degenerate elliptic systems. Am. J. Math. **115**, 1107–1134 (1993)
20. Dolzmann, G., Hungerbühler, N., Müller, S.: Uniqueness and maximal regularity for nonlinear elliptic systems of n-Laplace type with measure valued right hand side. J. Reine Angew. Math. **520**, 1–35 (2000)
21. Duzaar, F., Mingione, G.: Gradient estimates via non-linear potentials. Am. J. Math. **133**, 1093–1149 (2011)
22. Duzaar, F., Mingione, G.: Gradient estimates via linear and nonlinear potentials. J. Funct. Anal. **259**, 2961–2998 (2010)
23. Duzaar, F., Mingione, G.: Gradient continuity estimates. Calc. Var. PDE **39**, 379–418 (2010)
24. Evans, L.C.: A new proof of local $C^{1,\alpha}$ regularity for solutions of certain degenerate elliptic p.d.e. J. Differ. Equ. **45**, 356–373 (1982)
25. Giusti, E.: Direct Methods in the Calculus of Variations. World Scientific Publishing Co., Inc, River Edge (2003)
26. Grafakos, L.: Classical and Modern Fourier Analysis. Pearson Edu. Inc., Upper Saddle River (2004)
27. Hamburger, C.: Regularity of differential forms minimizing degenerate elliptic functionals. J. Reine Angew. Math. (Crelles J.) **431**, 7–64 (1992)
28. Havin, M., Maz'ja, V.G.: Non-linear potential theory. Russ. Math. Surveys **27**, 71–148 (1972)
29. Hedberg, L., Wolff, Th.H.: Thin sets in nonlinear potential theory. Ann. Inst. Fourier (Grenoble) **33**, 161–187 (1983)
30. Heinonen, J., Kilpeläinen, T., Martio, O.: Nonlinear Potential Theory of Degenerate Elliptic Equations. Oxford Mathematical Monographs, New York (1993)
31. Iwaniec, T.: Projections onto gradient fields and L^p-estimates for degenerated elliptic operators. Studia Math. **75**, 293–312 (1983)
32. Jaye, B., Verbitsky, I.: Local and global behaviour of solutions to nonlinear equations with natural growth terms. Arch. Ration. Mech. Anal. **204**, 627–681 (2012)
33. Jin, T., Mazya, V., Van Schaftingen, J.: Pathological solutions to elliptic problems in divergence form with continuous coefficients. Comptes Rendus Mathematique **347**, 773–778 (2009)
34. John, F., Nirenberg, L.: On functions of bounded mean oscillation. Commun. Pure Appl. Math. **14**, 415–426 (1961)
35. Kichenassamy, S., Verón, L.: Singular solutions to the p-Laplace equation. Math. Ann. **275**, 599–615 (1986)
36. Kilpeläinen, T., Kuusi, T., Tuhola-Kujanpää, A.: Superharmonic functions are locally renormalized solutions. Ann. Inst. H. Poincaré, Anal. Non Linèaire **28**, 775–795 (2011)
37. Kilpeläinen, T., Malý, J.: Degenerate elliptic equations with measure data and nonlinear potentials. Ann. Scuola Norm. Sup. Pisa Cl. Sci. (IV) **19**, 591–613 (1992)
38. Kilpeläinen, T., Malý, J.: The Wiener test and potential estimates for quasilinear elliptic equations. Acta Math. **172**, 137–161 (1994)
39. Korte, R., Kuusi, T.: A note on the Wolff potential estimate for solutions to elliptic equations involving measures. Adv. Calc. Var. **3**, 99–113 (2010)
40. Kuusi, T., Mingione, G.: Universal potential estimates. J. Funct. Anal. **262**, 4205–4269 (2012)
41. Kuusi, T., Mingione, G.: Linear potentials in nonlinear potential theory. Arch. Ration. Mech. Anal. **207**, 215–246 (2013)
42. Kuusi, T., Mingione, G.: The Wolff gradient bound for degenerate parabolic equations. J. Eur. Math. Soc. (2014, to appear)
43. Kuusi, T., Mingione, G.: Gradient regularity for nonlinear parabolic equations. Ann Scu. Norm. Sup. Cl. Sci. (Ser. V) **12**, 755–822 (2013)
44. Kuusi, T., Mingione, G.: Riesz potentials and nonlinear parabolic equations. Arch. Ration. Mech. Anal.
45. Kuusi, T., Mingione, G.: A nonlinear Stein theorem. Calc. Var. PDE.

46. Lewis, J.L.: Regularity of the derivatives of solutions to certain degenerate elliptic equations. Indiana Univ. Math. J. **32**, 849–858 (1983)
47. Lieberman, G.M.: The natural generalization of the natural conditions of Ladyzhenskaya and Ural'tseva for elliptic equations. Commun. PDE **16**, 311–361 (1991)
48. Lindqvist, P.: On the definition and properties of p-superharmonic functions. J. Reine Angew. Math. (Crelles J.) **365**, 67–79 (1986)
49. Lindqvist, P.: Notes on the p-Laplace equation. Univ. Jyväskylä, Report 102 (2006)
50. Malý, J.: Wolff potential estimates of superminimizers of Orlicz type Dirichlet integrals. Manuscr. Math. **110**, 513–525 (2003)
51. Manfredi, J.J.: Regularity for minima of functionals with p-growth. J. Differ. Equ. **76**, 203–212 (1988)
52. Maz'ya, V., Shaposhnikova, T.: On the Bourgain, Brezis, and Mironescu theorem concerning limiting embeddings of fractional Sobolev spaces. J. Funct. Anal. **195**, 230–238 (2002)
53. Mikkonen, P.: On the Wolff potential and quasilinear elliptic equations involving measures. Ann. Acad. Sci. Fenn. Math. Diss. **104** (1996)
54. Mingione, G.: The singular set of solutions to non-differentiable elliptic systems. Arch. Ration. Mech. Anal. **166**, 287–301 (2003)
55. Mingione, G.: The Calderón-Zygmund theory for elliptic problems with measure data. Ann Scu. Norm. Sup. Pisa Cl. Sci. (V) **6**, 195–261 (2007)
56. Mingione, G.: Gradient estimates below the duality exponent. Math. Ann. **346**, 571–627 (2010)
57. Mingione, G.: Gradient potential estimates. J. Eur. Math. Soc. **13**, 459–486 (2011)
58. Mingione, G.: Nonlinear measure data problems. Milan J. Math. **79**, 429–496 (2011)
59. Phuc, N.C., Verbitsky, I.E.: Quasilinear and Hessian equations of Lane–Emden type. Ann. Math. (II) **168**, 859–914 (2008)
60. Phuc, N.C., Verbitsky, I.E.: Singular quasilinear and Hessian equations and inequalities. J. Funct. Anal. **256**, 1875–1906 (2009)
61. Serrin, J.: Pathological solutions of elliptic differential equations. Ann. Scuola Norm. Sup. Pisa (III) **18**, 385–387 (1964)
62. Serrin, J.: Local behavior of solutions of quasi-linear equations. Acta Math. **111**, 247–302 (1964)
63. Stein, E.M.: Editor's note: the differentiability of functions in $\mathbb{R}^n$. Ann. Math. (II) **113**, 383–385 (1981)
64. Teixeira, E.: Sharp regularity for general Poisson equations with borderline sources. J. Math. Pures Appl. **99**, 150–164 (2013)
65. Trudinger, N.S., Wang, X.J.: On the weak continuity of elliptic operators and applications to potential theory. Am. J. Math. **124**, 369–410 (2002)
66. Trudinger, N.S., Wang, X.J.: Quasilinear elliptic equations with signed measure data. Disc. Contin. Dyn. Syst. A **23**, 477–494 (2009)
67. Ural'tseva, N.N.: Degenerate quasilinear elliptic systems. Zap. Na. Sem. Leningrad. Otdel. Mat. Inst. Steklov. (LOMI) **7**, 184–222 (1968)

4

Diagonalizations of two classes of unbounded Hankel operators

D. R. Yafaev

Abstract We show that every Hankel operator H is unitarily equivalent to a pseudo-differential operator A of a special structure acting in the space $L^2(\mathbb{R})$. As an example, we consider integral operators H in the space $L^2(\mathbb{R}_+)$ with kernels $P(\ln(t+s))$ $(t+s)^{-1}$ where $P(x)$ is an arbitrary real polynomial of degree K. In this case, A is a differential operator of the same order K. This allows us to study spectral properties of Hankel operators H with such kernels. In particular, we show that the essential spectrum of H coincides with the whole axis for K odd, and it coincides with the positive half-axis for K even. In the latter case we additionally find necessary and sufficient conditions for the positivity of H. We also consider Hankel operators whose kernels have a strong singularity at some positive point. We show that spectra of such operators consist of the zero eigenvalue of infinite multiplicity and eigenvalues accumulating to $+\infty$ and $-\infty$. We find the asymptotics of these eigenvalues.

Keywords Hankel operators · Necessary and sufficient conditions for the positivity · Essential spectrum · Quasi-Carleman operators · Discontinuous kernels · Asymptotics of eigenvalues

Mathematics Subject Classification (2000) 47A40 · 47B25

Communicated by Ari Laptev.

D. R. Yafaev (✉)
IRMAR, Université de Rennes I, Campus de Beaulieu, 35042 Rennes Cedex, France
e-mail: yafaev@univ-rennes1.fr

1 Introduction

Hankel operators can be defined as integral operators

$$(Hf)(t) = \int_0^\infty h(t+s)f(s)ds \tag{1.1}$$

in the space $L^2(\mathbb{R}_+)$ with kernels h that depend on the sum of variables only. We refer to the books [4,5] for basic information on Hankel operators. Of course H is symmetric if $h(t) = \overline{h(t)}$. There are very few cases when Hankel operators can be explicitly diagonalized. The most simple and important case $h(t) = t^{-1}$ was considered by T. Carleman in [2].

Here we study a class of Hankel operators generalizing the Carleman operator. The corresponding kernels are given by the formula

$$h(t) = P(\ln t)t^{-1} \tag{1.2}$$

where $P(x)$ is an arbitrary polynomial. Hankel operators H with such kernels are not bounded unless $P(x) = \text{const}$, but, for real $P(x)$, they can be uniquely defined as self-adjoint operators. We show that the Hankel operator with kernel (1.2) is unitarily equivalent to the differential operator

$$A = vQ(D)v, \quad D = id/d\xi, \tag{1.3}$$

in the space $L^2(\mathbb{R})$. Here v is the operator of multiplication by the universal function

$$v(\xi) = \frac{\sqrt{\pi}}{\sqrt{\cosh(\pi\xi)}} \tag{1.4}$$

and the polynomial $Q(x)$ is determined by $P(x)$. The polynomials $P(x)$ and $Q(x)$ have the same degree, and their coefficients are linked by an explicit formula (see subs. 3.2). In particular, $Q(x) = 1$ if $P(x) = 1$ which yields the familiar diagonalization of the Carleman operator.

Thus the spectral analysis of Hankel operators with kernels (1.2) reduces to the spectral analysis of differential operators which in principle is very well developed. However operators (1.3) are somewhat unusual because the function $v(\xi)$ tends to zero exponentially as $|\xi| \to \infty$ so that there is a strong degeneracy at infinity. Nevertheless we describe completely the essential spectrum of differential operators (1.3) under rather general assumptions on the function $v(\xi)$. We show that $\text{spec}_{\text{ess}}(A) = \mathbb{R}$ if $K := \deg P$ is odd, and $\text{spec}_{\text{ess}}(A) = [0, \infty)$ if K is even. Moreover, it turns out that zero is never an eigenvalue of A. In the case of even K we also find necessary and sufficient conditions for the positivity of A and for the infinitude of its negative spectrum. Of course the same spectral results are true for Hankel operators H with kernels (1.2). For real polynomials $P(x)$ of first order, our approach yields the explicit diagonalization of Hankel operators H. In particular, we show that in this case the spectrum of H is absolutely continuous, has multiplicity 1 and covers the whole real line.

Actually, the unitary equivalence of the operators H and A is quite explicit. Let $M : L^2(\mathbb{R}_+) \to L^2(\mathbb{R})$ be the Mellin transform; it is a unitary mapping. Set

$$(Ff)(\xi) = \frac{\Gamma(1/2 + i\xi)}{|\Gamma(1/2 + i\xi)|}(Mf)(\xi) \tag{1.5}$$

where $\Gamma(\cdot)$ is the gamma function. We show that

$$H = F^* A F. \tag{1.6}$$

Our proof of this identity follows the approach of [6] where general Hankel operators H were considered. For an arbitrary H, the function $Q(x)$ in formula (1.3) is a distribution which may be (for example, for finite rank H) very singular. However it is a polynomial for Hankel operators with kernels (1.2) so that A is the explicit differential operator in this case.

1.2. Kernels (1.2) are singular at the points $t = 0$ and $t = \infty$. We also consider another class of kernels which are singular at some point $t_0 > 0$. We assume that

$$h(t) = \sum_{k=0}^{K} h_k \delta^{(k)}(t - t_0), \quad h_k = \bar{h}_k, \quad h_K \neq 0, \tag{1.7}$$

where $\delta(\cdot)$ is the Dirac delta function. It turns out that Hankel operators with such kernels reduce to "differential" operators with the reflection and a shift of the argument.

Spectral properties of Hankel operators with kernels (1.2) and (1.7) are completely different. As discussed in [6], Hankel operators can be sign-definite only for $h \in C^\infty(\mathbb{R}_+)$ which is of course not true for kernels (1.7). If $K = 0$, then, as shown in [6], the spectrum of H consists of three eigenvalues 0, h_0 and $-h_0$ of infinite multiplicity each. We shall prove here that for $K \geq 1$ the spectrum of the operator H with kernel (1.7) consists of the zero eigenvalue of infinite multiplicity and an infinite number of eigenvalues of finite multiplicities accumulating both at $+\infty$ and $-\infty$. Moreover, we shall find the leading term of asymptotics of these eigenvalues.

Recall that a symbol of a Hankel operator with kernel $h(t)$ is defined as a function $\omega(\lambda)$, $\lambda \in \mathbb{R}$, such that $(2\pi)^{-1/2}(\Phi\omega)(t) = h(t)$ where Φ is the Fourier transform. Since by the Nehari theorem, Hankel operators with bounded symbols are bounded, the symbols of operators with kernels (1.1) and (1.7) are necessarily unbounded functions. For kernels (1.1) symbols can be constructed by the formula

$$\omega(\lambda) = 2i \int_0^\infty \sin(\lambda t) h(t) dt = 2i \int_0^\infty \sin t \, P(\ln t - \ln \lambda) \, t^{-1} dt$$

for $\lambda > 0$ and $\omega(-\lambda) = -\omega(\lambda)$. Thus $\omega(\lambda)$ is a C^∞ function for $\lambda \neq 0$ with logarithmic singularities at $\lambda = 0$ and $\lambda = \infty$. For kernels (1.7), the symbol equals

$$\omega(\lambda) = \sum_{k=0}^{K} h_k(-i\lambda)^k e^{i\lambda t_0},$$

so that it is a C^∞ function with a power growth and an oscillation as $|\lambda| \to \infty$.

1.3. Let us introduce some standard notation. We denote by Φ,

$$(\Phi u)(\xi) = (2\pi)^{-1/2} \int_{-\infty}^{\infty} u(x)e^{-ix\xi}dx,$$

the Fourier transform. The space $\mathcal{Z} = \mathcal{Z}(\mathbb{R})$ of test functions is defined as the subset of the Schwartz space $\mathcal{S} = \mathcal{S}(\mathbb{R})$ which consists of functions u admitting the analytic continuation to entire functions in the complex plane $\mathbb{C}$ and satisfying bounds

$$|u(z)| \leq C_n(1 + |z|)^{-n} e^{r|\mathrm{Im} z|}, \quad z \in \mathbb{C},$$

for some $r = r(u) > 0$ and all n. We recall (see, e.g., the book [3]) that the Fourier transform $\Phi : \mathcal{Z} \to C_0^\infty(\mathbb{R})$ and $\Phi^* : C_0^\infty(\mathbb{R}) \to \mathcal{Z}$. The dual classes of distributions (continuous antilinear functionals on $\mathcal{S}$, $C_0^\infty(\mathbb{R})$ and $\mathcal{Z}$) are denoted $\mathcal{S}'$, $C_0^\infty(\mathbb{R})'$ and $\mathcal{Z}'$, respectively. We use the notation $\langle \cdot, \cdot \rangle$ and $\langle \cdot, \cdot \rangle$ for the duality symbols in $L^2(\mathbb{R}_+)$ and $L^2(\mathbb{R})$, respectively. They are linear in the first argument and antilinear in the second argument.

We denote by $\mathsf{H}^K(\mathcal{J})$ the Sobolev space of functions defined on an interval $\mathcal{J} \subset \mathbb{R}$; $C_0^K(\mathcal{J})$ is the class of k-times continuously differentiable functions with compact supports in $\mathcal{J}$. We often use the same notation for a function and the operator of multiplication by this function. The letters c and C (sometimes with indices) denote various positive constants whose precise values are inessential.

Let us briefly describe the structure of the paper. We collect necessary results of [6] in Sect. 2. In Sect. 3 we establish the unitary equivalence of the Hankel operator H with kernel (1.2) and differential operator (1.3). Spectral properties of the operators A and hence of H are studied in Sect. (4). We emphasize that our results on the operator A do not require specific assumption (1.4). Finally, Hankel operators H with kernels (1.7) are studied in Sect. 5. Our presentation in this section is independent of general results of Sect. 2. On the other hand, it is rather similar to that in Section 6 of [6] where Hankel operators with discontinuous kernels (but not as singular as kernels (1.7) were considered.

2 Hankel and pseudo-differential operators

In this section we show that an arbitrary Hankel operator H is unitarily equivalent to a pseudo-differential operator A defined by formula (1.3) with a distribution $Q(x)$. Our presentation is close to [6], but we here insist upon the unitary equivalence of the operators H and A.

2.1. Let us consider a Hankel operator H defined by equality (1.1) in the space $L^2(\mathbb{R}_+)$. Actually, it is more convenient to work with sesquilinear forms instead of operators.

Let us introduce the Laplace convolution

$$(\bar{f_1} \star f_2)(t) = \int_0^t \overline{f_1(s)} f_2(t-s) ds$$

of functions $\bar{f_1}$ and f_2. Then

$$(Hf_1, f_2) = \langle h, \bar{f_1} \star f_2 \rangle \tag{2.1}$$

where we write $\langle \cdot, \cdot \rangle$ instead of $(\cdot, \cdot)$ because h may be a distribution.

We consider form (2.1) on elements $f_1, f_2 \in \mathcal{D}$ where the set $\mathcal{D}$ is defined as follows. Put

$$(Uf)(x) = e^{x/2} f(e^x).$$

Then $U : L^2(\mathbb{R}_+) \to L^2(\mathbb{R})$ is the unitary operator, and $f \in \mathcal{D}$ if and only if $Uf \in \mathcal{Z}$. Since $f(t) = t^{-1/2}(Uf)(\ln t)$ and $\mathcal{Z} \subset \mathcal{S}$, we see that functions $f \in \mathcal{D}$ and their derivatives satisfy the estimates

$$|f^{(m)}(t)| = C_{n,m} t^{-1/2-m} (1 + |\ln t|)^{-n} \tag{2.2}$$

for all n and m. Of course, the set $\mathcal{D}$ is dense in the space $L^2(\mathbb{R}_+)$. It is shown in [6] that if $f_1, f_2 \in \mathcal{D}$, then the function

$$\Omega(x) = (\bar{f_1} \star f_2)(e^x)$$

belongs to the space $\mathcal{Z}$.

With respect to h, we assume that the distribution

$$\theta(x) = e^x h(e^x) \tag{2.3}$$

is an element of the space $\mathcal{Z}'$. The set of all such h will be denoted $\mathcal{Z}'_+$, that is,

$$h \in \mathcal{Z}'_+ \Longleftrightarrow \theta \in \mathcal{Z}'.$$

It is shown in [6] that *this condition is satisfied for all bounded Hankel operators* H. Since $\Omega \in \mathcal{Z}$, the form

$$\langle h, \bar{f_1} \star f_2 \rangle = \int_0^\infty h(t)(f_1 \star \bar{f_2})(t) dt = \int_{-\infty}^\infty \theta(x) \overline{\Omega(x)} dx =: \langle \theta, \Omega \rangle$$

is correctly defined for all $f_1, f_2 \in \mathcal{D}$.

Note that $h \in \mathcal{Z}'_+$ if $h \in L^1_{\text{loc}}(\mathbb{R}_+)$ and the integral

$$\int_0^\infty |h(t)|(1 + |\ln t|)^{-\kappa} dt < \infty \tag{2.4}$$

converges for some κ. In this case the corresponding function (2.3) satisfies the condition

$$\int_{-\infty}^\infty |\theta(x)|(1 + |x|)^{-\kappa} dx < \infty,$$

and hence $\theta \in \mathcal{S}' \subset \mathcal{Z}'$.

2.2. Let us now give the definitions of the b- and s-functions of a Hankel operator H. We formally define

$$b(\xi) = \frac{1}{2\pi} \frac{\int_0^\infty h(t) t^{-i\xi} dt}{\int_0^\infty e^{-t} t^{-i\xi} dt}. \tag{2.5}$$

Of course $b(-\xi) = \overline{b(\xi)}$ if $h(t) = \overline{h(t)}$. We call $b(\xi)$ the b-function of a Hankel operator H. Formula (2.5) can be rewritten as

$$b(\xi) = (2\pi)^{-1/2} a(\xi) \Gamma(1 - i\xi)^{-1} \tag{2.6}$$

where

$$a(\xi) = (\Phi\theta)(\xi) = (2\pi)^{-1/2} \int_0^\infty h(t) t^{-i\xi} dt \tag{2.7}$$

is the Fourier transform of function (2.3).

Recall that the gamma function $\Gamma(z)$ is a holomorphic function in the right half-plane and $\Gamma(z) \neq 0$ for all $z \in \mathbb{C}$. According to the Stirling formula the gamma function $\Gamma(z)$ tends to zero exponentially as $|z| \to \infty$ parallel with the imaginary axis. To be more precise, we have

$$\Gamma(\alpha + i\xi) = e^{\pi i(2\alpha-1)/4} (2\pi/e)^{1/2} \xi^{\alpha-1/2} e^{i\xi(\ln\xi - 1)} e^{-\pi\xi/2} \big(1 + O(\xi^{-1})\big)$$

for a fixed $\alpha > 0$ and $\xi \to +\infty$. We also note that $\Gamma(\alpha - i\xi) = \overline{\Gamma(\alpha + i\xi)}$ and

$$|\Gamma(1/2 + i\xi)|^2 = \frac{\pi}{\cosh(\pi\xi)}.$$

Since the denominator in (2.6) tends to zero exponentially as $|\xi| \to \infty$, $b(\xi)$ is a "nice" function only under very stringent assumptions on $a(\xi)$ and hence on

$h(t)$. Therefore we are obliged to work with distributions which turn out to be very convenient. The Schwartz class is too restrictive for our purposes because of the exponential decay of $\Gamma(1-i\xi)$. Therefore we assume that $a \in C_0^\infty(\mathbb{R})'$; in this case $b(\xi)$ belongs to the same class. Our assumption on a means that $\theta \in \mathcal{Z}'$ or equivalently $h \in \mathcal{Z}'_+$.

Thus we are led to the following

Definition 2.1 Let $h \in \mathcal{Z}'_+$. The distribution $b \in C_0^\infty(\mathbb{R})'$ defined by formulas (2.3), (2.6) and (2.7) is called the b-function of the Hankel operator H (or of its kernel $h(t)$). Its Fourier transform $s = \sqrt{2\pi}\,\Phi^* b \in \mathcal{Z}'$ is called the s-function or the *sign-function* of H.

Let the unitary mapping $F : L^2(\mathbb{R}_+) \to L^2(\mathbb{R})$ be defined by formula (1.5) where $M = \Phi U$ is the Mellin transform. If $f \in \mathcal{D}$, then $Uf \in \mathcal{Z}$ and hence the function $Ff \in C_0^\infty(\mathbb{R})$. We recall that the function $v(\xi)$ was defined by formula (1.4) and set $(\mathcal{J}g)(\xi) = g(-\xi)$.

The following result was obtained in [6].

Theorem 2.2 *Suppose that $h \in \mathcal{Z}'_+$, and let $b \in C_0^\infty(\mathbb{R})'$ be the corresponding b-function. Let $f_j \in \mathcal{D}$, $j = 1, 2$. Then $g_j = Ff_j \in C_0^\infty(\mathbb{R})$ and the representation*

$$\langle h, \bar{f}_1 \star f_2 \rangle = \langle b, (v\mathcal{J}\bar{g}_1) * (vg_2) \rangle \tag{2.8}$$

holds.

Passing in the right-hand side of (2.8) to the Fourier transforms and using that

$$\Phi^*\big((v\mathcal{J}\bar{g}_1) * (vg_2)\big) = (2\pi)^{1/2}\overline{\Phi^*(vg_1)}\Phi^*(vg_2),$$

we obtain

Corollary 2.3 *Let $s \in \mathcal{Z}'$ be the sign-function of h, and let $u_j = \Phi^*(vFf_j) \in \mathcal{Z}$. Then*

$$\langle h, \bar{f}_1 \star f_2 \rangle = \langle s, \bar{u}_1 u_2 \rangle.$$

We note that, formally, the identity (2.8) can be rewritten as relation (1.6) where A is the "integral operator" with kernel $v(\xi)b(\xi - \eta)v(\eta)$. To put it differently,

$$A = v\Phi s \Phi^* v, \tag{2.9}$$

that is, A is the pseudo-differential operator defined by the amplitude $v(\xi)s(x)v(\eta)$. We emphasize that in general $s(x)$ is a distribution so that formula (2.9) has only a formal meaning. According to relation (1.6) a study of the operator H reduces to that of the operator A.

2.3. For an arbitrary distribution $h \in \mathcal{Z}'_+$, we have constructed in the previous subsection its sign-function $s \in \mathcal{Z}'$. It turns out that, conversely, the kernel $h(t)$ can be recovered from its sign-function $s(x)$. It is convenient to introduce the distribution

$$h^{\natural}(\lambda) = \lambda^{-1} s(-\ln \lambda). \tag{2.10}$$

Note that the inclusions $s \in \mathcal{Z}'$ and $h^{\natural} \in \mathcal{Z}'_{+}$ are equivalent. The proof of the following result can be found in [6].

Theorem 2.4 *Let $h \in \mathcal{Z}'_{+}$, and let $s \in \mathcal{Z}'$ be the corresponding sign-function (see Definition* (2.1)). *Then h can be recovered from function* (2.10) *by the formula*

$$h(t) = \int_0^\infty e^{-t\lambda} h^{\natural}(\lambda) \lambda d\lambda. \tag{2.11}$$

Formula (2.11) is understood of course in the sense of distributions. We emphasize the mappings $h \mapsto h^{\natural}$ as well as its inverse $h^{\natural} \mapsto h$ are one-to-one continuous mappings of the space $\mathcal{Z}'_{+}$ onto itself.

3 Quasi-Carleman and differential operators

3.1. Now we are in a position to consider Hankel operators with kernels defined by formula (1.2) where

$$P(x) = \sum_{k=0}^{K} p_k x^k, \quad p_K \neq 0, \tag{3.1}$$

is a polynomial with coefficients p_k, $k = 0, 1, \ldots, K$. It is easy to see that such operators (they will be denoted by H_0) are well defined on the set $\mathcal{D}_0$ of functions $f(t)$ satisfying estimate (2.2) for $m = 0$ and some $n > K + 1$. Indeed, by the Schwarz inequality for an arbitrary $\varepsilon > 0$, we have the estimate

$$\int_0^\infty dt \Big| \int_0^\infty ds \frac{|\ln(t+s)|^k}{t+s} s^{-1/2} (1 + |\ln s|)^{-n} \Big|^2$$
$$\leq C \int_0^\infty dt \int_0^\infty ds \frac{|\ln(t+s)|^{2k}}{(t+s)^2} (1 + |\ln s|)^{-2n+1+\varepsilon}. \tag{3.2}$$

Let us make the change of variables $(t, s) \mapsto (\tau, s) = (t + s, s)$ in the right-hand side and integrate first over $\tau \geq s$. Then using inequality

$$\int_s^\infty |\ln \tau|^{2k} \tau^{-2} d\tau \leq C_1 (1 + |\ln s|)^{2k} s^{-1},$$

we see that expression (3.2) is bounded by the integral

$$C_2 \int_0^\infty (1 + |\ln s|)^{-2n+2k+1+\varepsilon} s^{-1} ds$$

which converges if $n > k + 1 + \varepsilon/2$. It follows that $H_0 f \in L^2(\mathbb{R})$ for $f \in \mathcal{D}_0$. Moreover, using the Fubini theorem, we see that $(H_0 f_1, f_2) = (f_1, H_0 f_2)$ for f_1, $f_2 \in \mathcal{D}_0$ if all coefficients p_k are real.

Let us formulate the results obtained.

Lemma 3.1 *Let the kernel $h(t)$ of a Hankel operator H_0 be given by formulas* (1.2) *and* (3.1)*. Then H_0 is well defined on the set $\mathcal{D}_0$, and it is symmetric on $\mathcal{D}_0$ if all coefficients p_k are real.*

As we shall see below, the operator H_0 is essentially self-adjoint (see, e.g., the book [1], for background information on the theory of self-adjoint extensions of symmetric operators). The proof of this result as well as our study of spectral properties of the closure $\bar{H}_0 =: H$ of H_0 rely on the identity (1.6). We emphasize however that the proof of (1.6) does not require the assumption $p_k = \bar{p}_k, k = 0, 1, \dots, K$. The symmetricity of H_0 on the domain $\mathcal{D}$ is also a consequence of (1.6) so that the direct proof of Lemma (3.1) could be avoided.

3.2. Since kernels (1.2) satisfy condition (2.4) with any $\kappa > K + 1$, Theorem 2.2 can be directly applied in this case. We only have to calculate the corresponding b- and s-functions. If $h_k(t) = t^{-1} \ln^k t$, then the function (2.3) equals $\theta_k(x) = x^k$ and its Fourier transform equals

$$a_k(\xi) = (\Phi\theta_k)(\xi) = (2\pi)^{1/2} i^k \delta^{(k)}(\xi).$$

To simplify notation, we set$\omega(z) = \Gamma(1-z)^{-1}$. Then function (2.6) equals

$$b_k(\xi) = i^k \omega(i\xi)\delta^{(k)}(\xi) = \sum_{\ell=0}^{k} i^\ell C_k^\ell \omega^{(k-\ell)}(0)\delta^{(\ell)}(\xi)$$

where C_k^ℓ are the binomial coefficients.

It follows that the b-function of kernel (1.2), (3.1) is given by the formula

$$b(\xi) = \sum_{k=0}^{K} q_k i^k \delta^{(k)}(\xi)$$

where

$$q_k = \sum_{\ell=k}^{K} C_\ell^k \omega^{(\ell-k)}(0) p_\ell, \quad k = 0, \dots, K, \quad \omega(z) = \Gamma(1-z)^{-1}. \tag{3.3}$$

It means that the operator A acts by formula (1.3) where

$$Q(x) = \sum_{k=0}^{K} q_k x^k. \tag{3.4}$$

Thus for kernels (1.2) the sign-function $s(x) = Q(x)$ is the polynomial. Note that according to general formula (2.11), $P(x)$ can be recovered from $Q(x)$ by the equality

$$P(\ln t) = t \int_0^\infty Q(-\ln \lambda) e^{-t\lambda} d\lambda. \tag{3.5}$$

Observe that $q_K = p_K$ for all K. Recall that $\Gamma'(1) = -\gamma$ (the Euler constant) and $\Gamma''(1) = \gamma^2 + \pi^2/6$. Therefore we have

$$q_0 = p_0 - \gamma p_1, \quad \text{if} \quad K = 1, \tag{3.6}$$

and

$$q_0 = p_0 - \gamma p_1 + (\gamma^2 - \pi^2/6) p_2, \quad q_1 = p_1 - 2\gamma p_2, \quad \text{if} \quad K = 2. \tag{3.7}$$

The following assertion is a particular case of Theorem 2.2.

Theorem 3.2 *Let a kernel* $h(t)$ *be defined by formulas* (1.2) *and* (3.1). *Let* $Q(x)$ *be polynomial* (3.4) *with coefficients* (3.3), *and let* A *be differential operator* (1.3). *Then for all* $f_j \in \mathcal{D}$, $j = 1, 2$, *the identity*

$$(Hf_1, f_2) = (AFf_1, Ff_2) \tag{3.8}$$

holds.

Note that, for $h(t) = t^{-1}$, the identity (3.8) yields the familiar diagonalization of the Carleman operator. Indeed, in this case we have

$$\theta(x) = 1, \quad a(\xi) = (2\pi)^{1/2} \delta(\xi), \quad b(\xi) = \delta(\xi), \quad s(x) = 1.$$

Therefore the identity (3.8) reads as

$$(Hf_1, f_2) = \int_{-\infty}^{\infty} \frac{\pi}{\cosh(\pi\xi)} \tilde{f}_1(\xi) \overline{\tilde{f}_2(\xi)} d\xi$$

where $\tilde{f}_j = Mf_j = \Phi U f_j$, $j = 1, 2$, is the Mellin transform of f_j.

We emphasize that Theorem 3.2 does not require that the coefficients of $P(x)$ be real.

3.3. In view of Theorem 3.2 spectral properties of the Hankel operator H are the same as those of the differential operator A. Therefore we forget for a while Hankel operators and study differential operators A defined by formula (1.3), but we not assume that the function $v(\xi)$ has special form (1.4). *We suppose that* $v = \bar{v} \in C^K(\mathbb{R})$ *and that the coefficients of the polynomial* $Q(x)$ *of degree* K *are real and* $q_K \neq 0$. Then the operator A_0 defined by formula (1.3) on the domain $C_0^K(\mathbb{R})$ is symmetric in $L^2(\mathbb{R})$. We emphasize that operators (1.3) require a special study because the function $v(\xi)$ may tend to zero as $|\xi| \to \infty$.

Let us start with the case $K = 1$ when $Q(x) = q_0 + q_1 x$ and A_0 can be standardly reduced by a change of variables and a gauge transformation to the differential operator $q_1 D$. We suppose that $v(\xi) > 0$ and

$$\int_{\mathbb{R}_+} v(\xi)^{-2} d\xi = \int_{\mathbb{R}_-} v(\xi)^{-2} d\xi = \infty. \tag{3.9}$$

Under this assumption the operator $T : L^2(\mathbb{R}) \to L^2(\mathbb{R})$ defined by the relation

$$(Tg)(\xi) = v(\xi)^{-1} e^{i q_0 q_1^{-1} \xi} g\left(\int_0^\xi v(\eta)^{-2} d\eta\right) \tag{3.10}$$

is unitary and

$$A_0 T g = i q_1 T g', \quad g \in C_0^1(\mathbb{R}).$$

Recall that $D = id/d\xi$. Let the set $\mathcal{D}_* \subset L^2(\mathbb{R})$ consist of functions $g \in \mathsf{H}^1_{\rm loc}(\mathbb{R})$ such that $vQ(D)(vg) \in L^2(\mathbb{R})$. It is easy to see that $\mathcal{D}_* = T\mathsf{H}^1(\mathbb{R})$. Thus we are led to the following assertion.

Lemma 3.3 *Suppose that* $v \in C^1(\mathbb{R})$, $v(\xi) > 0$ *and condition* (3.9) *is satisfied. Then the operator* $A_0 = v(q_0 + q_1 D)v$ *is essentially self-adjoint on* $C_0^1(\mathbb{R})$, *and its closure* $\bar{A}_0 =: A$ *is self-adjoint on the domain* $\mathcal{D}_* =: \mathcal{D}(A)$. *The spectrum of the operator* A *is simple, absolutely continuous, and it coincides with* $\mathbb{R}$.

Remark 3.4 If both integrals (3.9) are finite, then A_0 reduces to the operator $q_1 D$ on a finite interval. Its deficiency indices equal $(1, 1)$. If only one of integrals (3.9) is finite, then A_0 reduces to the operator $q_1 D$ on a half-axis. Its deficiency indices equal $(0, 1)$ or $(1, 0)$.

As a by-product of our considerations, we obtain the following result. It is simple but perhaps was never explicitly mentioned.

Proposition 3.5 *Suppose that $v \in C^1(\mathbb{R})$ and $v(\xi) > 0$. Let the space $\mathcal{K} = \mathcal{K}_v$ consist of functions $g \in \mathsf{H}^1_{\mathrm{loc}}(\mathbb{R})$ with the norm*

$$\|g\|_v^2 = \int_{-\infty}^{\infty} \big(v^2(\xi)|g'(\xi)|^2 + v^{-2}(\xi)|g(\xi)|^2\big)d\xi.$$

Then the set $C_0^1(\mathbb{R})$ is dense in $\mathcal{K}$ if and only if condition (3.9) is satisfied.

Proof Let us make the change of variables

$$\tilde{g}(\tilde{\xi}) = g(\xi) \quad \text{where} \quad \tilde{\xi} = \int_0^{\xi} v(\eta)^{-2}d\eta$$

and put

$$v_\pm = \pm \int_{\mathbb{R}_\pm} v(\xi)^{-2}d\xi.$$

Then

$$\|g\|_v^2 = \int_{v_-}^{v_+} \big(|\tilde{g}'(\tilde{\xi})|^2 + |\tilde{g}(\tilde{\xi})|^2\big)d\tilde{\xi}.$$

Since $g \in C_0^1(\mathbb{R})$ if and only if $\tilde{g} \in C_0^1(v_-, v_+)$, it remains to use that the set $C_0^1(v_-, v_+)$ is dense in the Sobole space $\mathsf{H}^1(v_-, v_+)$ if and only if $v_- = -\infty$ and $v_+ = +\infty$. □

Returning to the Hankel operator H with kernel (1.2) and using Theorem 3.2 and equality (3.6), we obtain the following result.

Theorem 3.6 *Suppose that $P(x) = p_0 + p_1 x$ where $p_0 = \bar{p}_0$ and $p_1 = \bar{p}_1 \neq 0$. Then*

$$H = q_1 F^* T D T^* F$$

where T is defined by formula (3.10) with $q_0 = p_0 - \gamma p_1$, $q_1 = p_1$ and $v(\xi)$ is function (1.4). The operator H is essentially self-adjoint on the set $\mathcal{D}_0$, and it is self-adjoint on the set $F^*\mathcal{D}_*$. The spectrum of the operator H is simple, absolutely continuous, and it coincides with $\mathbb{R}$.

3.4. Let us pass to the case $K \geq 2$. We recall that the operator $A_0 = vQ(D)v$ is symmetric in $L^2(\mathbb{R})$ on $C_0^K(\mathbb{R})$. Let us use the notation $\mathcal{A}_0$ for the same operator considered as a mapping $\mathcal{A}_0 : C_0^K(\mathbb{R}) \to L^2(\mathbb{R})$. The operator $\mathcal{A}_0^* : L^2(\mathbb{R}) \to C_0^K(\mathbb{R})'$ is defined by the relation

$$(\mathcal{A}_0 g, y) = \langle g, \mathcal{A}_0^* y \rangle, \quad g \in C_0^K(\mathbb{R}), \ y \in L^2(\mathbb{R}), \tag{3.11}$$

and is given by the same differential expression (1.3) where derivatives are understood in the sense of distributions.

It is also quite easy to construct the operator A_0^* adjoint to A_0 in the space $L^2(\mathbb{R})$. Let the domain $\mathcal{D}_* \subset L^2(\mathbb{R})$ consist of y such that $\mathcal{A}_0^* y \in L^2(\mathbb{R})$. The following assertion is rather standard.

Lemma 3.7 *The operator A_0 is symmetric on $C_0^\infty(\mathbb{R})$ and its adjoint A_0^* is defined on the domain $\mathcal{D}(A_0^*) = \mathcal{D}_*$. For $y \in \mathcal{D}(A_0^*)$, we have $A_0^* y = \mathcal{A}_0^* y$.*

Proof By definition, $\mathcal{D}(A_0^*)$ consists of $g \in L^2(\mathbb{R})$ such that

$$(A_0 g, y) = (g, y_*) \tag{3.12}$$

for all $g \in C_0^K(\mathbb{R})$ and some $y_* \in L^2(\mathbb{R})$; in this case $y_* = A_0^* y$. Observe that the left-hand sides of (3.11) and (3.12) coincide. If $\mathcal{A}_0^* y \in L^2(\mathbb{R})$, then (3.12) is satisfied with $y_* = \mathcal{A}_0^* y$. Conversely, if (3.12) is satisfied, then

$$\langle g, \mathcal{A}_0^* y \rangle = (g, y_*), \quad \forall g \in C_0^K(\mathbb{R}),$$

and hence $y_* = \mathcal{A}_0^* y$ so that $\mathcal{A}_0^* y \in L^2(\mathbb{R})$. □

Under additional assumptions on $v(\xi)$ the operator A_0^* is symmetric. The proof of this result requires the following auxiliary assertion. Recall that $D = id/d\xi$.

Lemma 3.8 *Suppose that $v \in C^1(\mathbb{R})$, $v \in L^\infty(\mathbb{R})$, $v(\xi) > 0$ and*

$$|v'(\xi)| \le Cv(\xi). \tag{3.13}$$

Let $z \in \mathbb{C}$ and let $|Imz|$ be sufficiently large. Then the operator $v(D - z)^{-1} v^{-1}$ in $L^2(\mathbb{R})$ defined on functions with compact supports extends to a bounded operator.

Proof Let $z = a + ib$. Since

$$((D - z)^{-1} y_a)(\xi) = e^{-ia\xi}((D - ib)^{-1} y)(\xi), \quad \text{where} \quad y_a(\xi) = e^{-ia\xi} y(\xi),$$

we can suppose that $z = ib$ where $b \in \mathbb{R}$. We have to check the inequality

$$\|v(D - ib)^{-1} v^{-1} w\| \le C\|w\| \tag{3.14}$$

on a dense in $L^2(\mathbb{R})$ set of elements w with compact supports. Consider $w = v(D - ib)u$ where $u \in C_0^\infty(\mathbb{R})$ is arbitrary; then $w \in C_0^1(\mathbb{R})$. The set of such elements w is dense in $L^2(\mathbb{R})$. Indeed, suppose that

$$(v(D - ib)u, g_0) = 0$$

for some $g_0 \in L^2(\mathbb{R})$ and all $u \in C_0^\infty(\mathbb{R})$. Then $(D + ib)(vg_0) = 0$ and hence

$$v(\xi)g_0(\xi) = ce^{-b\xi}.$$

Since $v \in L^\infty(\mathbb{R})$, it implies that $c = 0$, and the equality $v(\xi)g_0(\xi) = 0$ implies that $g_0(\xi) = 0$ because $v(\xi) \neq 0$.

For $w = v(D - ib)u$, (3.14) is equivalent to the inequality

$$\begin{aligned}\int_{-\infty}^{\infty} v^2(\xi)|u(\xi)|^2 d\xi &\leq C \int_{-\infty}^{\infty} v^2(\xi)|u'(\xi) - bu(\xi)|^2 d\xi \\ &= C\left(\int_{-\infty}^{\infty} v^2(\xi)|u'(\xi)|^2 d\xi - 2b\mathrm{Re} \int_{-\infty}^{\infty} v^2(\xi)u'(\xi)\bar{u}(\xi)d\xi \right. \\ &\quad \left. + b^2 \int_{-\infty}^{\infty} v^2(\xi)|u(\xi)|^2 d\xi \right). \end{aligned} \tag{3.15}$$

Integrating in the second term in the right-hand side by parts and using condition (3.13), we see that

$$\mathrm{Re} \int_{-\infty}^{\infty} v^2(\xi)u'(\xi)\bar{u}(\xi)d\xi = -\int_{-\infty}^{\infty} v'(\xi)v(\xi)|u(\xi)|^2 d\xi$$

is bounded by

$$C \int_{-\infty}^{\infty} v^2(\xi)|u(\xi)|^2 d\xi.$$

This proves inequality (3.15) if b is large enough. □

Corollary 3.9 *Let $b \in \mathbb{R}$ be sufficiently large. Then for all $k = 0, 1, \ldots, K$, the operator $vD^k(Q(D) - ib)^{-1}v^{-1}$ in $L^2(\mathbb{R})$ defined on functions with compact supports extends to a bounded operator.*

Proof The equation $Q(x) = ib$ has K solutions $x_1(b), \ldots, x_K(b)$ which for large b are close to the solutions of the equation $q_K x^K = ib$. Therefore the roots $x_\ell(b)$ are simple and $|\mathrm{Im} x_\ell(b)| \to \infty$ as $b \to \infty$ for all $\ell = 1, \ldots K$. Let us expand the function $x^k(Q(x) - ib)^{-1}$ in a linear combination of the functions $(x - x_\ell)^{-1}$ and of the constant term 1 (for $k = K$). We can apply Lemma 3.8 to every term $v(D - x_\ell(b))^{-1}v^{-1}$. The contribution of 1 gives the identity operator. □

Recall that $\mathcal{D}(A_0^*) = \mathcal{D}_*$ according to Lemma 3.7. Below we need additional information on this set. Let us accept the following

Assumption 3.10 The function $v \in C^K(\mathbb{R})$, $v \in L^\infty(\mathbb{R})$, $v(\xi) > 0$ and estimate (3.13) holds.

Lemma 3.11 *Let Assumption* 3.10 *be satisfied. If* $g \in \mathcal{D}(A_0^*)$, *then* $vD^k(vg) \in L^2(\mathbb{R})$ *for all* $k = 1, \dots, K$ *and, in particular,* $g \in \mathsf{H}^K_{\mathrm{loc}}(\mathbb{R})$. *Moreover, the coercive estimates hold:*

$$\|vD^k(vg)\| \le C\big(\|vQ(D)(vg)\| + \|g\|\big), \quad k = 1, \dots, K.$$

Proof By definition of $\mathcal{D}_*$, we have $vQ(D)(vg) \in L^2(\mathbb{R})$ and hence $w := v(Q(D) - ib)(vg) \in L^2(\mathbb{R})$ for all b. Observe that $vD^k(vg) = (vD^k(Q(D) - ib)^{-1}v^{-1})w$. Thus it remains to use Corollary 3.9. □

This lemma shows that the set $\mathcal{D}_* \subset L^2(\mathbb{R})$ consists of functions $g \in \mathsf{H}^K_{\mathrm{loc}}(\mathbb{R})$ such that $vD^k(vg) \in L^2(\mathbb{R})$ for all $k = 1, \dots, K$. Now it is easy to check the following assertion.

Lemma 3.12 *Under Assumption* 3.10 *the set* $C_0^K(\mathbb{R})$ *is dense in* $\mathcal{D}(A_0^*)$ *in the graph-norm* $\|g\| + \|A_0^* g\|$.

Proof Let $\varphi \in C_0^\infty(\mathbb{R})$ and $\varphi(\xi) = 1$ for $|\xi| \le 1$. Set $\varphi_n(\xi) = \varphi(\xi/n)$. For an arbitrary $g \in \mathcal{D}(A_0^*)$, we put $g_n = g\varphi_n$. Of course $\|g - g_n\| \to 0$ as $n \to \infty$. Set $u = vg$, $u_n = vg_n$. We have to show that $\|vQ(D)(u - u_n)\| \to 0$ as $n \to \infty$ or that

$$\lim_{n\to\infty} \|vD^k(u - u_n)\| = 0, \quad k = 0, 1, \dots, K. \tag{3.16}$$

Recall that $vu^{(k)} \in L^2(\mathbb{R})$ by Lemma 3.11. Therefore

$$\lim_{n\to\infty} \|vu^{(k)}(1 - \varphi_n)\| = 0 \quad \text{and} \quad \lim_{n\to\infty} \|vu^{(k)}\varphi_n^{(l)}\| = 0$$

for all $k = 0, 1, \dots, K$ and $l \ge 1$. These relations imply (3.16). □

Lemma 3.12 shows that the operator A_0^* coincides with the closure $\bar{A}_0$ of the operator A_0. This yields the following assertion.

Theorem 3.13 *Let Assumption* 3.10 *be satisfied. Then the operator* A_0 *defined by formula* (1.3) *on* $C_0^K(\mathbb{R})$ *is essentially self-adjoint. Its closure* $\bar{A}_0 =: A$ *is self-adjoint on the set* $\mathcal{D}_* =: \mathcal{D}(A)$ *and* $Ag = vQ(D)(vg)$ *for* $g \in \mathcal{D}_*$.

For K even, it is also possible to define A in terms of the quadratic form

$$(Ag, g) = \int_{-\infty}^{\infty} \big(Q(D)(vg)\big)v\bar{g}\,d\xi. \tag{3.17}$$

We suppose that $q_K > 0$; then the form $(Ag, g) + c\|g\|^2$ is positive-definite for a sufficiently large $c > 0$. Similarly to Theorem 3.13, it can be verified that this form defined on $C_0^K(\mathbb{R})$ admits the closure, and it is closed on the set $\widetilde{\mathcal{D}}_*$ of functions $g \in L^2(\mathbb{R})$ such that $D^k(vg) \in L^2(\mathbb{R})$ for all $k = 1, \dots, K/2$. Then the operator $A + cI$ can be defined as a self-adjoint operator corresponding to this closed form. Note that $\widetilde{\mathcal{D}}_* = \mathcal{D}(\sqrt{A + cI})$.

3.5. Let us return to Hankel operators. We recall that according to Theorem 3.2 the Hankel operator H with kernel (1.2) is unitarily equivalent to differential operator (1.3) where v is defined by formula (1.4) and $Q(x)$ is polynomial (3.4) with the coefficients defined by formula (3.3). To be more precise, the operators H and A are linked by relation (1.6) where F is operator (1.5). In particular, we have

$$\mathcal{D}(H) = F^*\mathcal{D}(A) \quad \text{and} \quad \mathcal{D}(\sqrt{H + cI}) = F^*\mathcal{D}(\sqrt{A + cI}) \quad \text{for} \quad K \quad \text{even}.$$

Therefore the following result is a direct consequence of Theorem 3.13. Recall that the set $\mathcal{D}_0$ consists of functions $f(t)$ satisfying estimate (2.2) for $m = 0$ and some $n > K + 1$.

Theorem 3.14 *Let kernel $h(t)$ be defined by formulas* (1.2) *and* (3.1) *where $p_k = \bar{p}_k$ for $k = 0, 1, \dots, K$. The Hankel operator H with kernel $h(t)$ is essentially self-adjoint on the domain $\mathcal{D}_0$, and its closure is self-adjoint on the domain $F^*\mathcal{D}_*$.*

4 Spectral results

Here we study spectral properties of the operators A and H.

4.1. We recall that the precise definition of the operator A was given in Theorem 3.13. The following result relies on a construction of trial functions.

Theorem 4.1 *Let Assumption* 3.10 *be satisfied. Suppose additionally that*

$$v^{(k)}v^{-1} \in L^\infty(\mathbb{R}), \quad k = 1, \dots, K - 1, \tag{4.1}$$

and that, for some $\delta > 0$,

$$\left(\int_{-n(1+\delta)}^{n(1+\delta)} v(\xi)^{-2+4/K} d\xi \right) \left(\int_{-n}^{n} v(\xi)^{-2} d\xi \right)^{-1} \to 0 \tag{4.2}$$

as $n \to \infty$. If K is odd, then $\operatorname{spec}(A) = \mathbb{R}$. *If K is even and $q_K > 0$, then* $[0, \infty) \subset \operatorname{spec}(A)$.

Proof We shall construct Weyl sequences for all $\lambda \in \mathbb{R}$ in the case of odd K and for all $\lambda \in [0, \infty)$ in the case of even K. Let $\varphi = \bar{\varphi} \in C_0^\infty(\mathbb{R})$, $\varphi(\xi) = 1$ for $|\xi| \le 1$ and $\varphi(\xi) = 0$ for $|\xi| \ge 1 + \delta$. We set

$$G(\xi; \lambda) = \lambda^{1/K} \int_0^\xi v(\eta)^{-2/K} d\eta \tag{4.3}$$

and

$$g_n(\xi) = v(\xi)^{-1} e^{iG(\xi;\lambda)} \varphi_n(\xi) \quad \text{where} \quad \varphi_n(\xi) = \varphi(\xi/n).$$

Obviously, we have

$$\|g_n\|^2 \geq \int_{-n}^{n} v(\xi)^{-2} d\xi. \tag{4.4}$$

Let us calculate

$$Ag_n - (-1)^K q_K \lambda g_n = vQ(D)(e^{iG}\varphi_n) - (-1)^K q_K \lambda v^{-1} e^{iG} \varphi_n. \tag{4.5}$$

Differentiating exponentials and using definition (4.3) and conditions (4.1), we see that

$$D^k(e^{iG(\xi;\lambda)}) = O(v(\xi)^{-2k/K}), \quad k = 0, 1, \dots, K-1,$$

and

$$D^K(e^{iG(\xi;\lambda)}) = (-1)^K \lambda v(\xi)^{-2} e^{iG(\xi;\lambda)} + O(v(\xi)^{-2(K-1)/K})$$

as $|\xi| \to \infty$. Estimating the functions $\varphi_n(\xi)$ and their derivatives by constants, we find that

$$\begin{aligned} v(\xi)Q(D)(e^{iG(\xi;\lambda)}\varphi_n(\xi)) &= (-1)^K q_K \lambda v(\xi)^{-1} e^{iG(\xi;\lambda)} \varphi_n(\xi) \\ &+ O(v(\xi)^{1-2(K-1)/K}). \end{aligned} \tag{4.6}$$

Substituting this expression into (4.5), we see that the first term in the right-hand side of (4.6) is cancelled with the second term in the right-hand side of (4.5). This yields the estimate

$$\|Ag_n - (-1)^K q_K \lambda g_n\|^2 \leq C \int_{-n(1+\delta)}^{n(1+\delta)} v(\xi)^{2-4(K-1)/K} d\xi. \tag{4.7}$$

By virtue of condition (4.2), it follows from (4.4) and (4.7) that

$$\|Ag_n - (-1)^K q_K \lambda g_n\| \|g_n\|^{-1} \to 0$$

as $n \to \infty$ so that $(-1)^K q_K \lambda \in \operatorname{spec}(A)$. □

Let us discuss condition (4.2). If $K = 2$, it means that

$$n^{-1} \int_{-n}^{n} v(\xi)^{-2} d\xi \to \infty$$

as $n \to \infty$. Since the integral here can be estimated from below by $n \min_{|\xi| \geq n/2} v(\xi)^{-2}$, this condition is automatically satisfied provided $v(\xi) \to 0$ as $|\xi| \to \infty$.

Let $K > 2$. If

$$c|\xi|^{-\rho} \leq v(\xi) \leq C|\xi|^{-\rho}, \quad 0 < c < C < \infty, \quad \rho > 0,$$

then expression (4.2) is estimated by $C(\delta)n^{-4\rho/K}$. Hence condition (4.2) is satisfied in this case (for all δ). If

$$ce^{-\rho|\xi|} \leq v(\xi) \leq Ce^{-\rho|\xi|}, \quad 0 < c < C < \infty, \quad \rho > 0,$$

then expression (4.2) is estimated by

$$C \exp\Big(2\rho n\big((1 - 2K^{-1})(1 + \delta) - 1\big)\Big).$$

This expression tends to zero if $\delta < 2(K - 2)^{-1}$ so that condition (4.2) is again satisfied for such δ. On the other hand, condition (4.2) can be violated for $K > 2$ if $v(\xi)$ tends to zero very rapidly (as $e^{-e^{|\xi|}}$, for example).

For the next result, assumptions (4.1) and (4.2) are not necessary.

Proposition 4.2 *Let Assumption* 3.10 *be satisfied. Then* 0 *is not an eigenvalue of the operator* A.

Proof Let $Ag = 0$ for some $g \in \mathcal{D}(A)$. Put $u = vg$. Then $u \in L^2(\mathbb{R})$ because $v \in L^\infty(\mathbb{R})$. Since $v(\xi) > 0$, we have $Q(D)u = 0$. Denote by $x_1, \ldots, x_{K_0} \in \mathbb{C}$ different roots of the equation $Q(x) = 0$. Then

$$u(\xi) = \sum_{k=1}^{K_0} P_k(\xi) e^{-ix_k\xi} \tag{4.8}$$

for some polynomials $P_k(\xi)$. Observe that all exponentials do not decay at least at one of the infinities. Therefore function (4.8) does not belong to $L^2(\mathbb{R})$ unless all polynomials $P_k(\xi)$, $k = 1, \ldots, K_0$, are zeros. It follows that $u = 0$ whence $g = 0$. □

4.2. Let K be even and $q_K > 0$; then A is semi-bounded from below and according to (3.17) we have

$$(Ag, g) = \int_{-\infty}^{\infty} Q(x)|(\Phi^*(vg))(x)|^2 dx, \quad \forall g \in \mathcal{D}(A). \tag{4.9}$$

Clearly, $A \geq 0$ if $Q(x) \geq 0$. On the other hand, if $Q(x_0) < 0$ for some $x_0 \in \mathbb{R}$, then $Q(x) < 0$ for some interval Δ centered at the point x_0. For every N, we choose functions $\psi_n \in C_0^\infty(\mathbb{R})$ with $\operatorname{supp}\psi_n \subset \Delta$ for all $n = 1, \dots, N$ such that $\operatorname{supp}\psi_n \cap \operatorname{supp}\psi_m = \varnothing$ if $n \neq m$. The functions $g_n = v^{-1}\Phi\psi_n \in \mathcal{D}(A)$ and according to (4.9) the form

$$(Ag, g) = \sum_{n=1}^{N} |\alpha_n|^2 \int_{-\infty}^{\infty} Q(x)|\psi_n(x)|^2 dx < 0$$

on all non-trivial linear combinations $g = \sum_{n=1}^{N} \alpha_n g_n$ of the functions $g_1, \dots, g_N$. This leads to the following result.

Theorem 4.3 *Let Assumption* 3.10 *be satisfied. Suppose that K is even and $q_K > 0$. Then the operator A is positive if and only if $Q(x) \geq 0$ for all $x \in \mathbb{R}$. Moreover, if $Q(x_0) < 0$ for some $x_0 \in \mathbb{R}$, then the negative spectrum of A is infinite.*

Let $K \geq 2$ be even and $q_K > 0$. According to Theorem 4.1, $[0, \infty) \subset \operatorname{spec}_{\mathrm{ess}}(A)$. Let us show that actually we have the equality here. It follows from Theorem 4.3 that for sufficiently large ν the operator $A_\nu = v(Q(D) + \nu)v \geq 0$. Thus we have to check that adding the operator νv^2 does not change the essential spectrum of A.

Lemma 4.4 *In addition to the assumptions of Theorem* 4.3 *suppose that $v(\xi) \to 0$ as $|\xi| \to \infty$. Then the operator $v^2(A + i)^{-1}$ is compact.*

Proof Let a set of functions g_n be bounded in the graph-norm $\|Ag\| + \|g\|$. We have to check that it is compact in the norm $\|v^2 g\|$. Put $u_n = v g_n$. Lemma 3.11 implies that

$$\|v u_n'\| + \|v^{-1} u_n\| \leq C < \infty. \tag{4.10}$$

We have to show that the set u_n is compact in the norm $\|v u_n\|$ or in $L^2(\mathbb{R})$ because the function $v(\xi)$ is bounded. Since $v(\xi) \to 0$ as $|\xi| \to \infty$, the boundedness of the second term in (4.10) shows that the norms of u_n in $L^2(-\infty, -R)$ and $L^2(R, \infty)$ can be made arbitrary small uniformly in n if R is sufficiently large. Observe that $v(\xi) \geq c > 0$ on every compact interval, and hence the boundedness of the first term in (4.10) shows that the set u_n is bounded in the Sobolev space $\mathsf{H}^1(-R, R)$. It follows that this set is compact in $L^2(-R, R)$ for all $R < \infty$. □

Corollary 4.5 *For an arbitrary ν, we have*

$$\operatorname{spec}_{\mathrm{ess}}(A + \nu v^2) = \operatorname{spec}_{\mathrm{ess}}(A).$$

Putting together this result with Theorem 4.1, we obtain the following assertion.

Theorem 4.6 *In addition to the assumptions of Theorem* 4.1 *suppose that $v(\xi) \to 0$ as $|\xi| \to \infty$. If K is even and $q_K > 0$, then*

$$\operatorname{spec}_{\mathrm{ess}}(A) = [0, \infty). \tag{4.11}$$

We emphasize that equality (4.11) is due to the condition $v(\xi) \to 0$ as $|\xi| \to \infty$. If $v(\xi) = 1$, then of course $\mathrm{spec}(A) = \mathrm{spec}_{\mathrm{ess}}(A) = [P_{\min}, \infty)$ where $P_{\min} = \min P(x)$ for $x \in \mathbb{R}$.

4.3. Theorem 3.2 allows us to reformulate the results of the previous subsections in terms of Hankel operators. We recall that the precise definition of the operator H was given in Theorem 3.14. Since function (1.4) satisfies Assumption 3.10 and conditions (4.1), (4.2), the following result is a consequence of Theorems 4.1 and 4.6.

Theorem 4.7 *Let kernel $h(t)$ be defined by formulas* (1.2) *and* (3.1) *where $p_k = \bar{p}_k$ for $k = 0, 1, \dots, K$. Then:*

1^0 *The point 0 is not an eigenvalue of H.*

2^0 *If K is odd, then* $\mathrm{spec}(H) = \mathbb{R}$.

3^0 *If K is even and $p_K > 0$, then* $\mathrm{spec}_{\mathrm{ess}}(H) = [0, \infty)$.

We emphasize that, for $K = 1$, Theorem 3.6 yields a much stronger result.

Apparently the theory of Weyl-Titchmarsh-Kodaira does not apply to operators (1.3) because $v(\xi) \to 0$ as $|\xi| \to \infty$. Nevertheless we conjecture that the spectrum of A is absolutely continuous up to perhaps a discrete set of eigenvalues. Moreover, we expect that the spectrum of A is simple for odd K and it has multiplicity 2 for even K.

Let K be even and $p_K > 0$; then H is semi-bounded from below. Let us find conditions of the positivity of H. Since the operators H and A are unitarily equivalent, the following result is a direct consequence of Theorem 4.3.

Theorem 4.8 *Let kernel $h(t)$ be defined by formulas* (1.2) *and* (3.1) *where $p_k = \bar{p}_k$ for $k = 0, 1, \dots, K$. Suppose that K is even and $p_K > 0$. Let $Q(x)$ be polynomial* (3.4) *with coefficients* (3.3)*. Then the Hankel operator H is positive if and only if $Q(x) \geq 0$ for all $x \in \mathbb{R}$. Moreover, if $Q(x_0) < 0$ for some $x_0 \in \mathbb{R}$, then the negative spectrum of H is infinite.*

Theorem 4.8 shows that the positivity of the Hankel operator with kernel (1.2) defined by a polynomial $P(x)$ is determined by another polynomial $Q(x)$ defined by formula (3.4). Of course the condition $Q(x) \geq 0$ is stronger than $P(x) \geq 0$. This follows, for example, from representation (3.5).

In the case $K = 2$, the condition $Q(x) \geq 0$ reads as $q_1^2 \leq 4q_0q_2$. By virtue of (3.7) it can be rewritten as

$$p_1^2 + 2\pi^2 p_2^2/3 \leq 4p_0p_2. \tag{4.12}$$

Obviously, this condition is stronger than the condition $p_1^2 \leq 4p_0p_2$ guaranteeing that $h(t) \geq 0$.

The following assertion is a particular case of Theorem 4.8.

Proposition 4.9 *The Hankel operator H with kernel*

$$h(t) = (p_0 + p_1 \ln t + p_2 \ln^2 t)t^{-1}, \quad p_0 = \bar{p}_0, \quad p_1 = \bar{p}_1, \quad p_2 > 0,$$

is positive if and only if condition (4.12) *is satisfied. Moreover, if $p_1^2 + 2\pi^2 p_2^2/3 > 4p_0p_2$, then the negative spectrum of H is infinite.*

5 Hankel operators with discontinuous kernels

5.1. We here consider Hankel operators with singular kernels defined by formula (1.7). Hankel operators with such kernels are formally symmetric, and we shall see later that they are essentially self-adjoint on $C_0^\infty(\mathbb{R}_+)$. According to (1.1) we have

$$(Hf)(t) = \sum_{k=0}^{K} (-1)^k h_k f^{(k)}(t_0 - t), \quad t \in (0, t_0), \tag{5.1}$$

and $(Hf)(t) = 0$ for $t > t_0$. Formula (5.1) gives us the precise definition of the Hankel operator with distributional kernel (1.7).

Since $L^2(t_0, \infty) \subset \operatorname{Ker} H$, it suffices to study the restriction of H on the subspace $L^2(0, t_0)$. It is again given by differential expression (5.1) on C^∞ functions vanishing in a neighborhood of the point $t = 0$. Let us denote by H_0 the operator (5.1) in $L^2(0, t_0)$ with such domain $\mathcal{D}(H_0)$. Recall that $\mathsf{H}^K(0, t_0)$ is the Sobolev class. Let the set $\mathsf{D}_* \subset \mathsf{H}^K(0, t_0)$ consist of functions satisfying the boundary conditions

$$f(0) = f'(0) = \cdots = f^{(K-1)}(0) = 0. \tag{5.2}$$

The following assertion defines H as a self-adjoint operator.

Lemma 5.1 *The operator H_0 is symmetric and essentially self-adjoint. Its closure $\bar{H}_0 =: H$ is self-adjoint in $L^2(0, t_0)$ on the domain $\mathcal{D}(H) = \mathsf{D}_*$, and it acts by formula* (5.1).

Proof Let us denote by H_* differential operator (5.1) considered as a mapping $H_* : L^2(0, t_0) \to C_0^\infty(0, t_0)'$. Notice that $H_* : \mathsf{H}^K(0, t_0) \to L^2(0, t_0)$. Integrating by parts and using that $h_k = \bar{h}_k$, we see that

$$(H_* f, z) = -\sum_{k=1}^{K} (-1)^k h_k \sum_{l=0}^{k-1} f^{(k-1-l)}(t_0 - t)\overline{z^{(l)}(t)}\Big|_{t=0}^{t=t_0} + (f, H_* z) \tag{5.3}$$

for all $f, z \in \mathsf{H}^K(0, t_0)$. Observe that the non-integral terms here vanish if both functions f and z satisfy boundary conditions (5.2). Since $H_0 f = H_* f$ for $f \in \mathcal{D}(H_0)$, it follows that $(H_0 f, z) = (f, H_0 z)$ if $f, z \in \mathcal{D}(H_0)$.

Let us construct the adjoint operator H_0^*. Suppose that $(H_0 f, z) = (f, z_*)$ for all $f \in \mathcal{D}(H_0)$ and some $z, z_* \in L^2(0, t_0)$. Choosing first $f \in C_0^\infty(0, t_0)$ and using again (5.3), we see that $(H_0 f, z) = (f, H_* z)$ and hence $z_* = H_* z$. Since $z_* \in L^2(0, t_0)$, we find that $z \in \mathsf{H}^K(0, t_0)$.

For an arbitrary $f \in \mathcal{D}(H_0)$, only the nonintegral terms in (5.3) corresponding to $t = t_0$ are equal to zero. Therefore it follows from (5.3) that the sum of terms corresponding to $t = 0$ is also zero, that is,

$$\sum_{k=1}^{K}(-1)^k h_k \sum_{l=0}^{k-1} f^{(k-1-l)}(t_0)\overline{z^{(l)}(0)}$$
$$= -\sum_{p=0}^{K-1}(-1)^p f^{(p)}(t_0) \sum_{l=0}^{K-p-1}(-1)^l h_{p+l+1}\overline{z^{(l)}(0)} = 0.$$

Since the numbers $f^{(p)}(t_0)$ are arbitrary, we obtain a system of K equations

$$\sum_{l=0}^{K-p-1}(-1)^l h_{p+l+1}\overline{z^{(l)}(0)} = 0, \quad p = 0, \ldots, K-1, \tag{5.4}$$

for K numbers $\overline{z(0)}, \overline{z'(0)}, \ldots, \overline{z^{(K-1)}(0)}$. The matrix corresponding to this system consists of elements $a_{p,l}$ parametrized by indices $p, l = 0, \ldots, K-1$. We have $a_{p,l} = (-1)^l h_{p+l+1}$ for $p+l \leq K-1$ and $a_{p,l} = 0$ for $p+l > K-1$. The determinant of this matrix is the product of skew diagonal elements $a_{p,l}$ where $p + l = K - 1$ times the factor $(-1)^{(K-1)K/2}$. Thus it equals h_K^K which is not zero. Therefore it follows from (5.4) that necessarily $z(0) = z'(0) = \cdots = z^{(K-1)}(0) = 0$. It means that $\mathcal{D}(H_0^*) \subset \mathsf{D}_*$ and $H_0^* z = H_* z$ for $z \in \mathcal{D}(H_0^*)$.

Conversely, using again (5.3), we see that $(H_* f, z) = (f, H_* z)$ for all $f, z \in \mathsf{D}_*$. It follows that $\mathcal{D}(H_0^*) = \mathsf{D}_*$ and that the operator H_0^* is symmetric. Hence the operator $H_0^{**} = \bar{H}_0$ is self-adjoint. □

We note that zero is not an eigenvalue of the operator H. Indeed, after the change of variables $t \mapsto t_0 - t$ the equation $(Hf)(t) = 0$ reduces to the differential equation of order K. Therefore the unique solution of the equation $(Hf)(t) = 0$ satisfying conditions (5.2) is zero.

5.2. Clearly, H^2 is the differential operator of order $2K$ defined by the formula

$$(H^2 f)(t) = \sum_{k,l=0}^{K}(-1)^k h_k h_l f^{(k+l)}(t) \tag{5.5}$$

on functions in $\mathsf{H}^{2K}(0, t_0)$ satisfying boundary conditions (5.2) and

$$\sum_{l=0}^{K}(-1)^l h_l f^{(k+l)}(t_0) = 0, \quad k = 0, 1, \ldots, K-1.$$

Of course the spectrum of the operator H^2 consists of positive eigenvalues of multiplicity not exceeding K (because the differential equation $H^2 f = \lambda f$ together with conditions (5.2) has K linearly independent solutions). These eigenvalues accumulate to $+\infty$ and their asymptotics is given by the Weyl formula. However, to find the asymptotics of eigenvalues of the operator H, we have to distinguish its positive and negative eigenvalues. For this reason, it is convenient to introduce an auxiliary operator $\widetilde{H}$ with symmetric (with respect to the point 0) spectrum having the same asymptotics of eigenvalues as H.

We define the operator $\widetilde{H}$ by the same formula (5.1) as H but consider it on functions in $\mathsf{H}^K(0, t_0/2) \oplus \mathsf{H}^K(t_0/2, t_0)$ satisfying the boundary conditions

$$f^{(k)}(0) = f^{(k)}(t_0/2 - 0), \quad f^{(k)}(t_0/2 + 0) = f^{(k)}(t_0), \\ k = 0, \dots, K - 1, \tag{5.6}$$

for K odd or

$$f^{(k)}(0) = f^{(k)}(t_0/2 - 0) = 0, \quad f^{(k)}(t_0/2 + 0) = f^{(k)}(t_0) = 0, \\ k = 0, \dots, K/2 - 1, \tag{5.7}$$

for K even. The operator $\widetilde{H}$ is self-adjoint in the space $L^2(0, t_0/2) \oplus L^2(t_0/2, t_0)$, and it is determined by the matrix

$$\widetilde{H} = \begin{pmatrix} 0 & H_{1,2} \\ H_{2,1} & 0 \end{pmatrix}, \quad H_{1,2} = H_{2,1}^*, \tag{5.8}$$

where $H_{2,1} : L^2(0, t_0/2) \to L^2(t_0/2, t_0)$. The operator $H_{2,1}$ is again given by formula (5.1 on functions in $\mathsf{H}^K(0, t_0/2)$ satisfying conditions (5.6) for K odd or (5.7) for K even at the points 0 and $t_0/2 - 0$. It follows from representation (5.8) that the spectrum of the operator $\widetilde{H}$ is symmetric with respect to the point 0 and consists of eigenvalues $\pm\sqrt{\mu_n}$ where μ_n are eigenvalues of the operator $H_{2,1}^* H_{2,1} =: \mathbf{H}$.

Obviously, the operator $H_{2,1}^*$ is again given by formula (5.1) on functions in $\mathsf{H}^K(t_0/2, t_0)$ satisfying conditions (5.6) for K odd or (5.7) for K even at the points $t_0/2+0$ and t_0. The operator $\mathbf{H}$ acts in the space $L^2(0, t_0/2)$ according to equality (5.5) and its domain $\mathcal{D}(\mathbf{H})$ consists of functions $f \in \mathcal{D}(H_{2,1})$ such that $H_{2,1} f \in \mathcal{D}(H_{2,1}^*)$; in particular, $\mathcal{D}(\mathbf{H}) \subset \mathsf{H}^{2K}(0, t_0/2)$. If K is odd, we have the boundary conditions $f^{(k)}(0) = f^{(k)}(t_0/2)$ for $k = 0, \dots, 2K - 1$. If K is even, then equalities (5.7) should be complemented by the boundary conditions

$$\sum_{l=K/2-k}^{K} (-1)^l h_l f^{(l+k)}(0) = \sum_{l=K/2-k}^{K} (-1)^l h_l f^{(l+k)}(t_0/2) = 0 \tag{5.9}$$

for $k = 0, \dots, K/2 - 1$. Note that conditions (5.7) and (5.9) at the point 0 as well as at the point $t_0/2$ are linearly independent because $h_K \neq 0$.

Let μ_n be eigenvalues of the operator $\mathbf{H}$ enumerated in increasing order with multiplicities taken into account. According to the Weyl formula we have

$$\mu_n = h_K^2 (2\pi t_0^{-1} n)^{2K} (1 + O(n^{-1})), \quad n \to \infty.$$

This yields the asymptotics of eigenvalues $\pm\sqrt{\mu_n}$ of the operator $\widetilde{H}$.

Let us now observe that the operators H and $\widetilde{H}$ are self-adjoint extensions of the same symmetric operator H_{00} with finite deficiency indices $(2K, 2K)$. The operator H_{00} can be defined by formula (5.1) on C^∞ functions vanishing in some neighbourhoods of the points 0, $t_0/2$ and t_0. Therefore the operators H and $\widetilde{H}$ have the same asymptotics of spectra. Thus we have obtained the following result.

Theorem 5.2 *Let H be the self-adjoint Hankel operator with kernel* (1.7). *Then* Ker $H = L^2(t_0, \infty)$. *The non-zero spectrum of H consists of infinite number of eigenvalues $\lambda_n^{(\pm)}$ of multiplicities not exceeding K such that $0 < \lambda_1^{(+)} \le \lambda_2^{(+)} \le \cdots \le \lambda_n^{(+)} \le \cdots$ and $0 > \lambda_1^{(-)} \ge \lambda_2^{(-)} \ge \cdots \ge \lambda_n^{(-)} \ge \cdots$. Eigenvalues $\lambda_n^{(\pm)}$ accumulate to $\pm\infty$ as $n \to \infty$ and have the asymptotics*

$$\lambda_n^{(\pm)} = \pm|h_K|(2\pi t_0^{-1} n)^K (1 + O(n^{-1}))$$

as $n \to \infty$. The corresponding eigenfunctions $f_n^{(\pm)}(t)$ satisfy the equation

$$\sum_{k=0}^{K} (-1)^k h_k \frac{d^k f_n^{(\pm)}(t)}{dt^k} = \lambda_n^{(\pm)} f_n^{(\pm)}(t_0 - t), \quad t \in (0, t_0),$$

and boundary conditions (5.2).

Remark 5.3 In the case $h(t) = \delta'(t - t_0)$ we have the explicit formulas

$$\lambda_n^{(+)} = 2\pi t_0^{-1}(n - 1/4), \quad \lambda_n^{(-)} = -2\pi t_0^{-1}(n - 3/4).$$

References

1. Birman, M.Sh., Solomyak, M.Z.: Spectral Theory of Selfadjoint Operators in Hilbert Space. Reidel, Doldrecht (1987)
2. Carleman, T.: Sur les équations intégrales singulières à noyau réel et symétrique. Almqvist and Wiksell (1923)
3. Gel'fand, I.M., Shilov, G.E.: Generalized Functions. vol. 1. Academic Press, New York (1964)
4. Peller, V.V.: Hankel Operators and their Applications. Springer Verlag, Berlin (2002)
5. Power, S.R.: Hankel Operators on Hilbert Space. Pitnam, Boston (1982)
6. Yafaev, D.R.: Criteria for Hankel operators to be sign-definite, ArXiv: 1303.4040 (2013)

5

Recurrence relations and outer relative asymptotics of orthogonal polynomials with respect to a discrete Sobolev type inner product

Francisco Marcellán · M. Francisca Pérez-Valero · Yamilet Quintana · Alejandro Urieles

Abstract We investigate algebraic and analytic properties of sequences of polynomials orthogonal with respect to the Sobolev type inner product

$$\langle f, g\rangle = \int f(x)g(x)d\mu(x) + \sum_{k=1}^{K}\sum_{i=0}^{N_k} M_{k,i} f^{(i)}(b_k)g^{(i)}(b_k),$$

where μ is a finite positive Borel measure belonging to the Nevai class, the mass points b_k are located outside the support of μ, and $M_{k,i}$ are complex numbers such that $M_{k,N_k} \neq 0$. First, we study the existence as well as recurrence relations for such

Communicated by S. K. Jain.

The second author was partially supported by the Research Fellowship Program from Ministerio de Ciencia e Innovación (MTM 2009-12740-C03-01), Spain. The first, second and third authors were partially supported by a grant from Ministerio de Economía y Competitividad. Dirección General de Investigación Científica y Técnica (MTM 2012-36732-C03-01), Spain.

F. Marcellán · M. F. Pérez-Valero
Departamento de Matemáticas, Universidad Carlos III de Madrid,
Avenida de la Universidad 30, 28911 Leganés, Madrid, Spain
e-mail: pacomarc@ing.uc3m.es

M. F. Pérez-Valero
e-mail: mpvalero@math.uc3m.es

Y. Quintana (✉) · A. Urieles
Departamento de Matemáticas Puras y Aplicadas, Edificio Matemáticas y Sistemas (MYS),
Universidad Simón Bolívar, Apartado Postal: 89000, Caracas 1080 A, Venezuela
e-mail: yquintana@usb.ve

A. Urieles
e-mail: aurielesg@gmail.com

polynomials. When the values $M_{k,i}$ are nonnegative real numbers, we can deduce the coefficients of the recurrence relation in terms of the connection coefficients for the sequences of polynomials orthogonal with respect to the Sobolev type inner product and those orthogonal with respect to the measure μ. The matrix of a symmetric multiplication operator in terms of the above sequence of Sobolev type orthogonal polynomials is obtained from the Jacobi matrix associated with the measure μ. Finally, we focus our attention on some outer relative asymptotics of such polynomials, which are deduced by using the above connection formulas.

Keywords Nevai class · Sobolev inner products · Orthogonal polynomials · Recurrence relations · Connection coefficients · Outer relative asymptotics.

Mathematics Subject Classification (2000) 42C10 · 33C45

1 Introduction

The study of asymptotic properties for general orthogonal polynomials is an important challenge in approximation theory and their applications permeate many fields in science and engineering [16,25,26]. Although it may seem an old subject from the point of view of the standard orthogonality [26], this is not the case neither in the general setting (cf. [9,10,16,21–23,25]) nor from the point of view of Sobolev orthogonality, where it remains like a partially explored subject [3]. In fact, new results continue to appear in some recent publications [4,6–8,11–15,18–20].

A discrete Sobolev-type inner product is defined by

$$\langle f, g\rangle = \int f(x)g(x)d\mu(x) + \sum_{k=1}^{K}\sum_{i=0}^{N_k} M_{k,i} f^{(i)}(b_k)g^{(i)}(b_k), \tag{1}$$

where the functions f and g belong to $L^2(\mu)\cap C^\infty$ and we assume that their derivatives exist at b_k. Here μ is a finite positive Borel measure supported on an infinite subset of the real line, $M_{k,i}$ are real or complex numbers such that $M_{k,N_k} \neq 0$, and b_k, $k = 1, \ldots, K$, are real mass points. Let $(\hat{B}_n)_{n=0}^{\infty}$ be the sequence of polynomials such that

$$deg(\hat{B}_n) \leq n, \quad \langle \hat{B}_n, p\rangle = 0, \quad p \in \mathbb{P}_{n-1}, \quad \int |\hat{B}_n(x)|^2 d\mu(x) = 1.$$

The study of the asymptotic properties of the sequence $(\hat{B}_n)_{n=0}^{\infty}$ has been analyzed considering the cases 'mass points inside' or 'mass points outside' of supp μ, respectively, being supp μ a compact subset of $\mathbb{R}$ or, more recently, an unbounded interval of the real line (see, for instance [7,14]). The first results in the literature about asymptotic properties for orthogonal polynomials with respect to a Sobolev type inner product like (1) appear in [15], where the authors studied the case when there is only one mass point supporting the derivatives either inside or outside $[-1, 1]$ and μ is a measure in the Nevai class $M(0, 1)$. Let us remind that the support of a measure of the Nevai class consists of $[-1, 1]$ and at most a denumerable set of real mass points located outside $[-1, 1]$ which, when infinitely many, may accumulate only at the end points of the above interval. In particular, if $\mu' > 0$, a.e. in $[-1, 1]$, then $\mu \in M(0, 1)$. In [11], using an approach based on the theory of Padé approximants, the authors obtain the outer relative asymptotics for orthogonal polynomials with respect to the Sobolev-type inner product (1) provided that μ belongs to Nevai class $M(0, 1)$ and the mass points b_k belong to $\mathbb{C}\backslash\text{supp}\mu$. The same problem with the mass points in supp $\mu = [-1, 1]$ was solved in [24], provided that $\mu'(x) > 0$ a.e. $x \in [-1, 1]$ and $M_{k,i}$ being nonnegative constants. The asymptotics for orthogonal polynomials with respect to the Sobolev-type inner product (1) with $\mu \in M(0, 1)$, some b_k belonging supp $\mu\backslash[-1, 1]$ and $M_{k,i}$ complex numbers such that $M_{k,N_k} \neq 0$, was solved in [2].

Other results corresponding to the study of asymptotics for orthogonal polynomials with respect to diagonal (resp. non-diagonal) Sobolev inner products with respect to measures supported on the complex plane can be found in [1,4,6,13]. Results related to asymptotics for extremal polynomials associated to non-diagonal Sobolev norms may be seen in [12,18–20].

In this paper, we emphasize the role of the recurrence relations that the Sobolev orthogonal polynomials satisfy as well as their matrix interpretation in terms of a multiplication operator which is related to the evaluation of such a polynomial in the Jacobi matrix associated with the measure μ. Based on a matrix approach, we find the connection coefficients for such sequences. Thus, an efficient algorithm is given in order to deduce those recurrence coefficients.

Notice that the main idea of [2,24] and our contribution is to compare the polynomials $(\hat{B}_n)_{n=0}^{\infty}$ with the polynomials $(p_n)_{n=0}^{\infty}$ orthonormal with respect to the measure μ. In order to obtain estimates of the Sobolev orthogonal polynomials $\hat{B}_n$ we follow, essentially, the same techniques used in these papers and we present an alternative proof for a special case of [11, Theorem 4] (see the proof of Theorem 1) based on the connection formulas.

The paper is structured as follows. Section 2 has an auxiliary character and provides some background as well as some technical lemmas in order to simplify the proof of Theorem 1 about the outer relative asymptotics of the polynomials $(\hat{B}_n)_{n=0}^{\infty}$. In Sect. 3 we deduce a new matrix interpretation of the recurrence relation satisfied by the polynomial sequence $(\hat{B}_n)_{n=0}^{\infty}$ in terms of a matrix polynomial of the Jacobi matrix associated with the sequence of orthonormal polynomials $(p_n)_{n=0}^{\infty}$. The analysis of the connection coefficients for such sequences constitutes a basic tool for such an approach. In Sect. 4 we give the proof of the Theorem concerning the outer relative asymptotics explained above.

2 Background and previous results

Let μ be a finite positive Borel measure supported on the interval $[-1, 1]$ with infinitely many points at the support and let b_k, $k = 1, \dots, K$, be real numbers located outside $[-1, 1]$ with a period. For f and g in $L^2(\mu) \cap C^{\infty}$ such that there exist the derivatives at b_k, we can introduce the Sobolev-type inner product

$$\langle f, g\rangle = (f, g) + \sum_{k=1}^{K}\sum_{i=0}^{N_k} M_{k,i} f^{(i)}(b_k) g^{(i)}(b_k), \tag{2}$$

with $(f, g) = \int f(x)g(x)d\mu(x)$, $M_{k,i} \in \mathbb{C}$, and $M_{k,N_k} \neq 0$. For convenience, we work with another normalization of the Sobolev orthogonal polynomials. Let $(\tilde{B}_n)_{n=0}^{\infty}$ be the sequence of monic polynomials of least degree such that

$$\langle \tilde{B}_n, p\rangle = 0, \quad p \in \mathbb{P}_{n-1},$$

where $\mathbb{P}_{n-1}$ is the linear space of all polynomials with complex coefficients of degree less than or equal to $n-1$. The existence of $\tilde{B}_n \in \mathbb{P}_n$ for each $n \in \mathbb{Z}_+$ follows from solving a linear system of n homogeneous equations and $n+1$ unknowns. Since $\int |\tilde{B}_n(x)|^2 d\mu(x) = 1/\tau_n^2 > 0$, we can define $\hat{B}_n(x) = \tau_n \tilde{B}_n(x)$ and we have a sequence $(\hat{B}_n)_{n=0}^{\infty}$ such that

$$deg(\hat{B}_n) \leq n, \quad \langle \hat{B}_n, p\rangle = 0, \quad p \in \mathbb{P}_{n-1}, \quad \int |\hat{B}_n(x)|^2 d\mu(x) = 1. \tag{3}$$

It is clear that the polynomials $\hat{B}_n$ are not orthonormal with respect to (2), but it is possible to prove that, for μ belonging to the Nevai class $M(0, 1)$ and n large enough, they are equal up to constant factors α_n, with $\lim_{n\to\infty} \alpha_n = 1$. More precisely,

Lemma 1 [2, Lemma 2.2] *For $\mu \in M(0, 1)$, the polynomials $\hat{B}_n$ satisfy the conditions*

(i) *If $M_{k,i} \neq 0$ then* $\lim_{n\to\infty} \hat{B}_n^{(i)}(b_k) = 0$.
(ii) $\lim_{n\to\infty} \langle \hat{B}_n, \hat{B}_n\rangle = 1$.
(iii) *There exists a positive integer n_0 such that $deg(\hat{B}_n) = n$ for all $n \geq n_0$.*

Let $(p_n)_{n=0}^{\infty}$ be the sequence of orthonormal polynomials with respect to μ. In what follows, we assume that either $\mu' > 0$ a.e. on the support of μ or $\mu \in M(0, 1)$ with the additional assumption that none of the mass points b_k belong to the support of the measure μ (cf. [16,21,22]), and we denote by $\kappa(\Pi_n)$ the leading coefficient of $\Pi_n \in \mathbb{P}_n$ with $\deg(\Pi_n) = n$. Let us consider $N = \sum_{k=1}^{K}(N_k + 1)$ and the polynomial

$$\omega_N(x) := \prod_{k=1}^{K}(x - b_k)^{N_k+1}. \tag{4}$$

Let I_k be the number of coefficients $M_{k,j}$, $j = 0, \ldots, N_k$, different from 0 in (1) and let J_k be such that $I_k + J_k = N_k + 1$.

In order to deduce asymptotic properties for the polynomials $\hat{B}_n$, a successful strategy has certainly been to find orthogonality relations involving the polynomials $\hat{B}_n\, p_n$, ω_N, and the monomials $(x - b_k)^m$, for $m \in \{1, \ldots, I_k\} \cup \{1, \ldots, J_k\}$. Concerning this issue, we have the following result.

Lemma 2 (i) *For $m = 1, \ldots, I_k$, we have*

$$\lim_{n\to\infty} \int \frac{\hat{B}_n(x)p_n(x)}{(x-b_k)^m} d\mu(x) = 0.$$

(ii) *If $J_k > 0$ then*

$$\lim_{n\to\infty} \int \frac{\omega_N(x)\hat{B}_n(x)p_{n-N}(x)}{(x-b_k)^m} d\mu(x) = 0, \quad m = 1, \ldots, J_k$$

Proof It suffices to follow the proof given in [2, Lemma 3.1], making the corresponding modifications. □

Lemma 3 *For $n \geq n_0$, the polynomial $\omega_N \hat{B}_n$ has the following representation in terms of the sequence $(p_n)_{n=0}^{\infty}$ of orthonormal polynomials with respect to μ.*

$$\omega_N(x)\hat{B}_n(x) = \sum_{j=0}^{2N} A_{n,j}\, p_{n+N-j}(x), \quad A_{n,0} \neq 0. \tag{5}$$

Moreover, $A_{n,j}$ are bounded and $A_{n,2N} = \frac{\kappa(p_{n-N})}{\kappa(p_{n+N})}\frac{1}{A_{n,0}}\langle \hat{B}_n, \hat{B}_n\rangle \neq 0$.

Proof (5) is an immediate consequence of

$$\int \omega_N(x)\hat{B}_n(x)p_i(x)d\mu(x) = \langle \hat{B}_n, \omega_N p_i\rangle.$$

Thus,

$$\sum_{i=0}^{2N} |A_{n,i}|^2 = \int \omega_N^2(x)|\hat{B}_n(x)|^2 d\mu(x) \leq \max_{x\in \operatorname{supp}\mu} \omega_N^2(x),$$

$$A_{n,0} = \int \omega_N(x)\hat{B}_n(x)p_{n+N}(x)d\mu(x) = \frac{\kappa(\hat{B}_n)}{\kappa(p_{n+N})},$$

$$\begin{aligned} A_{n,2N} &= \int \omega_N(x)\hat{B}_n(x)p_{n-N}(x)d\mu(x) = \langle \hat{B}_n, \omega_N p_{n-N}\rangle \\ &= \frac{\kappa(p_{n-N})}{\kappa(\hat{B}_n)}\langle \hat{B}_n, \hat{B}_n\rangle, \quad n \geq n_0. \end{aligned}$$

Therefore, the coefficients $A_{n,j}$ are bounded, $0 \leq j \leq 2N$, and $A_{n,0}, A_{n,2N}$ satisfy

$$A_{n,0}A_{n,2N} = \frac{\kappa(p_{n-N})}{\kappa(p_{n+N})}\langle \hat{B}_n, \hat{B}_n\rangle.$$

□

3 Recurrence relations and connection coefficients

Next, we will assume that the values $M_{k,i}$ in the inner product (2) are nonnegative real numbers. In such a way, $n_0 = 0$ in Lemma 1 and we can define the sequence of orthogonal polynomials $(\hat{B}_n)_{n=0}^{\infty}$ with $deg(\hat{B}_n) = n$. Thus, it constitutes a basis of the linear space $\mathbb{P}_n$.

Lemma 4 *The polynomial $\hat{B}_n$ has the following representation in terms of the sequence $(p_n)_{n=0}^{\infty}$ of orthonormal polynomials with respect to μ.*

$$\hat{B}_n(x) = \sum_{j=0}^{n} \alpha_{n,j} p_j(x) \tag{6}$$

where

$$\alpha_{n,n} = \frac{\tau_n}{\kappa(p_n)},$$
$$\alpha_{n,j} = -\sum_{k=1}^{K}\sum_{i=0}^{N_k} M_{k,i} \hat{B}_n^{(i)}(b_k) p_j^{(i)}(b_k), \quad \textit{for } 0 \le j \le n-1.$$

Proof By orthonormality of p_n, we have

$$\alpha_{n,n} = \int \hat{B}_n(x) p_n(x) d\mu(x) = \tau_n \int \tilde{B}_n(x) p_n(x) d\mu(x) = \frac{\tau_n}{\kappa(p_n)}.$$

For $0 \le j \le n-1$, using (3) we have

$$\begin{aligned}
\alpha_{n,j} &= \int \hat{B}_n(x) p_j(x) d\mu(x) \\
&= \langle \hat{B}_n, p_j \rangle - \sum_{k=1}^{K}\sum_{i=0}^{N_k} M_{k,i} \hat{B}_n^{(i)}(b_k) p_j^{(i)}(b_k) \\
&= -\sum_{k=1}^{K}\sum_{i=0}^{N_k} M_{k,i} \hat{B}_n^{(i)}(b_k) p_j^{(i)}(b_k).
\end{aligned}$$

□

Lemma 5 *The polynomial p_n has the following representation in terms of the sequence $(\hat{B}_n)_{n=0}^{\infty}$.*

$$p_n(x) = \sum_{j=0}^{n} \beta_{n,j} \hat{B}_j(x), \tag{7}$$

where

$$\beta_{n,n} = \frac{\tau_n}{\langle \hat{B}_n, \hat{B}_n \rangle \kappa(p_n)} + \frac{1}{\langle \hat{B}_n, \hat{B}_n \rangle} \sum_{k=1}^{K} \sum_{i=0}^{N_k} M_{k,i} p_n^{(i)}(b_k) \hat{B}_n^{(i)}(b_k),$$

$$\beta_{n,j} = \frac{1}{\langle \hat{B}_j, \hat{B}_j \rangle} \sum_{k=1}^{K} \sum_{i=0}^{N_k} M_{k,i} p_n^{(i)}(b_k) \hat{B}_j^{(i)}(b_k), \quad \textit{for } 0 \le j \le n-1.$$

Proof This is a straightforward result that follows by using the same arguments as in the previous lemma. □

In order to write in matrix form, we introduce the following notation.

$$\hat{\mathbb{B}} = \left(\hat{B}_0(x), \ldots, \hat{B}_n(x), \ldots\right)^T,$$
$$\mathcal{P} = \left(p_0(x), \ldots, p_n(x), \ldots\right)^T,$$

and Λ being the following lower triangular infinite matrix

$$\Lambda = \begin{pmatrix} \alpha_{0,0} & 0 & 0 & \cdots \\ \alpha_{1,0} & \alpha_{1,1} & 0 & \cdots \\ \alpha_{2,0} & \alpha_{2,1} & \alpha_{2,2} & \cdots \\ \vdots & \vdots & \vdots & \ddots \end{pmatrix}.$$

Then $\hat{\mathbb{B}} = \Lambda \mathcal{P}$ and (6), (7) and (5), respectively, can be written in matrix form as follows.

$$\hat{B}_n(x) = \left(\alpha_{n,0}, \ldots, \alpha_{n,n}\right) \left(p_0(x), \ldots, p_n(x)\right)^T,$$
$$p_n(x) = \left(\beta_{n,0}, \ldots, \beta_{n,n}\right) \left(\hat{B}_0(x), \ldots, \hat{B}_n(x)\right)^T,$$

and

$$\begin{aligned} \omega_N(x)\hat{B}_n(x) &= \left(A_{n,2N}, \ldots, A_{n,0}\right) \left(p_{n-N}(x), \ldots, p_{n+N}(x)\right)^T \\ &= \left(A_{n,2N}, \ldots, A_{n,0}\right) \tilde{H} \left(\hat{B}_0(x), \ldots, \hat{B}_{n+N}(x)\right)^T, \end{aligned}$$

where $\tilde{H} \in \mathbb{R}^{(2N+1)\times(n+N+1)}$ is the Hessenberg matrix given by

$$\tilde{H} = \begin{pmatrix} \beta_{n-N,0} & \cdots & \beta_{n-N,n-N} & \cdots & 0 & 0 & \cdots & 0 \\ \vdots & \vdots & \vdots & \vdots & \vdots & \vdots & \vdots & \vdots \\ \beta_{n,0} & \cdots & \beta_{n,n-N} & \cdots & \beta_{n,n} & 0 & \cdots & 0 \\ \vdots & \vdots & \vdots & \vdots & \vdots & \vdots & \ddots & \vdots \\ \beta_{n+N,0} & \cdots & \beta_{n+N,n-N} & \cdots & \beta_{n+N,n} & \beta_{n+N,n+1} & \cdots & \beta_{n+N,n+N} \end{pmatrix}.$$

Now, we decompose

$$\tilde{H}\left[\begin{pmatrix} \hat{B}_0(x) \\ \vdots \\ \hat{B}_{n-N-1}(x) \\ 0 \\ \vdots \\ 0 \end{pmatrix} + \begin{pmatrix} 0 \\ \vdots \\ 0 \\ \hat{B}_{n-N}(x) \\ \vdots \\ \hat{B}_{n+N}(x) \end{pmatrix}\right]$$

$$= \begin{pmatrix} \beta_{n-N,0} & \cdots & \beta_{n-N,n-N-1} \\ \vdots & & \vdots \\ \beta_{n+N,0} & \cdots & \beta_{n+N,n-N-1} \end{pmatrix}\begin{pmatrix} \hat{B}_0(x) \\ \vdots \\ \hat{B}_{n-N-1}(x) \end{pmatrix} + \hat{H}\begin{pmatrix} \hat{B}_{n-N}(x) \\ \vdots \\ \hat{B}_{n+N}(x) \end{pmatrix},$$

where $\hat{H}$ is the matrix obtained from $\tilde{H}$ deleting its firts column, i.e.

$$\hat{H} = \begin{pmatrix} \beta_{n-N,n-N} & \cdots & 0 \\ \vdots & \ddots & \vdots \\ \beta_{n+N,n-N} & \cdots & \beta_{n+N,n+N} \end{pmatrix}.$$

Having in mind that $\omega_N \hat{B}_n(x) \in \text{span}(\hat{B}_{n-N}, \ldots, \hat{B}_{n+N})$ it follows that

$$\left(A_{n,2N}, \ldots, A_{n,0}\right)\begin{pmatrix} \beta_{n-N,0} & \cdots & \beta_{n-N,n-N-1} \\ \vdots & & \vdots \\ \beta_{n+N,0} & \cdots & \beta_{n+N,n-N-1} \end{pmatrix} = (0, \ldots, 0).$$

Finally, we obtain

$$\omega_N(x)\hat{B}_n(x) = \left(A_{n,2N}, \ldots, A_{n,0}\right)\hat{H}\left(\hat{B}_{n-N}(x), \ldots, \hat{B}_{n+N}(x)\right)^T.$$

These remarks can be summarized as follows.

Proposition 1 *The sequence of polynomials* $(\hat{B}_n)_{n=0}^{\infty}$ *satisfies the following recurrence relation.*

$$\omega_N(x)\hat{B}_n(x) = \sum_{j=0}^{2N} c_{n,j}\hat{B}_{n+N-j}(x),$$

where

$$\left(c_{n,2N}, \dots, c_{n,0}\right) = \left(A_{n,2N}, \dots, A_{n,0}\right) \begin{pmatrix} \beta_{n-N,n-N} & \cdots & 0 \\ \vdots & \ddots & \vdots \\ \beta_{n+N,n-N} & \cdots & \beta_{n+N,n+N} \end{pmatrix}.$$

On the other hand, according to Lemma 3 and Lemma 4

$$\begin{aligned} A_{n,j} &= \int \omega_N(x)\hat{B}_n(x)p_{n+N-j}(x)d\mu(x) \\ &= \langle \omega_N \hat{B}_n, p_{n+N-j}\rangle = \langle \hat{B}_n, \omega_N p_{n+N-j}\rangle \\ &= \left\langle \sum_{l=0}^{n} \alpha_{n,l} p_l, \omega_N p_{n+N-j} \right\rangle \\ &= \sum_{l=0}^{n} \alpha_{n,l} \langle p_l, \omega_N p_{n+N-j}\rangle. \end{aligned}$$

But,

$$\langle \omega_N p_{n+N-j}, p_l\rangle = [\omega_N(J)]_{n+N-j,l},$$

i.e., $\langle \omega_N p_{n+N-j}, p_l\rangle$ is the entry $(n+N-j, l)$ in the $(2N+1)$-diagonal matrix $\omega_N(J)$, where J is the Jacobi matrix associated with the measure μ, i.e. the matrix associated with the multiplication operator in terms of the orthonormal basis $(p_n)_{n=0}^{\infty}$.

Notice that

$$\begin{aligned} A_{n,2N} &= \sum_{l=0}^{n} \alpha_{n,l}[\omega_N(J)]_{n-N,l} = \sum_{l=n-2N}^{n} \alpha_{n,l}[\omega_N(J)]_{n-N,l} \\ A_{n,2N-1} &= \sum_{l=0}^{n} \alpha_{n,l}[\omega_N(J)]_{n-N+1,l} = \sum_{l=n-2N+1}^{n} \alpha_{n,l}[\omega_N(J)]_{n-N+1,l} \\ &\vdots \\ A_{n,0} &= \sum_{l=0}^{n} \alpha_{n,l}[\omega_N(J)]_{n+N,l} = \alpha_{n,n}[\omega_N(J)]_{n+N,n}. \end{aligned}$$

Thus,

$$\begin{aligned} &\left(A_{n,2N}, \dots, A_{n,0}\right) \\ &= \left(\alpha_{n,n-2N}, \dots, \alpha_{n,n}\right) \begin{pmatrix} [\omega_N(J)]_{n-N,n-2N} & \cdots & 0 \\ \vdots & \ddots & \vdots \\ [\omega_N(J)]_{n-N,n} & \cdots & [\omega_N(J)]_{n+N,n} \end{pmatrix}. \end{aligned}$$

As a conclusion, from Proposition 1 we obtain

$$\left(c_{n,2N},\ \ldots,\ c_{n,0}\right)$$

$$= \left(\alpha_{n,n-2N},\ \ldots,\ \alpha_{n,n}\right)\begin{pmatrix} [\omega_N(J)]_{n-N,n-2N} & \cdots & 0 \\ \vdots & \ddots & \vdots \\ [\omega_N(J)]_{n-N,n} & \cdots & [\omega_N(J)]_{n+N,n} \end{pmatrix}$$

$$\times \begin{pmatrix} \beta_{n-N,n-N} & \cdots & 0 \\ \vdots & \ddots & \vdots \\ \beta_{n+N,n-N} & \cdots & \beta_{n+N,n+N} \end{pmatrix}.$$

This yields the relation between the parameters of the recurrence formula for $(\hat{B}_n)$ in terms of $(\alpha_{n,j})_{j=n-2N}^{n}$ and $(\beta_{n-N+k,j})_{k=0}^{2N}$.

On the other hand, from Lemma 3

$$\omega_N \hat{\mathbb{B}} = H\mathcal{P}.$$

Here H denotes the $2N+1$ banded infinite matrix with entries $h_{k,j} = A_{k,k+N-j}$ $k-N \le j \le k+N$, and 0 otherwise.

Given $\mathcal{C} = (c_0(x), \ldots, c_n(x), \ldots)^T$ and $\mathcal{D} = (d_0(x), \ldots, d_n(x), \ldots)^T$, we will denote by $(\mathcal{C}, \mathcal{D}^T)$ and $\langle \mathcal{C}, \mathcal{D}^T \rangle$ the infinite matrices whose entries are $(c_i(x), d_j(x))$ and $\langle c_i(x), d_j(x) \rangle$, respectively.

Since $(\mathcal{P}, \mathcal{P}^T) = I$, we have

$$\begin{aligned} H &= (\omega_N \hat{\mathbb{B}}, \mathcal{P}^T) = (\hat{\mathbb{B}}, \omega_N \mathcal{P}^T) = (\hat{\mathbb{B}}, \mathcal{P}^T \omega_N(J)) \\ &= (\Lambda \mathcal{P}, \mathcal{P}^T)\omega_N(J) = \Lambda \omega_N(J). \end{aligned}$$

From Proposition 1 we get

$$\omega_N \hat{\mathbb{B}} = S\hat{\mathbb{B}},$$

hence $S\langle \hat{\mathbb{B}}, \hat{\mathbb{B}}^T \rangle = \langle \omega_N \hat{\mathbb{B}}, \hat{\mathbb{B}}^T \rangle$. Here S is the $2N+1$ banded infinite matrix with entries $s_{k,j} = c_{k,k+N-j}$ $k-N \le j \le k+N$, and 0 otherwise. But $\langle \omega_N \hat{\mathbb{B}}, \hat{\mathbb{B}}^T \rangle = (\omega_N \hat{\mathbb{B}}, \hat{\mathbb{B}}^T) = (H\mathcal{P}, \mathcal{P}^T \Lambda^T) = H\Lambda^T$. Taking into account that $\langle \hat{\mathbb{B}}, \hat{\mathbb{B}}^T \rangle = D$ is a diagonal matrix according to the orthogonality of the polynomial sequence $(\hat{B}_n)_{n=0}^{\infty}$ with respect to our Sobolev inner product, then $SD = H\Lambda^T$ and we get the following result.

Proposition 2 *The matrices S and $\omega_N(J)$ satisfy the following connection relation*

$$SD = \Lambda \omega_N(J) \Lambda^T. \tag{8}$$

4 Outer relative asymptotics

We can already provide an alternative (and more simple) proof of a well-known result about the outer relative asymptotics for the polynomials $\hat{B}_n$, which is a special case of [11, Theorem 4]. We denote by T_j the jth Chebyshev polynomial of the first kind and by $\varphi^{\pm}(x) := x \pm \sqrt{x^2 - 1}$ with the assumption that the square root is positive for $x > 1$.

Theorem 1 [11, formula (1.10)] *Let μ be a finite positive Borel measure in the Nevai class $M(0, 1)$, such that all the mass points $b_k \notin \operatorname{supp}\mu$. Then the polynomials $(\hat{B}_n)_{n=0}^{\infty}$ satisfy*

$$\lim_{n\to\infty} \frac{\hat{B}_n(x)}{p_n(x)} = \prod_{k=1}^{K} \left(\frac{1}{|\varphi^+(b_k)|} \frac{(\varphi^+(x) - \varphi^+(b_k))^2}{2\varphi^+(x)(x - b_k)} \right)^{I_k}, \tag{9}$$

uniformly on compact sets of $\bar{\mathbb{C}} \backslash \operatorname{supp}\mu$.

Proof Since $\mu \in M(0, 1)$,

$$\lim_{n\in\Lambda} \frac{\omega_N(x)\hat{B}_n(x)}{p_{n+N}(x)} = \sum_{j=0}^{2N} A_j(\varphi^-(x))^j \tag{10}$$

uniformly on compact sets of $\bar{\mathbb{C}} \backslash \operatorname{supp}\mu$.

Now, we are going to show that the A_j's are completely determined for any sequence of nonnegative integers Λ. In order to do it, we need to obtain a factorization of the polynomial $\sum_{j=0}^{2N} A_j z^j$.

Since $b_k \notin \operatorname{supp}\mu$ $1/(x - b_k)^i$, are continuous functions on supp μ. Hence, for a fixed $k \in \{1, \ldots, K\}$, by orthogonality, (2.4), and the weak asymptotic property (see [2, formula (2)]) we get

$$\lim_{n\in\Lambda} \int \frac{\omega_N(x)}{(x - b_k)^i} \hat{B}_n(x) p_{n+N}(x) d\mu(x) = \frac{1}{\pi} \int_{-1}^{1} \frac{\sum_{j=0}^{2N} A_j T_j(x)}{(x - b_k)^i} \frac{dx}{\sqrt{1 - x^2}} = 0,$$

$$i = 1, \ldots, N_k + 1. \tag{11}$$

According to the residue's theorem, expression (11) means that the polynomial $\sum_{j=0}^{2N} A_j z^j$ has a zero of multiplicity at least $N_k + 1$ at $\varphi^-(b_k)$.

On the other hand,

$$\begin{aligned}
&\lim_{n\to\infty} \int \frac{\omega_N(x)}{(x - b_k)^{N_k+1}} \hat{B}_n(x) \frac{p_{n+N}(x)}{(x - b_k)^m} d\mu(x) \\
&= \lim_{n\to\infty} \int \left\{ \sum_{j=0}^{m-1} \frac{1}{j!} \left(\frac{\omega_N}{(x - b_k)^{N_k+1}} \hat{B}_n \right)^{(j)} (b_k)(x - b_k)^j \right. \\
&\quad \left. + (x - b_k)^m \pi_{n+N-N_k-1-m}(x) \right\} \frac{p_{n+N}(x)}{(x - b_k)^m} d\mu(x) \\
&= \lim_{n\to\infty} \sum_{j=0}^{m-1} \frac{1}{j!} \left(\frac{\omega_N}{(x - b_k)^{N_k+1}} \hat{B}_n \right)^{(j)} (b_k) \int \frac{p_{n+N}(x)}{(x - b_k)^{m-j}} d\mu(x) = 0,
\end{aligned}$$

where π_{n+N-N_k-1-m} is some polynomial of degree $n+N-N_k-1-m$ and the last equality holds as a consequence of the following two facts (cf. [2, the proof of statement (i) of Lemma 3.1. and Lemma 2.3]).

(i)

$$\lim_{n\to\infty} \frac{n^{I_k-1-j_1}}{j_1!p_{n+N}(b_k)} \left(\frac{\omega_N}{(x-b_k)^{N_k+1}}\hat{B}_n\right)^{(j_1)}(b_k)$$
$$= \cdots = \lim_{n\to\infty} \frac{n^{I_k-1-j_{J_k}}}{j_{J_k}!p_{n+N}(b_k)} \left(\frac{\omega_N}{(x-b_k)^{N_k+1}}\hat{B}_n\right)^{(j_{J_k})}(b_k) = 0, \qquad (12)$$

where $j_1 < j_2 < \cdots < j_{J_k}$ are the nonnegative integers corresponding to the masses $M_{k,j} = 0$.

(ii)

$$\lim_{n\to\infty} \frac{1}{n^{m-j-1}} \int \frac{p_{n+N}(x)p_{n+N}(b_k)}{(x-b_k)^{m-j}} d\mu(x) = \frac{(-1)^{m-j}}{(m-j-1)!} \left(\frac{1}{\sqrt{b_k^2-1}}\right)^{m-j}, \qquad (13)$$

for each b_k and $m-j>0$.

Then $\varphi^-(b_k)$ is a zero of the polynomial $\sum_{j=0}^{2N} A_j z^j$ of multiplicity at least N_k+1+I_k.

From statement (ii) of Lemma 2, we have

$$\lim_{n\to\infty} \int \frac{\omega_N(x)\hat{B}_n(x)p_{n-N}(x)}{(x-b_k)^m} d\mu(x) = 0, \quad m = 1,\ldots,N_k+1-I_k.$$

As a consequence,

$$\lim_{n\in\Lambda} \int \frac{\omega_N(x)\hat{B}_n(x)p_{n-N}(x)}{(x-b_k)^m} d\mu(x)$$
$$= \frac{1}{\pi}\int_{-1}^{1} \frac{\sum_{j=0}^{2N} A_j T_{2N-j}(x)}{(x-b_k)^m} \frac{dx}{\sqrt{1-x^2}} = 0, \quad m = 1,\ldots,N_k+1-I_k. \qquad (14)$$

Hence, (14) means that $\varphi^+(b_k)$ is a zero of the polynomial $\sum_{j=0}^{2N} A_j z^j$ of multiplicity at least N_k+1-I_k.

Therefore, $\sum_{j=0}^{2N} A_j z^j$ has the following factorization

$$\sum_{j=0}^{2N} A_j z^j = A_{2N} \prod_{k=1}^{K} (z-\varphi^-(b_k))^{N_k+1+I_k}(z-\varphi^+(b_k))^{N_k+1-I_k},$$

and for $z = \varphi^-(x)$ we obtain

$$\sum_{j=0}^{2N} A_j(\varphi^-(x))^j = A_{2N} \prod_{k=1}^{K} (\varphi^-(x) - \varphi^-(b_k))^{N_k+1+I_k} (\varphi^-(x) - \varphi^+(b_k))^{N_k+1-I_k}. \tag{15}$$

If x tends to infinity, then we find $A_0 = A_{2N} \prod_{k=1}^{K} (\varphi^-(b_k))^{2I_k}$, and having in mind that $A_0 A_{2N} = \frac{1}{2^{2N}}$, we can deduce that

$$A_{2N} = \frac{\prod_{k=1}^{K} |\varphi^+(b_k)|^{I_k}}{2^N},$$

and this last equation determines completely A_j for any sequence of nonnegative integers Λ. Also, from (10) and (15) we can deduce that

$$\begin{aligned}
&\lim_{n\to\infty} \frac{\omega_N(x)\hat{B}_n(x)}{p_{n+N}(x)} \\
&= \frac{\prod_{k=1}^{K} |\varphi^+(b_k)|^{I_k}}{2^N} \prod_{k=1}^{K} (\varphi^-(x) - \varphi^-(b_k))^{N_k+1+I_k} (\varphi^-(x) - \varphi^+(b_k))^{N_k+1-I_k} \\
&= \frac{1}{2^N} \prod_{k=1}^{K} |\varphi^+(b_k)|^{I_k} (\varphi^-(x) - \varphi^-(b_k))^{N_k+1+I_k} (\varphi^-(x) - \varphi^+(b_k))^{N_k+1-I_k} \\
&= \frac{1}{2^N} \prod_{k=1}^{K} \frac{|\varphi^+(b_k)|^{I_k} (\varphi^-(x) - \varphi^-(b_k))^{I_k}}{(\varphi^-(x) - \varphi^+(b_k))^{I_k}} ((\varphi^-(x) - \varphi^-(b_k))(\varphi^-(x) - \varphi^+(b_k)))^{N_k+1}.
\end{aligned}$$

Finally, taking into account the outer ratio asymptotics for orthonormal polynomials associated with measures in the Nevai class, as well as the fact that

$$\omega_N(x) = \frac{1}{2^N} (\varphi^+(x))^N \prod_{k=1}^{K} ((\varphi^-(x))^2 - 2b_k \varphi^-(x) + 1)^{N_k+1}, \tag{16}$$

on compact subsets of $\bar{\mathbb{C}} \setminus \operatorname{supp}\mu$ we get

$$\lim_{n\to\infty} \frac{\hat{B}_n(x)}{p_n(x)} = \prod_{k=1}^{K} \left(\frac{1}{|\varphi^+(b_k)|} \frac{(\varphi^+(x) - \varphi^+(b_k))^2}{2\varphi^+(x)(x - b_k)} \right)^{I_k}.$$

□

Remark 1 When $b_k \in \operatorname{supp}\mu$, $k = 1, \ldots, K$, for instance, $\{a_k\}_{k=1}^{K_1}$ and $\{b'_j\}_{j=1}^{K_2}$ are the mass points of μ on $[-1, 1]$ and the mass points of μ on supp $\mu \setminus [-1, 1]$, respectively, for $k = 1, \ldots, K_1$ and $j = 1, \ldots, K_2$, $K = K_1 + K_2$, it was proved in [2, Theorem

3.1, (i)] that for $\mu \in M(0,1)$ the following outer relative asymptotics holds uniformly on compact subsets of $\bar{\mathbb{C}} \backslash \operatorname{supp}\mu$.

$$\lim_{n\to\infty} \frac{\hat{B}_n(x)}{q_n(x)} = \prod_{k=1}^{K_2} \left(\frac{1}{|\varphi^+(b'_k)|} \frac{(\varphi^+(x) - \varphi^+(b'_k))^2}{2\varphi^+(x)(x-b'_k)} \right)^{I_k}, \tag{17}$$

where $(q_n)_{n=0}^{\infty}$ is the sequence of orthonormal polynomials with respect to the measure $\nu \in M(0,1)$ defined by $\nu = \mu - \sum_{k=1}^{K_1} \mu(\{a_k\})\delta_{a_k} - \sum_{k=1}^{K_2} \mu(\{b'_k\})\delta_{b'_k}$.

Even though in the previous proof we follow the ideas of the proof given in [2, Theorem 3.1], it is worthwhile to point out that the existing difference between the arguments of both proofs is the use of Lemma 2, which is necessary for us since we do not consider the mass points of μ inside supp μ.

Remark 2 It would be interesting to analyze two open problems concerning the inner strong asymptotics (resp. the outer relative asymptotics) for orthogonal polynomials with respect to Sobolev-type inner products when $K = \infty$ and μ belongs to the Szegő (resp. Nevai) class. Notice that when you consider standard inner products associated with measures μ belonging to the Szegő class $\mathcal{S}$, i.e. $\frac{ln\mu'}{(1-x^2)^{1/2}} \in L^1([-1,1])$, and you add a denumerable set of mass points outside the interval $[-1,1]$ which accumulate at the ends of the interval, assuming that these mass points satisfy Blaschke's condition, then inner strong asymptotics for the corresponding sequence of standard orthonormal polynomials has been obtained in [17]. On the other hand, if $\mu' > 0$ a.e. in $[-1,1]$ and you add an infinite number of mass points outside the interval $[-1,1]$, then the new measure belongs to the Nevai class, but nothing is said about the outer relative asymptotics for the new sequence of standard orthonormal polynomials (see [5]). Thus, this last situation needs a deep analysis. Of course, in the Sobolev case, when you add a denumerable set of mass points at the non standard component of the inner product, the asymptotic behavior (inner strong asymptotics for measures in the Szegő class and outer relative asymptotics in the Nevai class, respectively) of the corresponding orthonormal polynomials constitute more intriguing problems than those pointed out above.

Acknowledgments The authors thank the referees for their careful revision of the manuscript. Their helpful comments and suggestions contributed to improve substantially the presentation of the manuscript.

References

1. Alfaro, M., Marcellán, F., Rezola, M.L., Ronveaux, A.: Sobolev-type orthogonal polynomials: the non-diagonal case. J. Approx. Theory **83**, 266–287 (1995)
2. Alvarez Rocha, I., Salto, L.: Asymptotics of polynomials orthogonal with respect to a discrete-complex Sobolev inner product. J. Comput. Appl. Math. **178**, 1–19 (2005)

3. Baratchart, I., Martínez-Finkelshtein, A., Jiménez, D., Lubinsky, D.S., Mhaskar, H.N., Pritsker, I., Putinar, M., Stylianopoulus, M., Totik, V., Varju, P., Xu, Y.: Open problems in constructive function theory. Electron. Trans. Numer. Anal. **25**, 511–525 (2006)
4. Branquinho, A., Foulquié, A., Marcellán, F.: Asymptotic behavior of Sobolev type orthogonal polynomials on a rectifiable Jordan curve or arc. Constr. Approx. **18**, 161–182 (2002)
5. Denisov, S.A.: On Rakhmanov's theorem for Jacobi matrices. Proc. Am. Math. Soc. **132**, 847–852 (2004)
6. Dueñas, H., Marcellán, F.: Asymptotic behaviour of Laguerre-Sobolev type orthogonal polynomials: a nondiagonal case. J. Comput. Appl. Math. **235**, 998–1007 (2010)
7. Dueñas, H., Huertas, E.J., Marcellán, F.: Asymptotic properties of Laguerre-Sobolev type orthogonal polynomials. Numer. Algorithms **60**, 51–73 (2011)
8. Foulquié, A., Marcellán, F., Pan, K.: Asymptotic behavior of Sobolev-type orthogonal polynomials on the unit circle. J. Approx. Theory **100**, 345–363 (1999)
9. López Lagomasino, G.: Relative asymptotics of orthogonal polynomials on the real axis, Mat. Sb. **137**(179), 266–287 (1988) (English translation: Math. USSR-Sb. **65**, 505–527 (1990))
10. López Lagomasino, G.: Asymptotics of polynomials orthogonal with respect to varying measures. Constr. Approx. **5**, 199–219 (1989)
11. López Lagomasino, G., Marcellán, F., Van Assche, W.: Relative symptotics for polynomials orthogonal with respect to a discrete Sobolev inner product. Constr. Approx. **11**, 107–137 (1995)
12. López Lagomasin, G., Pérez Izquierdo, I., Pijeira, H.: Asymptotic of extremal polynomials in the complex plane. J. Approx. Theory **137**, 226–237 (2005)
13. López Lagomasino, G., Pijeira, H.: Zero location and n-th root asymptotics of Sobolev orthogonal polynomials. J. Approx. Theory **99**, 30–43 (1999)
14. Marcellán, F., Moreno-Balcázar, J.J.: Asymptotics and zeros of Sobolev orthogonal polynomials on unbounded supports. Acta Appl. Math. **94**, 163–192 (2006)
15. Marcellán, F., Van Assche, W.: Relative asymptotics for orthogonal polynomials with a Sobolev inner product. J. Approx. Theory **72**, 193–209 (1992)
16. Nevai, P.: Orthogonal Polynomials. Members of American Mathematical Socirty, vol. 213. American Mathematical Society, Providence (1979)
17. Peherstorfer, F., Yuditskii, P.: Asymptotics of orthonormal polynomials in the presence of a denumerable set of mass points. Proc. Am. Math. Soc. **129**, 3213–3220 (2001)
18. Portilla, A., Rodríguez, J.M., Tourís, E.: The multiplication operator, zero location and asymptotic for non-diagonal Sobolev norms. Acta Appl. Math. **111**, 205–218 (2010)
19. Portilla, A., Quintana, Y., Rodríguez, J.M., Tourís, E.: Zero location and asymptotic behavior for extremal polynomials with non-diagonal Sobolev norms. J. Approx. Theory **162**, 2225–2242 (2010)
20. Portilla, A., Quintana, Y., Rodríguez, J.M., Tourís, E.: Concerning asymptotic behavior for extremal polynomials associated to non-diagonal Sobolev norms. J. Funct. Spaces Appl. 2013, 1–11 (2013, article ID 628031)
21. Rakhmanov, E.A.: On the asymptotics of the ratio of orthogonal polynomials. Mat. Sb. **103**(145), 271–291 (1977) (English translation: Math. USSR-Sb. **32**, 199–213 (1977))
22. Rakhmanov, E.A.: On the asymptotics of the ratio of orthogonal polynomials II. Mat. Sb. **118**(160), 271–291 (1982) (English translation: Math. USSR-Sb. **46**, 105–117 (1983))
23. Rakhmanov, E.A.: On the asymptotics of polynomials orthogonal on the circle with weights not satisfying Szegő's condition. Mat. Sb. **118**(172), 151–169 (1986) (English translation: Math. USSR-Sb. **58**, 149–167 (1987))
24. Rocha, I.A., Marcellán, F., Salto, L.: Relative asymptotics and Fourier Series of orthogonal polynomials with a discrete Sobolev inner product. J. Approx. Theory **121**, 336–356 (2003)
25. Stahl, H., Totik, V.: General Orthogonal Polynomials. Cambridge University Press, Cambridge (1992)
26. Szegő, G.: Orthogonal Polynomials. Colloquium Publications, American Mathematical Society, vol. 23, 4th edn. American Mathematical Society, Providence (1975)

6

Inverse problems: seeing the unseen

Gunther Uhlmann

Abstract This survey article deals mainly with two inverse problems and the relation between them. The first inverse problem we consider is whether one can determine the electrical conductivity of a medium by making voltage and current measurements at the boundary. This is called electrical impedance tomography and also Calderón's problem since the famous analyst proposed it in the mathematical literature (Calderón in On an inverse boundary value problem. Seminar on numerical analysis and its applications to continuum physics (Rio de Janeiro, 1980), Soc Brasil Mat. Rio de Janeiro, pp. 65–73, 1980). The second is on travel time tomography. The question is whether one can determine the anisotropic index of refraction of a medium by measuring the travel times of waves going through the medium. This can be recast as a geometry problem, the boundary rigidity problem. Can we determine a Riemannian metric of a compact Riemannian manifold with boundary by measuring the distance function between boundary points? These two inverse problems concern *visibility*, that is whether we can determine the internal properties of a medium by making measurements at the boundary. The last topic of this paper considers the opposite issue: *invisibility*: Can one make objects invisible to different types of waves, including light?

Communicated by Ari Laptev.

Gunther Uhlmann: Simons Fellowship.

G. Uhlmann (✉)
Department of Mathematics, University of Washington,
Seattle, WA 98195, USA
e-mail: gunther@math.washington.edu

1 Calderón's problem

1.1 Introduction

In 1980 Calderón published a short paper entitled "On an inverse boundary value problem" [37]. This pioneer contribution motivated many developments in inverse problems, in particular in the construction of "complex geometrical optics" solutions of partial differential equations to solve several inverse problems. We survey some these developments in this paper. In his talk at the ICM in Berlin in 1998 the author proposed 7 open problems [199] on this subject. This section is to a large extent a report on the progress made in solving these problems.

The problem that Calderón considered was whether one can determine the electrical conductivity of a medium by making voltage and current measurements at the boundary of the medium. This inverse method is known as *Electrical Impedance Tomography* (EIT). Calderón was motivated by oil prospection. In the 40's he worked as an engineer for Yacimientos Petroliferos Fiscales (YPF), the state oil company of Argentina, and he thought about this problem then although he did not publish his results until many years later. For use of electrical methods in geophysical prospection see [208]. Parenthetically Calderón said in his speech accepting the "Doctor Honoris Causa" of the Universidad Autónoma de Madrid that his work at YPF had been very interesting but he was not well treated there; he would have stayed at YPF otherwise [38]. It goes without saying that the bad treatment of Calderón by YPF was very fortunate for Mathematics! EIT also arises in medical imaging given that human organs and tissues have quite different conductivities [103]. One exciting potential application is the early diagnosis of breast cancer [210]. The conductivity of a malignant breast tumor is typically 0.2 mho which is significantly higher than normal tissue which has been typically measured at 0.03 mho. Another application is to monitor pulmonary functions [94]. See the book [79] and the issue of Physiological Measurement [80] for other medical imaging applications of EIT. This inverse method has also been used to detect leaks from buried pipes [102]. For other reviews see [29,43] and [74]. We now describe more precisely the mathematical problem. Let $\Omega \subseteq \mathbb{R}^n$ be a bounded domain with smooth boundary (many of the results we will describe are valid for domains with Lipschitz boundaries). The electrical conductivity of Ω is represented by a bounded and positive function $\gamma(x)$. In the absence of sinks or sources of current the equation for the potential is given by

$$\nabla \cdot (\gamma \nabla u) = 0 \text{ in } \Omega \tag{1}$$

since, by Ohm's law, $\gamma \nabla u$ represents the current flux. Given a potential $f \in H^{\frac{1}{2}}(\partial\Omega)$ on the boundary the induced potential $u \in H^1(\Omega)$ solves the Dirichlet problem

$$\begin{aligned} \nabla \cdot (\gamma \nabla u) &= 0 \text{ in } \Omega, \\ u\big|_{\partial\Omega} &= f. \end{aligned} \tag{2}$$

The Dirichlet to Neumann map, or voltage to current map, is given by

$$\Lambda_\gamma(f) = \left(\gamma \frac{\partial u}{\partial \nu}\right)\Big|_{\partial\Omega} \tag{3}$$

where ν denotes the unit outer normal to $\partial\Omega$. The inverse problem is to determine γ knowing Λ_γ. It is difficult to find a systematic way of prescribing voltage measurements at the boundary to be able to find the conductivity. Calderón took instead a different route. Using the divergence theorem we have

$$Q_\gamma(f) := \int_\Omega \gamma |\nabla u|^2 dx = \int_{\partial\Omega} \Lambda_\gamma(f) f \, dS \tag{4}$$

where dS denotes surface measure and u is the solution of (2). In other words $Q_\gamma(f)$ is the quadratic form associated to the linear map $\Lambda_\gamma(f)$, and to know $\Lambda_\gamma(f)$ or $Q_\gamma(f)$ for all $f \in H^{\frac{1}{2}}(\partial\Omega)$ is equivalent. $Q_\gamma(f)$ measures the energy needed to maintain the potential f at the boundary. Calderón's point of view is that if one looks at $Q_\gamma(f)$ the problem is changed to finding enough solutions $u \in H^1(\Omega)$ of the Eq. (1) in order to find γ in the interior. These are the complex geometrical optics (CGO) solutions considered in Sect. 1.3. A short summary of the contents of this section is as follows. In Sect. 1.2 we describe results about uniqueness, stability and reconstruction, for the boundary values of a conductivity and its normal derivative. In Sect. 1.3 we describe the construction by Sylvester and Uhlmann [190,191] of CGO solutions for the Schrödinger equation associated to a bounded potential. These solutions behave like Calderón's complex exponential solutions for large complex frequencies. In Sect. 1.4 we use these solutions to prove, in dimension $n \geq 3$, a global identifiability result, stability estimates and a reconstruction method for the inverse problem. We also describe an extension of the identifiability result to non-linear conductivities [184] and give other applications of CGO solutions. In Sect. 1.5 we consider the partial data problem, that is the case when the DN map is measured on a part of the boundary. We describe the results of [107] for the non-linear problem in dimension three or larger. This uses a larger class of CGO solutions, having a non-linear phase function that are constructed using Carleman estimates. We also review the article [49] for the linearized problem with partial data. In Sect. 1.6 we consider the two dimensional case. In particular we describe briefly the recent work of Astala and Päivärinta proving uniqueness for bounded measurable coefficients, and the work of Bukhgeim proving uniqueness for a potential from Cauchy data associated to the Schrödinger equation. Finally we consider the work of Imanuvilov, Uhlmann and Yamamoto on the partial data problem in two dimensions [87,88]. These sections deal with the case of *isotropic* conductivities. In Sect. 1.7 we consider the case of *anisotropic* conductivities, i.e. the conductivity depends also on direction. In two dimensions that there has been substantial progress in the understanding of anisotropic problems since one can usually reduce the problem to the isotropic case by using isothermal coordinates. In dimension three the problem as pointed out in [121] is of geometric nature. We review the results of [52,124].

1.2 Boundary determination

Kohn and Vogelius proved the following identifiability result at the boundary [114].

Theorem 1.1 *Let* $\gamma_i \in C^\infty(\overline{\Omega})$ *be strictly positive. Assume* $\Lambda_{\gamma_1} = \Lambda_{\gamma_2}$. *Then*

$$\partial^\alpha \gamma_1\Big|_{\partial\Omega} = \partial^\alpha \gamma_2\Big|_{\partial\Omega}, \qquad \forall|\alpha|.$$

This settled the identifiability question for the non-linear problem in the real-analytic category. They extended the identifiability result to piecewise real-analytic conductivities in [115].

Sketch of proof of Theorem 1.1. We outline an alternative proof to the one given by Kohn and Vogelius of 1.1. In the case $\gamma \in C^\infty(\overline{\Omega})$ we know, by another result of Calderón [39], that Λ_γ is a classical pseudodifferential operator of order 1. Let (x', x^n) be coordinates near a point $x_0 \in \partial\Omega$ so that the boundary is given by $x^n = 0$. The function $\lambda_\gamma(x', \xi')$ denotes the full symbol of Λ_γ in these coordinates. It was proved in [192] that

$$\lambda_\gamma(x', \xi') = \gamma(x', 0)|\xi'| + a_0(x', \xi') + r(x', \xi') \tag{5}$$

where $a_0(x', \xi')$ is homogeneous of degree 0 in ξ' and is determined by the normal derivative of γ at the boundary and tangential derivatives of γ at the boundary. The term $r(x', \xi')$ is a classical symbol of order -1. Then $\gamma\Big|_{\partial\Omega}$ is determined by the principal symbol of Λ_γ and $\frac{\partial\gamma}{\partial x^n}\Big|_{\partial\Omega}$ by the principal symbol and the term homogeneous of degree 0 in the expansion of the full symbol of Λ_γ. More generally the higher order normal derivatives of the conductivity at the boundary can be determined recursively. In Lee and Uhlmann [121] one can find a general approach to the calculation of the full symbol of the Dirichlet to Neumann map that applies to more general situations. We note that this gives also a reconstruction procedure. We first can reconstruct γ at the boundary since $\gamma\Big|_{\partial\Omega}|\xi'|$ is the principal symbol of Λ_γ [see (5)]. In other words in coordinates (x', x^n) so that $\partial\Omega$ is locally given by $x^n = 0$ we have

$$\gamma(x', 0)a(x') = \lim_{s\to\infty} e^{-is<x',\omega'>}\frac{1}{s}\Lambda_\gamma(e^{is<x',\omega'>}a(x'))$$

with $\omega' \in \mathbb{R}^{n-1}$ and $|\omega'| = 1$ and a a smooth and compactly supported function. In a similar fashion, using (5), one can find $\frac{\partial\gamma}{\partial\nu}\Big|_{\partial\Omega}$ by computing the principal symbol of $(\Lambda_\gamma - \gamma|_{\partial\Omega}\Lambda_1)$ where Λ_1 denotes the Dirichlet to Neumann map associated to the conductivity 1. The other terms can be reconstructed recursively in a similar fashion. We also observe, by taking an appropriate cut-off function a above, that this procedure is *local*, that is we only need to know the DN map in an open set of the boundary to determine the Taylor series of the conductivity in that open set. This method also leads to stability estimates at the boundary [192].

Theorem 1.2 *Suppose that* γ_1 *and* γ_2 *are* C^∞ *functions on* $\overline{\Omega} \subseteq \mathbb{R}^n$ *satisfying*

i) $0 < \frac{1}{E} \leq \gamma_i \leq E$

ii) $\|\gamma_i\|_{C^2(\overline{\Omega})} \leq E$

Given any $0 < \sigma < \frac{1}{n+1}$, *there exists* $C = C(\Omega, E, n, \sigma)$ *such that*

$$\|\gamma_1 - \gamma_2\|_{L^\infty(\partial\Omega)} \leq C\|\Lambda_{\gamma_1} - \Lambda_{\gamma_2}\|_{\frac{1}{2},\frac{-1}{2}} \tag{6}$$

and

$$\left\|\frac{\partial\gamma_1}{\partial\nu} - \frac{\partial\gamma_2}{\partial\nu}\right\|_{L^\infty(\partial\Omega)} \leq C\|\Lambda_{\gamma_1} - \Lambda_{\gamma_2}\|^{\sigma}_{\frac{1}{2},\frac{-1}{2}}. \tag{7}$$

This result implies that we don't need the conductivity to be smooth to determine the conductivity and its normal derivative at the boundary. In the case γ is continuous on $\overline{\Omega}$ we can determine γ at the boundary by using the stability estimate (6) and an approximation argument. In the case that $\gamma \in C^1(\overline{\Omega})$ we can determine, knowing the DN map, γ and its normal derivative at the boundary using the estimate (7) above and an approximation argument. For other results and approaches to boundary determination of the conductivity see [5,31,135,141]. In one way or another the boundary determination involves testing the DN map against highly oscillatory functions at the boundary.

1.3 Complex geometrical optics solutions with a linear phase

Motivated by Calderón exponential solutions used in [37] (see [92] for a numerical study) in the study of the linearized problem at a constant conductivity, Sylvester and Uhlmann [190, 191] constructed in dimension $n \geq 2$ complex geometrical optics (CGO) solutions of the conductivity equation for C^2 conductivities that behave like Calderón exponential solutions for large frequencies. This can be reduced to constructing solutions in the whole space (by extending $\gamma = 1$ outside a large ball containing Ω) for the Schrödinger equation with potential. We describe this more precisely below. Let $\gamma \in C^2(\mathbb{R}^n)$, γ strictly positive in $\mathbb{R}^n$ and $\gamma = 1$ for $|x| \geq R$, $R > 0$. Let $L_\gamma u = \nabla \cdot \gamma\nabla u$. Then we have

$$\gamma^{-\frac{1}{2}} L_\gamma(\gamma^{-\frac{1}{2}}) = \Delta - q \tag{8}$$

where

$$q = \frac{\Delta\sqrt{\gamma}}{\sqrt{\gamma}}. \tag{9}$$

Therefore, to construct solutions of $L_\gamma u = 0$ in $\mathbb{R}^n$ it is enough to construct solutions of the Schrödinger equation $(\Delta - q)u = 0$ with q of the form (9). The next result proven in [190, 191] states the existence of complex geometrical optics solutions for the Schrödinger equation associated to any bounded and compactly supported potential.

Theorem 1.3 *Let $q \in L^\infty(\mathbb{R}^n)$, $n \geq 2$, with $q(x) = 0$ for $|x| \geq R > 0$. Let $-1 < \delta < 0$. There exists $\epsilon(\delta)$ and such that for every $\rho \in \mathbb{C}^n$ satisfying*

$$\rho \cdot \rho = 0$$

and

$$\frac{\|(1+|x|^2)^{1/2} q\|_{L^\infty(\mathbb{R}^n)} + 1}{|\rho|} \leq \epsilon$$

there exists a unique solution to

$$(\Delta - q)u = 0$$

of the form

$$u = e^{x\cdot\rho}(1 + \psi_q(x, \rho)) \tag{10}$$

with $\psi_q(\cdot, \rho) \in L^2_\delta(\mathbb{R}^n)$. Moreover $\psi_q(\cdot, \rho) \in H^2_\delta(\mathbb{R}^n)$ and for $0 \leq s \leq 2$ there exists $C = C(n, s, \delta) > 0$ such that

$$\|\psi_q(\cdot, \rho)\|_{H^s_\delta} \leq \frac{C}{|\rho|^{1-s}}. \tag{11}$$

Here

$$L^2_\delta(\mathbb{R}^n) = \left\{ f;\ \int (1+|x|^2)^\delta |f(x)|^2 dx < \infty \right\}$$

with the norm given by $\|f\|^2_{L^2_\delta} = \int (1+|x|^2)^\delta |f(x)|^2 dx$ and $H^m_\delta(\mathbb{R}^n)$ denotes the corresponding Sobolev space. Note that for large $|\rho|$ these solutions behave like Calderón's exponential solutions $e^{x\cdot\rho}$. The equation for ψ_q is given by

$$(\Delta + 2\rho \cdot \nabla)\psi_q = q(1 + \psi_q). \tag{12}$$

The Eq. (12) is solved by constructing an inverse for $(\Delta + 2\rho \cdot \nabla)$ and solving the integral equation

$$\psi_q = (\Delta + 2\rho \cdot \nabla)^{-1}(q(1 + \psi_q)). \tag{13}$$

Lemma 1.4 *Let $-1 < \delta < 0$, $0 \leq s \leq 1$. Let $\rho \in \mathbb{C}^n - 0$, $\rho \cdot \rho = 0$. Let $f \in L^2_{\delta+1}(\mathbb{R}^n)$. Then there exists a unique solution $u_\rho \in L^2_\delta(\mathbb{R}^n)$ of the equation*

$$\Delta_\rho u_\rho := (\Delta + 2\rho \cdot \nabla)u_\rho = f. \tag{14}$$

Moreover $u_\rho \in H^2_\delta(\mathbb{R}^n)$ *and*

$$\|u_\rho\|_{H^s_\delta(\mathbb{R}^n)} \leq \frac{C_{s,\delta}\|f\|_{L^2_{\delta+1}}}{|\rho|^{1-s}}$$

for $0 \leq s \leq 2$ *and for some constant* $C_{s,\delta} > 0$.

The integral equation (12) can then be solved in $L^2_\delta(\mathbb{R}^n)$ for large $|\rho|$ since

$$(I - (\Delta + 2\rho \cdot \nabla)^{-1} q)\psi_q = (\Delta + 2\rho \cdot \nabla)^{-1} q$$

and $\|(\Delta + 2\rho \cdot \nabla)^{-1} q\|_{L^2_\delta \to L^2_\delta} \leq \frac{C}{|\rho|}$ for some $C > 0$ where $\|\cdot\|_{L^2_\delta \to L^2_\delta}$ denotes the operator norm between $L^2_\delta(\mathbb{R}^n)$ and $L^2_\delta(\mathbb{R}^n)$. We will not give details of the proof of Lemma 1.4 here. We refer to the papers [190,191] . We note that there has been other approaches to construct CGO solutions for the Schrödinger equation [73,97]. These constructions don't give uniqueness of the CGO solutions that are used in the reconstruction method of the conductivity from the DN map (see Sect. 1.4.5). If 0 is not a Dirichlet eigenvalue for the Schrödinger equation we can also define the DN map

$$\Lambda_q(f) = \frac{\partial u}{\partial \nu}|_{\partial\Omega}$$

where u solves

$$(\Delta - q)u = 0; \quad u|_{\partial\Omega} = f.$$

More generally we can define the set of Cauchy data for the Schrödinger equation. Let $q \in L^\infty(\Omega)$. We define the Cauchy data as the set

$$\mathcal{C}_q = \left\{ \left(u\Big|_{\partial\Omega}, \frac{\partial u}{\partial \nu}\Big|_{\partial\Omega} \right) \right\}, \tag{15}$$

where $u \in H^1(\Omega)$ is a solution of

$$(\Delta - q)u = 0 \text{ in } \Omega. \tag{16}$$

We have $\mathcal{C}_q \subseteq H^{\frac{1}{2}}(\partial\Omega) \times H^{-\frac{1}{2}}(\partial\Omega)$. If 0 is not a Dirichlet eigenvalue of $\Delta - q$, then in fact $\mathcal{C}_q$ is a graph, namely

$$\mathcal{C}_q = \{(f, \Lambda_q(f)) \in H^{\frac{1}{2}}(\partial\Omega) \times H^{-\frac{1}{2}}(\partial\Omega)\}.$$

Complex geometrical optics for first order equations and systems under different regularity assumptions of the coefficients have been constructed in [142,144,162,163,197]. For the case of the magnetic Schrödinger operator unique identifiability of the magnetic field and the electrical potential was shown in [120] assuming that both the electrical potential and magnetic potential are both just bounded. We refer to the article a more up to date developments on this topic and the references given there.

1.4 The Calderón problem in dimension $n \geq 3$

In this section we summarize some of the basic theoretical results for Calderón's problem in dimension three or higher.

1.4.1 Uniqueness

The identifiability question was resolved in [190] for smooth enough conductivities. The result is

Theorem 1.5 *Let $\gamma_i \in C^2(\overline{\Omega})$, γ_i strictly positive, $i = 1, 2$. If $\Lambda_{\gamma_1} = \Lambda_{\gamma_2}$ then $\gamma_1 = \gamma_2$ in $\overline{\Omega}$.*

In dimension $n \geq 3$ this result is a consequence of a more general result. Let $q \in L^\infty(\Omega)$.

Theorem 1.6 *Let $q_i \in L^\infty(\Omega)$, $i = 1, 2$. Assume $\mathcal{C}_{q_1} = \mathcal{C}_{q_2}$, then $q_1 = q_2$.*

We now show that Theorem 1.6 implies Theorem 1.5. Using (8) we have

$$\mathcal{C}_{q_i} = \left\{ \left(f, \left(\frac{1}{2}\gamma_i^{-1}\Big|_{\partial\Omega} \frac{\partial \gamma_i}{\partial \nu}\Big|_{\partial\Omega} \right) f + \gamma_i^{-\frac{1}{2}}\Big|_{\partial\Omega} \Lambda_{\gamma_i} \left(\gamma^{-\frac{1}{2}}\Big|_{\partial\Omega} f \right) \right), \quad f \in H^{\frac{1}{2}}(\partial\Omega) \right\}.$$

Then we conclude $\mathcal{C}_{q_1} = \mathcal{C}_{q_2}$ using the boundary identifiability result of Kohn and Vogelius [114] and its extension [192].

Proof of Theorem 1.6 Let $u_i \in H^1(\Omega)$ be a solution of

$$(\Delta - q_i)u_i = 0 \text{ in } \Omega, \qquad i = 1, 2.$$

Then using the divergence theorem we have

$$\int_\Omega (q_1 - q_2)u_1 u_2 dx = \int_{\partial\Omega} \left(\frac{\partial u_1}{\partial \nu} u_2 - u_1 \frac{\partial u_2}{\partial \nu} \right) dS. \tag{17}$$

Now it is easy to prove that if $\mathcal{C}_{q_1} = \mathcal{C}_{q_2}$ then the LHS of (17) is zero

$$\int_\Omega (q_1 - q_2)u_1 u_2 dx = 0. \tag{18}$$

Now we extend $q_i = 0$ in Ω^c. We take solutions of $(\Delta - q_i)u_i = 0$ in $\mathbb{R}^n$ of the form

$$u_i = e^{x\cdot\rho_i}(1 + \psi_{q_i}(x, \rho_i)), \qquad i = 1, 2 \tag{19}$$

with $|\rho_i|$ large, $i = 1, 2$, with

$$\begin{aligned}\rho_1 &= \frac{\eta}{2} + i\left(\frac{k+l}{2}\right)\\ \rho_2 &= -\frac{\eta}{2} + i\left(\frac{k-l}{2}\right)\end{aligned} \tag{20}$$

and $\eta, k, l \in \mathbb{R}^n$ such that

$$\begin{aligned}&\eta \cdot k = k \cdot l = \eta \cdot l = 0\\ &|\eta|^2 = |k|^2 + |l|^2.\end{aligned} \tag{21}$$

Condition (21) guarantees that $\rho_i \cdot \rho_i = 0, i = 1, 2$. Substituting (19) into (18) we conclude

$$\widehat{(q_1 - q_2)}(-k) = -\int_\Omega e^{ix\cdot k}(q_1 - q_2)(\psi_{q_1} + \psi_{q_2} + \psi_{q_1}\psi_{q_2})dx. \tag{22}$$

Now $\|\psi_{q_i}\|_{L^2(\Omega)} \leq \frac{C}{|\rho_i|}$. Therefore by taking $|l| \to \infty$ we obtain

$$\chi_\Omega \widehat{(q_1 - q_2)}(k) = 0 \quad \forall\, k \in \mathbb{R}^n$$

concluding the proof. Theorem 1.5 has been extended to conductivities having 3/2 derivatives in some sense in [32, 152]. Uniqueness for conormal conductivies in $C^{1+\epsilon}$ was shown in [63]. Recently Haberman and Tataru in a very nice article [75] have extended the uniqueness result to C^1 conductivities or small in the $W^{1,\infty}$ norm. It is an open problem whether uniqueness holds in dimension $n \geq 3$ for Lipschitz or less regular conductivities. Theorem 1.6 was extended to potentials in $L^{n/2}$ and small potentials in the Fefferman-Phong class in [41]. For conormal potentials with strong singularities so that the potential is not in $L^{n/2}$, for instance almost a delta function of an hypersurface, uniqueness was shown in [63].

Similar problems for higher order operators were considered in [97, 118].

1.4.2 Non-linear conductivities

We now give an extension of this result to conductivities that depend on the voltage. Let $\gamma(x, t)$ be a function with domain $\overline{\Omega} \times \mathbb{R}$. Let α be such that $0 < \alpha < 1$. We assume

$$\gamma \in C^{1,\alpha}(\overline{\Omega} \times [-T, T]), \quad \forall\, T, \tag{23}$$

$$\gamma(x, t) > 0, \ \forall\, (x, t) \in \overline{\Omega} \times \mathbb{R}. \tag{24}$$

Given $f \in C^{2,\alpha}(\partial\Omega)$, there exists a unique solution of the Dirichlet problem (see [59])

$$\nabla \cdot (\gamma(x,u)\nabla u) = 0 \text{ in } \Omega, \quad u\Big|_{\partial\Omega} = f. \tag{25}$$

Then the Dirichlet to Neumann map is defined by

$$\Lambda_\gamma(f) = \gamma(x,f)\Big|_{\partial\Omega} \frac{\partial u}{\partial \nu}\Big|_{\partial\Omega} \tag{26}$$

where u is a solution to (25). Sun [184] proved the following result.

Theorem 1.7 *Let $n \geq 3$. Assume $\gamma_i \in C^{1,1}(\overline{\Omega} \times [-T,T])\ \forall\ T > 0$, $i = 1, 2$, and $\Lambda_{\gamma_1} = \Lambda_{\gamma_2}$. Then $\gamma_1(x,t) = \gamma_2(x,t)$ on $\overline{\Omega} \times \mathbb{R}$.*

The main idea is to linearize the Dirichlet to Neumann map at constant boundary data equal to the parameter t (then the solution of (25) is equal to t). Isakov [96] was the first to use a linearization technique to study an inverse parabolic problems associated to non-linear equations. The case of the Dirichlet to Neumann map associated to the Schrödinger equation with a non-linear potential was considered in [99] under some assumptions on the potential. We note that, in contrast to the linear case, one cannot reduce the study of the inverse problem of the conductivity equation (25) to the Schrödinger equation with a non-linear potential. The main technical lemma in the proof of Theorem 1.7 is

Lemma 1.8 *Let $\gamma(x,t)$ be as in* (23) *and* (24). *Let $1 < p < \infty$, $0 < \alpha < 1$. Let us define*

$$\gamma^t(x) = \gamma(x,t). \tag{27}$$

Then for any $f \in C^{2,\alpha}(\partial\Omega)$, $t \in \mathbb{R}$

$$\lim_{s\to 0} \left\|\frac{1}{s}\Lambda_\gamma(t+sf) - \Lambda_{\gamma^t}(f)\right\|_{W^{1-\frac{1}{p},p}(\partial\Omega)} = 0. \tag{28}$$

The proof of Theorem 1.7 follows immediately from the lemma. Namely (28) and the hypotheses $\Lambda_{\gamma_1} = \Lambda_{\gamma_2} \Rightarrow \Lambda_{\gamma_1^t} = \Lambda_{\gamma_2^t}$ for all $t \in \mathbb{R}$. Then using the linear result, Theorem 1.5, we conclude that $\gamma_1^t = \gamma_2^t$ proving the theorem. We remark that the reduction from the non-linear problem to the linear is also valid in the two dimensional case [98]. Using the result of Astala and Päivärinta [11], which is reviewed in Sect. 1.6, one can extend Theorem 1.5 to $L^\infty(\Omega)$ conductivities in the two dimensional case. There are several open questions when the conductivity also depends on ∇u, see [185] for a survey of results and open problems in this direction.

1.4.3 Other applications

We give a short list of other applications to inverse problems using the CGO solutions described above for the Schrödinger equation.

- Quantum Scattering. In dimension $n \geq 3$ and in the case of a compactly supported electric potential, uniqueness for the fixed energy scattering problem was proven in [135,145,159]. In the earlier paper [146] this was done for small potentials. For compactly supported potentials, knowledge of the scattering amplitude at fixed energy is equivalent to knowing the Dirichlet-to-Neumann map for the Schrödinger equation measured on the boundary of a large ball containing the support of the potential (see [200,202] for an account). Then Theorem 1.6 implies the result. Melrose [128] suggested a related proof that uses the density of products of scattering solutions. Applications of CGO solutions to the 3-body problem were given in [203].
- Optics. The DN map associated to the Helmholtz equation $-\Delta + k^2 n(x)$ with an isotropic index of refraction n determines uniquely a bounded index of refraction in dimension $n \geq 3$.
- Optical tomography in the diffusion approximation. In this case we have $\nabla \cdot \mathcal{D}(x)\nabla u - \sigma_a(x)u - i\omega u = 0$ in Ω where u represents the density of photons, $\mathcal{D}$ the diffusion coefficient, and σ_a the optical absorption. Using the result of [190] one can show in dimension three or higher that if $\omega \neq 0$ one can recover both $\mathcal{D}$ and σ_a from the corresponding DN map. If $\omega = 0$ then one can recover one of the two parameters.
- Electromagnetics. For Maxwell's equations the analog of the DN map is the admittance map that maps the tangential component of the electric field to the tangential component of the magnetic field [175]. The admittance map for isotropic Maxwell's equations determines uniquely the isotropic electric permittivity, magnetic permeability and conductivity [147]. This system can in fact be reduced to the Schrödinger equation $\Delta - Q$ with Q an 8×8 system and Δ the Laplacian times the identity matrix [148].
- Elasticity. For the isotropic elasticity system the problem of determining the Lamé parameters from the analog of the DN map in this case which sends the displacement at the boundary to the traction of the boundary has been solved if the Lamé parameter μ is close to a constant [55,142,143].
- Determination of Inclusions and Obstacles. The CGO solutions constructed in Theorem 1.3 have been applied to determine inclusions for Helmholtz equations in [84] and Maxwell's equations in [209] using the enclosure method [83,84].
- Coupled-Physics Inverse Problems. In these problems one tries to combine the best features of two type of waves, one with high contrast and the other with high resolution to find the electromagnetic, optical or elastic properties of a medium. This combination is done through some physical principle. Examples are Photoacoustic and Thermoacoustic Tomography, Ultrasound Modulated Optical Tomography, Ultrasound Modulated Electrical Impedance Tomography, Magnetic Resonance Elastography and Transient Elastography among others. See [13,176] for a review of some of these inverse methods. CGO solutions have been used in these hybrid methods in [15–18,42,110]. Inverse transport see [14] for applications of 60 solutions to inverse

problem for the transport equation. The case where the electrical measurements are made on an unknown boundary was considered in [111].

1.4.4 Stability

The arguments used in the proofs of Theorems 1.5, 1.6, 1.1 can be pushed further to prove the following stability estimates proven in [4].

Theorem 1.9 *Let $n \geq 3$. Suppose that $s > \frac{n}{2}$ and that γ_1 and γ_2 are C^∞ conductivities on $\overline{\Omega} \subseteq \mathbb{R}^n$ satisfying*

i) $0 < \frac{1}{E} \leq \gamma_j \leq E,\ j = 1, 2.$
ii) $\|\gamma_j\|_{H^{s+2}(\Omega)} \leq E,\ j = 1, 2.$

Then there exist $C = C(\Omega, E, n, s)$ and $0 < \sigma < 1$ $(\sigma = \sigma(n, s))$ such that

$$\|\gamma_1 - \gamma_2\|_{L^\infty(\Omega)} \leq C\big(|\log \|\Lambda_{\gamma_1} - \Lambda_{\gamma_2}\|_{\frac{1}{2},\frac{-1}{2}}|^{-\sigma} + \|\Lambda_{\gamma_1} - \Lambda_{\gamma_2}\|_{\frac{1}{2},\frac{-1}{2}}\big) \quad (29)$$

where $\|\ \ \|_{\frac{1}{2},\frac{-1}{2}}$ denotes the operator norm as operators from $H^{\frac{1}{2}}(\partial\Omega)$ to $H^{-\frac{1}{2}}(\partial\Omega)$.

Notice that this logarithmic type stability estimates indicates that the problem is severely ill-posed. Mandache [127] has shown that this estimate is optimal up to the value of the exponent. There is the question of whether under some additional a-priori condition one can improve this logarithmic type stability estimate. Alessandrini and Vessella [7] have shown that this is indeed the case and one has a Lipschitz type stability estimate if the conductivity is piecewise constant with jumps on a finite number of domains. Rondi [160] has subsequently shown that the constant in the estimate grows exponentially with the number of domains. It is conjectured, and this is supported by numerical experiments, that the stability estimate should be "better" near the boundary and gets increasingly worse as one penetrated deeper into the domain (Theorem 1.2 shows that at the boundary we have Lipschitz type stability estimate.) This type of depth dependence stability estimate has been proved in [138] for the case of some electrical inclusions. For a recent review of stability issues in EIT see [6]. Theorem 1.9 is a consequence of Theorem 1.2 and the following result.

Theorem 1.10 *Assume 0 is not a Dirichlet eigenvalue of $\Delta - q_i,\ i = 1, 2$. Let $s > \frac{n}{2}$, $n \geq 3$ and*

$$\|q_j\|_{H^s(\Omega)} \leq M.$$

Then there exists $C = C(\Omega, M, n, s)$ and $0 < \sigma < 1$ $(\sigma = \sigma(n, s))$ such that

$$\|q_1 - q_2\|_{H^{-1}(\Omega)} \leq C\big(|\log \|\Lambda_{q_1} - \Lambda_{q_2}\|_{\frac{1}{2},\frac{-1}{2}}|^{-\sigma} + \|\Lambda_{q_1} - \Lambda_{q_2}\|_{\frac{1}{2},\frac{-1}{2}}\big). \quad (30)$$

It was shown in [140] for the acoustic equation that the stability improves with frequency.

1.4.5 Reconstruction

The complex geometrical optics solutions of Theorems 1.5 and 1.6 were also used by A. Nachman [135] and R. Novikov [145] to give a reconstruction procedure of the conductivity from Λ_γ. As we have already noticed in Sect. 1.2 we can reconstruct the conductivity at the boundary and its normal derivative from the DN map. Therefore if we know Λ_γ we can determine Λ_q. We will then show how to reconstruct q from Λ_q. Once this is done, to find $\sqrt{\gamma}$, we solve the problem

$$\begin{aligned} \Delta u - qu &= 0 \text{ in } \Omega, \\ u|_{\partial\Omega} &= \sqrt{\gamma}\Big|_{\partial\Omega}. \end{aligned} \tag{31}$$

Let $q_1 = q, q_2 = 0$ in formula (17). Then we have

$$\int_\Omega quvdx = \int_{\partial\Omega} (\Lambda_q - \Lambda_0)\left(v|_{\partial\Omega}\right) u|_{\partial\Omega} dS \tag{32}$$

where $u, v \in H^1(\Omega)$ solve$\Delta u - qu = 0$, $\Delta v = 0$ in Ω. Here Λ_0 denotes the Dirichlet to Neumann map associated to the potential $q = 0$. We choose $\rho_i, i = 1, 2$ as in (20). Take $v = e^{x\cdot\rho_1}$, $u := u_\rho = e^{x\cdot\rho_2}(1 + \psi_q(x, \rho_2))$ as in Theorem 1.3. By taking $|l| \to \infty$ in (32) we conclude

$$\widehat{q}(-k) = \lim_{|l|\to\infty} \int_{\partial\Omega} (\Lambda_q - \Lambda_0)(e^{x\cdot\rho_1}\Big|_{\partial\Omega}) u_\rho\Big|_{\partial\Omega} dS.$$

So the problem is then to recover the boundary values of the solutions u_ρ from Λ_q. The idea is to find $u_\rho\Big|_{\partial\Omega}$ by looking at the exterior problem. Namely by extending $q = 0$ outside Ω, u_ρ solves

$$\begin{aligned} \Delta u_\rho &= 0 \text{ in } \mathbb{R}^n - \Omega \\ \frac{\partial u_\rho}{\partial \nu}\Big|_{\partial\Omega} &= \Lambda_q(u_\rho\Big|_{\partial\Omega}). \end{aligned} \tag{33}$$

Also note that

$$e^{-x\cdot\rho_2} u_\rho - 1 \in L^2_\delta(\mathbb{R}^n). \tag{34}$$

Let $\rho \in \mathbb{C}^n - 0$ with $\rho \cdot \rho = 0$. Let $G_\rho(x, y) \in \mathcal{D}'(\mathbb{R}^n \times \mathbb{R}^n)$ denote the Schwartz kernel of the operator Δ_ρ^{-1}. Then we have that

$$g_\rho(x) = e^{x\cdot\rho} G_\rho(x) \tag{35}$$

is a Green's kernel for Δ, namely

$$\Delta g_\rho = \delta_0. \tag{36}$$

We write the solution of (33) and (34) in terms of single and double layer potentials using this Green's kernel. This is also called Faddeev Green's kernel [56] who considered it in the context of scattering theory. We define the single and double layer potentials

$$S_\rho f(x) = \int_{\partial\Omega} g_\rho(x-y) f(y) dS_y, \qquad x \in \mathbb{R}^n - \Omega, \tag{37}$$

$$D_\rho f(x) = \int_{\partial\Omega} \frac{\partial g_\rho}{\partial \nu}(x-y) f(y) dS_y, \qquad x \in \mathbb{R}^n - \Omega \tag{38}$$

$$B_\rho f(x) = p.v. \int_{\partial\Omega} \frac{\partial g_\rho}{\partial \nu}(x-y) f(y) dS_y, \qquad x \in \partial\Omega. \tag{39}$$

Nachman showed that $f_\rho = u_\rho\Big|_{\partial\Omega}$ is a solution of the integral equation

$$f_\rho = e^{x\cdot\rho} - (S_\rho \Lambda_q - B_\rho - \frac{1}{2}I) f_\rho. \tag{40}$$

Moreover (40) is an inhomogeneous integral equation of Fredholm type for f_ρ and it has a unique solution in $H^{\frac{3}{2}}(\partial\Omega)$. The uniqueness of the homogeneous equation follows from the uniqueness of the CGO solutions in Theorem 1.6.

1.5 The partial data problem

In several applications in EIT one can only measure currents and voltages on part of the boundary. Substantial progress has been made recently on the problem of whether one can determine the conductivity in the interior by measuring the DN map on part of the boundary. We review here the articles [49, 107]. The paper [35] used the method of Carleman estimates with a linear weight to prove that, roughly speaking, knowledge of the DN map in "half" of the boundary is enough to determine uniquely a C^2 conductivity. The regularity assumption on the conductivity was relaxed to $C^{1+\epsilon}, \epsilon > 0$ in [108]. Stability estimates for the uniqueness result of [35] were given in [76]. Stability estimates for the magnetic Schrödinger operator with partial data in the setting of [35] can be found in [198]. The result [35] was substantially improved in [107]. The latter paper contains a global identifiability result where it is assumed that the DN map is measured on any open subset of the boundary of a strictly convex domain for all functions supported, roughly, on the complement. We state the theorem more precisely below. The key new ingredient is the construction of a larger class of CGO solutions than the ones considered in Sect. 1.4. Let $x_0 \in \mathbf{R}^n \setminus \overline{\mathrm{ch}\,(\Omega)}$, where $\mathrm{ch}\,(\Omega)$ denotes the convex hull of Ω. Define the front and the back faces of $\partial\Omega$ by

$$F(x_0) = \{x \in \partial\Omega;\ (x - x_0)\cdot\nu \le 0\},\ B(x_0) = \{x \in \partial\Omega;\ (x - x_0)\cdot\nu > 0\}.$$

The main result of [107] is the following:

Theorem 1.11 *Let $n > 2$. With Ω, x_0, $F(x_0)$, $B(x_0)$ defined as above, let $q_1, q_2 \in L^\infty(\Omega)$ be two potentials and assume that there exist open neighborhoods $\widetilde{F}$, $\widetilde{B} \subset \partial\Omega$ of $F(x_0)$ and $B(x_0) \cup \{x \in \partial\Omega; (x - x_0)\cdot \nu = 0\}$ respectively, such that*

$$\Lambda_{q_1} u = \Lambda_{q_2} u \text{ in } \widetilde{F}, \text{ for all } u \in H^{\frac{1}{2}}(\partial\Omega) \cap \mathscr{E}'(\widetilde{B}). \tag{41}$$

Then $q_1 = q_2$.

Here $\mathscr{E}'(\widetilde{B})$ denotes the space of compactly supported distributions in $\widetilde{B}$. The proof of this result uses Carleman estimates for the Laplacian with limiting Carleman weights (LCW). The Carleman estimates allow one to construct, for large τ, a larger class of CGO solutions for the Schrödinger equation than previously used. These have the form

$$u = e^{\tau(\phi + i\psi)}(a + r), \tag{42}$$

where $\nabla\phi \cdot \nabla\psi = 0$, $|\nabla\phi|^2 = |\nabla\psi|^2$ and ϕ is the LCW. Moreover a is smooth and non-vanishing and $\|r\|_{L^2(\Omega)} = O(\frac{1}{\tau})$, $\|r\|_{H^1(\Omega)} = O(1)$. Examples of LCW are the linear phase $\phi(x) = x \cdot \omega$, $\omega \in S^{n-1}$, used previously, and the non-linear phase $\phi(x) = \ln|x - x_0|$, where $x_0 \in \mathbf{R}^n \backslash \overline{\text{ch}(\Omega)}$ which was used in [107]. Any conformal transformation of these would also be a LCW. Below we give a characterization of all the LCW in $\mathbb{R}^n$, $n > 2$, see [52]. In two dimensions any harmonic function is a LCW [205]. The CGO solutions used in [107] are of the form

$$u(x,\tau) = e^{\ln|x-x_0| + i d(\frac{x-x_0}{|x-x_0|}, \omega)}(a + r) \tag{43}$$

where x_0 is a point outside the convex hull of Ω, ω is a unit vector and $d(\frac{x-x_0}{|x-x_0|}, \omega)$ denote distance. We take directions ω so that the distance function is smooth for $x \in \overline{\Omega}$.

1.5.1 Limiting Carleman weights

We only recall here the main ideas in the construction of the CGO solutions. We will denote $\tau = \frac{1}{h}$ in order to use the standard semiclassical notation. Let $P_0 = -h^2\Delta$, where $h > 0$ is a small semi-classical parameter. The weighted L^2-estimate

$$\|e^{\phi/h} u\| \leq C\|e^{\phi/h} P_0 u\|$$

is of course equivalent to the unweighted estimate for a conjugated operator:

$$\|v\| \leq C\|e^{\phi/h} P_0 e^{-\phi/h} v\|.$$

The semi-classical principal symbol of P_0 is $p(x,\xi) = \xi^2$, and that of the conjugated operator $e^{\phi/h} P_0 e^{-\phi/h}$ is

$$p(x, \xi + i\phi'(x)) = a(x,\xi) + ib(x,\xi),$$

where

$$a(x,\xi) = \xi^2 - \phi'(x)^2, b(x,\xi) = 2\xi \cdot \phi'(x).$$

Here we denote by ϕ' the gradient of ϕ. Write the conjugated operator as $A + iB$, with A and B formally selfadjoint and with a and b as their associated principal symbols. Then

$$\|(A+iB)u\|^2 = \|Au\|^2 + \|Bu\|^2 + (i[A,B]u|u).$$

The principal symbol of $i[A,B]$ is $h\{a,b\}$, where $\{\cdot,\cdot\}$ denotes the Poisson bracket. In order to get enough negativity to satisfy Hörmander's solvability condition we require that

$$a(x,\xi) = b(x,\xi) = 0 \Rightarrow \{a,b\} \leq 0.$$

It is then indeed possible to get an a-priori estimate for the conjugated operator. We are led to the limiting case since we need to have CGO solutions for both ϕ and $-\phi$.

Definition 1.12 ϕ is a limiting Carleman weight (LCW) on some open set Ω if $\nabla\phi(x)$ is non-vanishing there and we have

$$a(x,\xi) = b(x,\xi) = 0 \Rightarrow \{a,b\}(x) = 0, x \in \Omega.$$

We remark that if ϕ is a LCW so is $-\phi$. In [52] we have classified locally all the LCW in Euclidean space.

Theorem 1.13 *Let Ω be an open subset of $\mathbb{R}^n$, $n \geq 3$. The limiting Carleman weights in Ω are locally of the form*

$$\phi(x) = a\phi_0(x - x_0) + b$$

where $a \in \mathbb{R}\backslash\{0\}$ and ϕ_0 is one of the following functions:

$$\begin{array}{llll} & \langle x,\xi\rangle, & \arg\langle x, \omega_1 + i\omega_2\rangle, & \\ \log|x|, & \dfrac{\langle x,\xi\rangle}{|x|^2}, & \arg\left(e^{i\theta}(x+i\xi)^2\right), & \log\dfrac{|x+\xi|^2}{|x-\xi|^2} \end{array}$$

with ω_1, ω_2 orthogonal unit vectors, $\theta \in [0, 2\pi)$ and $\xi \in \mathbb{R}^n\backslash\{0\}$.

As noted earlier, in two dimensions, any harmonic function with a non-vanishing gradient is a limiting Carleman weight.

1.5.2 Construction of CGO Solutions with a non-linear phase

A key ingredient in the construction of a richer family of CGO solutions is the following Carleman estimate.

Proposition 1.14 *Let* $\phi \in C^\infty(\text{neigh}(\overline{\Omega}))$ *be an LCW,* $P = -h^2\Delta + h^2 q$, $q \in L^\infty(\Omega)$. *Then, for* $u \in C^\infty(\overline{\Omega})$, *with* $u_{|\partial\Omega} = 0$, *we have*

$$-\frac{h^3}{C}((\phi'_x \cdot \nu)e^{\phi/h}\partial_\nu u|e^{\phi/h}\partial_\nu u)_{\partial\Omega_-} + \frac{h^2}{C}(\|e^{\phi/h}u\|^2 + \|e^{\phi/h}h\nabla u\|^2)$$
$$\leq Ch^3((\phi'_x \cdot \nu)e^{\phi/h}\partial_\nu u|e^{\phi/h}\partial_\nu u)_{\partial\Omega_+} + \|e^{\phi/h}Pu\|^2, \tag{44}$$

where norms and scalar products are in $L^2(\Omega)$ *unless a subscript* A *(like for instance* $A = \partial\Omega_-$*) indicates that they should be taken in* $L^2(A)$. *Here*

$$\partial\Omega_\pm = \{x \in \partial\Omega;\ \pm\nu(x)\cdot\phi'(x) \geq 0\}.$$

The proof of existence of solutions of the form (42) follows by using the Hahn-Banach theorem for the adjoint equation $e^{-\phi/h}Pe^{\phi/h}u = v$. Let ϕ be a LCW and write $p(x, \xi + \phi'(x)) = a(x,\xi) + ib(x,\xi)$. Then we know that a and b are in involution on their common zero set, and in this case it is well-known and exploited in [53] that we can find plenty of local solutions to the Hamilton-Jacobi system

$$a(x,\psi'(x)) = 0,\ b(x,\psi'(x)) = 0 \Leftrightarrow \psi'^2 = \phi'^2,\ \psi'\cdot\phi' = 0 \tag{45}$$

We need the following more global statement:

Proposition 1.15 *Let* $\phi \in C^\infty(\text{neigh}(\overline{\Omega}))$ *be a LCW, where* Ω *is a domain in* $\mathbb{R}^n$ *and define the hypersurface* $G = p^{-1}(C_0)$ *for some fixed value of* C_0. *Assume that each integral curve of* $\phi'\cdot\nabla_x$ *through a point in* Ω *also intersects* G *and that the corresponding projection map* $\Omega \to G$ *is proper. Then we get a solution of (45) in* $C^\infty(\Omega)$ *by solving first* $g'(x)^2 = \phi'(x)^2$ *on* G *and then defining* ψ *by* $\psi_{|G} = g$, $\phi'(x)\cdot\partial_x\psi = 0$. *The vector fields* $\phi'\cdot\partial_x$ *and* $\psi'\cdot\partial_x$ *commute.*

This result will be applied with a new domain Ω that contains the original one. Next consider the WKB-problem

$$P_0(e^{\frac{1}{h}(-\phi+i\psi)}a(x)) = e^{\frac{1}{h}(-\phi+i\psi)}\mathcal{O}(h^2). \tag{46}$$

The transport equation for a is of Cauchy-Riemann type along the two-dimensional integral leaves of $\{\phi'\cdot\partial_x, \psi'\cdot\partial_x\}$. We have solutions that are smooth and everywhere $\neq 0$. (See [53]). The existence result for $e^{\phi/h}Pe^{-\phi/h}$ mentioned in one of the remarks after Proposition 1.14 permits us to replace the right hand side of (46) by zero; more precisely, we can find $r = \mathcal{O}(h)$ in the semi-classical Sobolev space H^1 equipped with the norm $\|r\| = \|\langle hD\rangle r\|$, such that

$$P(e^{\frac{1}{h}(-\phi+i\psi)}(a+r)) = 0. \tag{47}$$

1.5.3 The uniqueness proof

We sketch the proof for the case that $\widetilde{B} = \partial\Omega$. All the arguments in this section are in dimension $n > 2$. The arguments are similar to those of [35] using the new CGO solutions. Let ϕ be an LCW for which the constructions of Sect. 1.5.2 are available. Let $q_1, q_2 \in L^\infty(\Omega)$ be as in Theorem 1.11 with

$$\Lambda_{q_1}(f) = \Lambda_{q_2}(f) \quad \text{in } \partial\Omega_{-,\epsilon_0}, \forall f \in H^{\frac{1}{2}}(\partial\Omega), \tag{48}$$

where

$$\partial\Omega_{-,\epsilon_0} = \{x \in \partial\Omega;\ \nu(x)\cdot\phi'(x) < \epsilon_0\}$$
$$\partial\Omega_{+,\epsilon_0} = \{x \in \partial\Omega;\ \nu(x)\cdot\phi'(x) \geq \epsilon_0\}.$$

Let

$$u_2 = e^{\frac{1}{h}(\phi+i\psi_2)}(a_2 + r_2)$$

solve

$$(\Delta - q_2)u_2 = 0 \text{ in } \Omega,$$

with $\|r_2\|_{H^1} = \mathcal{O}(h)$. Let u_1 solve

$$(\Delta - q_1)u_1 = 0 \quad \text{in } \Omega, u_{1|\partial\Omega} = u_{2|\partial\Omega}.$$

Then according to the assumptions in the theorem, we have $\partial_\nu u_1 = \partial_\nu u_2$ in $\partial\Omega_{-,\epsilon_0}$ if $\epsilon_0 > 0$ has been fixed sufficiently small and we choose $\phi(x) = \ln|x - x_0|$.

Put $u = u_1 - u_2$, $q = q_2 - q_1$, so that

$$(\Delta - q_1)u = qu_2, u_{|\partial\Omega} = 0, \text{supp}\,(\partial_\nu u_{|\partial\Omega}) \subset \partial\Omega_{+,\epsilon_0}. \tag{49}$$

For $v \in H^1(\Omega)$ with $\Delta v \in L^2$, we get from Green's formula

$$\begin{aligned}\int_\Omega q u_2 \overline{v} dx &= \int_\Omega (\Delta - q_1)u\overline{v}dx \\ &= \int_\Omega u\,\overline{(\Delta - \overline{q}_1)v}dx + \int_{\partial\Omega_{+,\epsilon_0}} \partial_\nu u \overline{v} dS.\end{aligned} \tag{50}$$

Similarly, we choose

$$v = e^{-\frac{1}{h}(\phi+i\psi_1)}(a_1 + r_1),$$

with

$$(\Delta - \overline{q}_1)v = 0.$$

Then

$$\int_{\Omega} q e^{\frac{i}{h}(\psi_1+\psi_2)}(a_2+r_2)(\overline{a_1+r_1})dx = \int_{\partial\Omega_{+,\epsilon_0}} \partial_\nu u\, e^{-\frac{1}{h}(\phi - i\psi_1)}(\overline{a_1+r_1})dS. \quad (51)$$

Assume that ψ_1, ψ_2 are slightly h-dependent with

$$\frac{1}{h}(\psi_1+\psi_2) \to f, h \to 0.$$

The left hand side of (51) tends to

$$\int_{\Omega} q e^{if} a_2 \bar{a}_1 dx,$$

when $h \to 0$. The modulus of the right hand side is

$$\leq \|a_1 + r_1\|_{L^2(\partial\Omega_{+,\epsilon_0})} \left(\int_{\partial\Omega_{+,\epsilon_0}} e^{-2\phi/h} |\partial_\nu u|^2 dS \right)^{\frac{1}{2}}.$$

Here the first factor is bounded when $h \to 0$. In the Carleman estimate (44) we can replace ϕ by $-\phi$ and make the corresponding permutation of $\partial\Omega_-$ and $\partial\Omega_+$. Applying this variant to the equation (49), we see that the second factor tends to 0, when $h \to 0$. Thus,

$$\int_{\Omega} e^{if(x)} a_2(x)\overline{a_1(x)}q(x)dx = 0.$$

Here we can arrange it so that f, a_2, a_1 are real-analytic and so that a_1, a_2 are non-vanishing. Moreover if f can be attained as a limit of $(\psi_1+\psi_2)/h$ when $h \to 0$, so can λf for any $\lambda > 0$. Thus we get the conclusion

$$\int_{\Omega} e^{i\lambda f(x)} a_2(x)\overline{a_1(x)}q(x)dx = 0. \quad (52)$$

To show that $q = 0$ one uses arguments of analytic microlocal analysis [107]. A reconstruction method based on the uniqueness proof has been proposed in [137]. In [9] it was shown that if the potential is known in a neighborhood of the boundary and the DN map is measured on any open subset with Dirichlet data supported in the same set, the potential can be reconstructed from this data. It is an open problem whether this is valid in the general case. Isakov [95] proved a uniqueness result in dimension three or higher when the DN map is given on an arbitrary part of the boundary assuming that the remaining part is an open subset of a plane or a sphere and the DN map is measured on the plane or sphere. The case of partial data on a slab was studied in [122]. The DN map with partial data for the magnetic Schrödinger operator was studied in [51,109,117,198]. The case of the polyharmonic operator was considered in [118]. The case of Helmholtz equation [139]. We also mention that in [66] (resp. [101]) CGO approximate solutions are concentrated near planes (resp. spheres) and provided some local results related to the local DN map. It would be very interesting to extend the partial data result to systems. See [164] for Dirac systems, [40] for Maxwell and [91] for elasticity. Using methods of hyperbolic geometry similar to [101] it is shown in [82] that one can reconstruct inclusions from the *local* DN map using CGO solutions that decay exponentially inside a ball and grow exponentially outside. These are called *complex spherical waves*. A numerical implementation of this method has been done in [82]. The construction of complex spherical waves can also be done using the CGO solutions constructed in [107]. This was done in [204] in order to detect elastic inclusions, in [205] to detect inclusions in the two dimensional case for a large class of systems with inhomogeneous background, and in [165] for the case of inclusions contained in a slab. Partial data for higher order elliptic operators has been studied in [119]. We mention that methods of hyperbolic geometry have been also studied earlier in the works [19,58,166].

1.5.4 The Linearized Calderón partial data problem

It is an open problem in dimension $n \geq 3$ that if the Dirichlet to Neumann map for the conductivity or potential is measured on an open non-empty subset of the boundary for Dirichlet data supported in that set we can determine uniquely the potential. In this section we consider the linearized version of this problem, generalizing Calderón's approach. We add the constraint that the restriction of the harmonic functions to the boundary vanishes on any fixed closed proper subset of the boundary. We show that the product of these harmonic functions is dense. More precisely

Theorem 1.16 *Let Ω be a connected bounded open set in $\mathbb{R}^n$, $n \geq 2$, with smooth boundary. The set of products of harmonic functions on Ω which vanish on a closed proper subset $\Gamma \subsetneq \partial\Omega$ of the boundary is dense in $L^1(\Omega)$.*

Sketch of the Proof. We take $f \in L^1(\Omega)$. Assume

$$\int_{\Omega} fuv dx = 0, \tag{53}$$

for all harmonic functions u, v with $u|_\Gamma = v|_\Gamma = 0$. First one proves a local result. Fix a point x_0 on the boundary. It is shown that if $f = 0$ in a neighborhood of x_0 then $f = 0$ in the whole domain. See [49] for the proof. We now extend $f = 0$ outside Ω. We reduce the problem to the case where the point x_0 has a hyperplane tangent to the boundary at the point x_0. We use Calderón's exponential solutions for all the possible complex frequencies ρ such that $\rho \cdot \rho = 0$ (previously we used in Sects. 1.4 and 1.5.3 the cancellation of the real parts when taking the products). Using these solutions and (53) one obtains a decay of the Bargmann-Segal transform of f (see [173])

$$Tf(z) = \int_{\mathbb{R}^n} e^{-\frac{1}{2h}(z-y)^2} f(y)dy \tag{54}$$

for certain complex directions. Using the watermelon approach [105, 174], one then shows that there is an exponential decay of this transform for other directions implying that the point (x_0, ν), where ν is the normal to the point x_0, is not in the analytic wave front set of f in contradiction to the microlocal version of Holmgren's uniqueness theorem [81, 173]. We explain below some of the details of the proof. One can assume that $\Omega \backslash \{x_0\}$ is on one side of the tangent hyperplane $T_{x_0}(\Omega)$ at x_0 by making a conformal transformation. Pick $a \in \mathbb{R}^n \backslash \overline{\Omega}$ which sits on the line segment in the direction of the outward normal to $\partial\Omega$ at x_0; there is a ball $B(a, r)$ such that $\partial B(a, r) \cap \overline{\Omega} = \{x_0\}$, and there is a conformal transformation

$$\psi : \mathbb{R}^n \backslash B(a, r) \to \overline{B(a, r)} \tag{55}$$

which fixes x_0 and exchanges the interior and the exterior of the ball $B(a, r)$. The hyperplane $H : (x - x_0) \cdot (a - x_0) = 0$ is tangent to $\psi(\Omega)$, and the image $\psi(\Omega)\backslash\{x_0\}$ by the conformal transformation lies inside the ball $B(a, r)$, therefore on one side of H. The fact that functions are supported on the boundary close to x_0 is left unchanged. Since a function is harmonic on Ω if and only if its Kelvin transform

$$u^* = r^{n-2}|x - a|^{-n+2} u \circ \psi$$

is harmonic on $\psi(\Omega)$, (1.53) becomes

$$0 = \int_{\Omega} fuv\, dx = \int_{\psi(\Omega)} r^4 |x - a|^{-4} f \circ \psi\, u^* v^*\, dx$$

for all harmonic functions u^*, v^* on $\psi(\Omega)$. If one can prove that if $|x - a|^{-4} f \circ \psi$ vanishes close to x_0 then so does f. Moreover, by scaling one can assume that Ω is contained in a ball of radius 1. Our setting will therefore be as follows: $x_0 = 0$, the tangent hyperplane at x_0 is given by $x_1 = 0$ and

$$\Omega \subset \left\{x \in \mathbb{R}^n : |x + e_1| < 1\right\}, \quad \Gamma = \left\{x \in \partial\Omega : x_1 \geq -2c\right\}. \tag{56}$$

The prime will be used to denote the last $n - 1$ variables so that $x = (x_1, x')$ for instance. The Laplacian on $\mathbb{R}^n$ has $p(\xi) = \xi^2$ as a principal symbol, if we still denote by $p(\zeta) = \zeta^2$ the continuation of this principal symbol on $\mathbb{C}^n$, we consider

$$p^{-1}(0) = \left\{\zeta \in \mathbb{C}^n : \zeta^2 = 0\right\}.$$

In dimension $n = 2$, this set is the union of two complex lines

$$p^{-1}(0) = \mathbb{C}\gamma \cup \mathbb{C}\overline{\gamma}$$

where $\gamma = ie_1 + e_2 = (i, 1) \in \mathbb{C}^2$. Note that $(\gamma, \overline{\gamma})$ is a basis of $\mathbb{C}^2$: the decomposition of a complex vector in this basis reads

$$\zeta = \zeta_1 e_1 + \zeta_2 e_2 = \frac{\zeta_2 - i\zeta_1}{2}\gamma + \frac{\zeta_2 + i\zeta_1}{2}\overline{\gamma}. \tag{57}$$

Similarly for $n \geq 2$, the differential of the map

$$\begin{aligned} s : p^{-1}(0) \times p^{-1}(0) &\to \mathbb{C}^n \\ (\zeta, \eta) &\mapsto \zeta + \eta \end{aligned}$$

at (ζ_0, η_0) is surjective

$$\begin{aligned} Ds(\zeta_0, \eta_0) : T_{\zeta_0} p^{-1}(0) \times T_{\eta_0} p^{-1}(0) &\to \mathbb{C}^n \\ (\zeta, \eta) &\mapsto \zeta + \eta \end{aligned}$$

provided $\mathbb{C}^n = T_{\zeta_0} p^{-1}(0) + T_{\eta_0} p^{-1}(0)$, i.e. provided ζ_0 and η_0 are linearly independent. In particular, this is the case if $\zeta_0 = \gamma$ and $\eta_0 = -\overline{\gamma}$; as a consequence all $z \in \mathbb{C}^n$, $|z - 2ie_1| < 2\varepsilon$ may be decomposed as a sum of the form

$$z = \zeta + \eta, \quad \text{with } \zeta, \eta \in p^{-1}(0),\ |\zeta - \gamma| < C\varepsilon,\ |\eta + \overline{\gamma}| < C\varepsilon \tag{58}$$

provided $\varepsilon > 0$ is small enough. The exponentials with linear weights

$$e^{-\frac{i}{h} x \cdot \zeta}, \quad \zeta \in p^{-1}(0)$$

are harmonic functions. We need to add a correction term in order to obtain harmonic functions u satisfying the boundary requirement $u|_\Gamma = 0$. Let $\chi \in \mathcal{D}(\mathbb{R}^n)$ be a cutoff function which equals 1 on Γ, and consider the solution w to the Dirichlet problem

$$\Delta w = 0 \quad \text{in } \Omega,\ w|_{\partial\Omega} = -(e^{-\frac{i}{h} x \cdot \zeta} \chi)|_{\partial\Omega}. \tag{59}$$

The function

$$u(x, \zeta) = e^{-\frac{i}{h} x \cdot \zeta} + w(x, \zeta)$$

is harmonic and satisfies $u|_\Gamma = 0$. We have the following bound on w:

$$\begin{aligned} \|w\|_{H^1(\Omega)} &\leq C_1 \|e^{-\frac{i}{h} x \cdot \zeta} \chi\|_{H^{\frac{1}{2}}(\partial\Omega)} \\ &\leq C_2 (1 + h^{-1}|\zeta|)^{\frac{1}{2}} e^{\frac{1}{h} H_K(\mathrm{Im}\zeta)} \end{aligned} \tag{60}$$

where H_K is the supporting function of the compact subset $K = \operatorname{supp} \chi \cap \partial\Omega$ of the boundary

$$H_K(\xi) = \sup_{x \in K} x \cdot \xi, \quad \xi \in \mathbb{R}^n.$$

In particular, if we take χ to be supported in $x_1 \leq -c$ and equal to 1 on $x_1 \leq -2c$ then the bound (60) becomes

$$\|w\|_{H^1(\Omega)} \leq C_2(1+h^{-1}|\zeta|)^{\frac{1}{2}} e^{-\frac{c}{h}\mathrm{Im}\zeta_1} e^{\frac{1}{h}|\mathrm{Im}\zeta'|} \quad \text{when } \mathrm{Im}\zeta_1 \geq 0. \tag{61}$$

The starting point is the cancellation of the integral

$$\int_\Omega f(x)u(x,\zeta)u(x,\eta)\,dx = 0, \quad \zeta,\eta \in p^{-1}(0) \tag{62}$$

which may be rewritten in the form

$$\int_\Omega f(x)e^{-\frac{i}{h}x\cdot(\zeta+\eta)}\,dx = -\int_\Omega f(x)e^{-\frac{i}{h}x\cdot\zeta}w(x,\eta)\,dx$$
$$-\int_\Omega f(x)e^{-\frac{i}{h}x\cdot\eta}w(x,\zeta)\,dx - \int_\Omega f(x)w(x,\zeta)w(x,\eta)\,dx.$$

This allows to give a bound on the left-hand side

$$\left|\int_\Omega f(x)e^{-\frac{i}{h}x\cdot(\zeta+\eta)}\,dx\right| \leq \|f\|_{L^\infty(\Omega)}\big(\|e^{-\frac{i}{h}x\cdot\zeta}\|_{L^2(\Omega)}\|w(x,\eta)\|_{L^2(\Omega)}$$
$$+\|e^{-\frac{i}{h}x\cdot\eta}\|_{L^2(\Omega)}\|w(x,\zeta)\|_{L^2(\Omega)} + \|w(x,\eta)\|_{L^2(\Omega)}\|w(x,\zeta)\|_{L^2(\Omega)}\big).$$

Thus using (61)

$$\left|\int_\Omega f(x)e^{-\frac{i}{h}x\cdot(\zeta+\eta)}\,dx\right| \leq C_3\|f\|_{L^\infty(\Omega)}(1+h^{-1}|\eta|)^{\frac{1}{2}}(1+h^{-1}|\zeta|)^{\frac{1}{2}}$$
$$\times e^{-\frac{c}{h}\min(\mathrm{Im}\zeta_1,\mathrm{Im}\eta_1)} e^{\frac{1}{h}(|\mathrm{Im}\zeta'|+|\mathrm{Im}\eta'|)}$$

when $\mathrm{Im}\zeta_1 \geq 0$, $\mathrm{Im}\eta_1 \geq 0$ and $\zeta,\eta \in p^{-1}(0)$. In particular, if $|\zeta - a\gamma| < C\varepsilon a$ and $|\eta + a\overline{\gamma}| < C\varepsilon a$ with $\varepsilon \leq 1/2C$ then

$$\left|\int_\Omega f(x)e^{-\frac{i}{h}x\cdot(\zeta+\eta)}\,dx\right| \leq C_4 h^{-1}\|f\|_{L^\infty(\Omega)} e^{-\frac{ca}{2h}} e^{\frac{2C\varepsilon a}{h}}.$$

Take $z \in \mathbb{C}^n$ with $|z - 2ae_1| < 2\varepsilon a$ and with ε small enough. Once rescaled the decomposition (58) gives

$$z = \zeta+\eta, \quad \zeta,\eta \in p^{-1}(0), \; |\zeta - a\gamma| < C\varepsilon a, \; |\eta + a\overline{\gamma}| < C\varepsilon a,$$

and we therefore get the estimate

$$\left|\int_{\Omega} f(x)e^{-\frac{i}{h}x\cdot z}\,dx\right| \leq C_4 h^{-1}\|f\|_{L^\infty(\Omega)}e^{-\frac{ca}{2h}}\,e^{\frac{2C\varepsilon a}{h}} \tag{63}$$

for all $z \in \mathbb{C}^n$ such that $|z - 2aie_1| < 2\varepsilon a$. This implies that the Bargmann-Segal transform of f satisfies

$$|Tf(z)| \leq C|f\|_{L^\infty(\Omega)}e^{\frac{1}{2h}(|Imz|^2-|Rez|^2-\frac{ca}{2})} \tag{64}$$

for some $\epsilon, a, c > 0$ and for all $z \in \mathbb{C}^n$ such that $|z - 2aie_1| < 2\varepsilon a$. By the definition of the analytic wave front set, the last estimate says that the point $(0, 2ae_1)$ is not in the analytic wave front set of f. By Kashiwara's watermelon theorem [105,174], since f is supported in the half space $x_1 \leq 0$, if 0 is in the support of f then $(0, \nu)$ with ν the unit normal to the boundary is also in the analytic wave front set but this is a contradiction since $2ae_1$ is also normal to $x_1 = 0$. Therefore 0 is not in the support of f and f vanishes in a neighborhood of 0.

1.6 The Calderón problem in two dimensions

Astala and Päivärinta [11], in a seminal contribution, have recently extended significantly the uniqueness result of [134] for conductivities having two derivatives in an appropriate sense and the result of [33] for conductivities having one derivative in appropriate sense, by proving that any L^∞ conductivity in two dimensions can be determined uniquely from the DN map. We remark that the method of [33,134] uses, besides CGO solutions, the $\overline{\partial}$ method introduced in one dimension by Beals and Coifman [21] and generalized to several dimensions in [1,22,136,196]. The $\overline{\partial}$ method has been used in numerical reconstruction procedures in two dimensions in [93,172] among others. The proof of [11] relies also on construction of CGO solutions for the conductivity equation with L^∞ coefficients and the $\overline{\partial}$ method. This is done by transforming the conductivity equation to a quasi-regular map. Let $\mathcal{D}$ be the unit disk in the plane. Then we have

Lemma 1.17 *Assume $u \in H^1(\mathcal{D})$ is real valued and satisfies the conductivity equation on $\mathcal{D}$. Then there exists a function $v \in H^1(\mathcal{D})$, unique up to a constant, such that $f = u + iv$ satisfies the Beltrami equation*

$$\overline{\partial} f = \mu\overline{\partial f}, \tag{65}$$

where $\mu = (1-\gamma)/(1+\gamma)$. Conversely, if $f \in H^1(\mathcal{D})$ satisfies (65) with a real-valued μ, then $u = Ref$ and $v = Imf$ satisfy

$$\nabla\cdot\gamma\nabla u = 0 \quad and \quad \nabla\cdot\frac{1}{\gamma}\nabla v = 0, \tag{66}$$

respectively, where $\gamma = (1-\mu)/(1+\mu)$.

Let us denote $\kappa = ||\mu||_{L^\infty} < 1$. Then (65) means that f is a quasi-regular map. The function v is called the γ-harmonic conjugate of u and it is unique up to a constant. Astala and Päivärinta consider the μ-Hilbert transform $\mathcal{H}_\mu : H^{1/2}(\partial\Omega) \to H^{1/2}(\partial\Omega)$ that is defined by

$$\mathcal{H}_\mu : u\big|_{\partial\Omega} \mapsto v\big|_{\partial\Omega}$$

and show that the DN map Λ_γ determines $\mathcal{H}_\mu$ and vice versa. Below we use the complex notation $z = x_1 + ix_2$. Moreover, for the Eq. (65), it is shown that for every $k \in \mathbb{C}$ there are complex geometrical optics solutions of the Beltrami equation that have the form

$$f_\mu(z, k) = e^{ikz} M_\mu(z, k), \tag{67}$$

where

$$M_\mu(z, k) = 1 + \mathcal{O}\left(\frac{1}{z}\right) \text{ as } |z| \to \infty. \tag{68}$$

In the case of non-smooth coefficients the function M_μ grows sub-exponentially in k. Astala and Päivärinta introduce the "transport matrix" to deal with this problem. Using a result of Bers connecting pseudoanalytic functions with quasi-regular maps they show that this matrix is determined by the Hilbert transform H_μ and therefore the DN map. Then they use the transport matrix to show that Λ_γ determines uniquely γ. See [11] for more details. Logarithmic type stability estimates for Hölder conductivities of positive exponent have been given in [20].

1.6.1 Bukhgeim's result

In a recent breakthrough, Bukhgeim [34] proved that a potential in $W^{2,p}(\Omega)$, $p > 2$ can be uniquely determined from the set of Cauchy data as defined in (15). An earlier result [187] gave this for a generic class of potentials. As before, if two potentials q_1, q_2 have the same set of Cauchy data, we have

$$\int_\Omega (q_1 - q_2) u_1 u_2 dx = 0 \tag{69}$$

where u_i, $i = 1, 2$, are solutions of the Schrödinger equation. Assume now that $0 \in \Omega$. Bukhgeim takes CGO solutions of the form

$$u_1(z, k) = e^{z^2 k}(1 + \psi_1(z, k)), \quad u_2(z, k) = e^{-\bar{z}^2 k}(1 + \psi_2(z, k)) \tag{70}$$

where $z, k \in \mathbb{C}$ and we have used the complex notation $z = x_1 + ix_2$. Moreover ψ_1 and ψ_2 decay uniformly in Ω, in an appropriate sense, for $|k|$ large. Note that the weight $z^2 k$ in the exponential is a limiting Carleman weight since it is a harmonic function but it is singular at 0 since its gradient vanishes there. Substituting (70) into (69) we obtain

$$\int_{\Omega} e^{2i\tau x_1 x_2}(q_1 - q_2)(1 + \psi_1 + \psi_2 + \psi_1\psi_2)dx = 0.$$

Now using the decay of ψ_i in τ, $i = 1, 2$, and applying stationary phase (the phase function $x_1 x_2$ that has a non-degenerate critical point at 0) we obtain $q_1(0) = q_2(0) = 0$ in Ω. Of course we can do this at any point of Ω proving the result. This result also shows that complex conductivities can be determined uniquely from the DN map. Francini has shown in [57] that this was the case for conductivities with small imaginary part. It also implies unique determination of a potential from the fixed energy scattering amplitude in two dimensions. A general result for first order systems is in [3], generalizing the results of [71] and [104].

1.6.2 Partial data problem in 2D

It is shown in [87] that for a two dimensional bounded domain the Cauchy data for the Schrödinger equation measured on an arbitrary open subset of the boundary determines uniquely the potential. This implies, for the conductivity equation, that if one measures the current fluxes at the boundary on an arbitrary open subset of the boundary produced by voltage potentials supported in the same subset, one can determine uniquely the conductivity. The paper [87] uses Carleman estimates with weights which are harmonic functions with non-degenerate critical points to construct appropriate complex geometrical optics solutions to prove the result. We describe this more precisely below. Let $\Omega \subset \mathbb{R}^2$ be a bounded domain which consists of N smooth closed curves γ_j, $\partial\Omega = \cup_{j=}^{N}\gamma_j$. As before we define the set of Cauchy data for a bounded potential q by:

$$\widehat{C_q} = \left\{ \left(u|_{\partial\Omega}, \left.\frac{\partial u}{\partial\nu}\right|_{\partial\Omega} \right) \mid (\Delta - q)u = 0 \text{ on } \Omega,\ u \in H^1(\Omega) \right\}. \tag{71}$$

Let $\Gamma \subset \partial\Omega$ be a non-empty open subset of the boundary. Denote $\Gamma_0 = \partial\Omega \backslash \overline{\Gamma}$. The main result of [87] gives global uniqueness by measuring the Cauchy data on Γ. Let $q_j \in C^{2+\alpha}(\overline{\Omega})$, $j = 1, 2$ for some $\alpha > 0$ and let q_j be complex-valued. Consider the following sets of Cauchy data on Γ:

$$\mathcal{C}_{q_j} = \left\{ \left(u|_{\Gamma}, \left.\frac{\partial u}{\partial\nu}\right|_{\Gamma} \right) \mid (\Delta - q_j)u = 0 \text{ in } \Omega,\ u|_{\Gamma_0} = 0,\ u \in H^1(\Omega) \right\},\quad j = 1, 2. \tag{72}$$

Theorem 1.18 *Assume $\mathcal{C}_{q_1} = \mathcal{C}_{q_2}$. Then $q_1 = q_2$.*

Using Theorem 1.18 one concludes immediately, as a corollary, the following global identifiability result for the conductivity Eq. (2). This result uses that knowledge of the Dirichlet-to-Neumann map on an open subset of the boundary determines γ and its first derivatives on Γ (see [113, 192]).

Corollary 1.19 *With some $\alpha > 0$, let $\gamma_j \in C^{4+\alpha}(\overline{\Omega})$, $j = 1, 2$, be non-vanishing functions. Assume that*

$$\Lambda_{\gamma_1}(f) = \Lambda_{\gamma_2}(f) \text{ on } \Gamma \text{ for all } f \in H^{\frac{1}{2}}(\Gamma), \ \text{supp}\, f \subset \Gamma.$$

Then $\gamma_1 = \gamma_2$.

It is easy to see that Theorem 1.18 implies the analogous result to [107] in the two dimensional case. Notice that Theorem 1.18 does not assume that Ω is simply connected. An interesting inverse problem is whether one can determine the potential or conductivity in a region of the plane with holes by measuring the Cauchy data only on the accessible boundary. This is also called the obstacle problem. Let Ω, D be domains in $\mathbb{R}^2$ with smooth boundaries such that $\overline{D} \subset \Omega$. Let $V \subset \partial\Omega$ be an open set. Let $q_j \in C^{2+\alpha}(\overline{\Omega \backslash D})$, for some $\alpha > 0$ and $j = 1, 2$. Let us consider the following set of partial Cauchy data

$$\widetilde{C}_{q_j} = \{(u|_V, \frac{\partial u}{\partial \nu}|_V) | (\Delta - q_j)u = 0 \text{ in } \Omega\backslash\overline{D},\ u|_{\partial D \cup \partial\Omega\backslash V} = 0, u \in H^1(\Omega\backslash\overline{D})\}.$$

Corollary 1.20 *Assume $\widetilde{C}_{q_1} = \widetilde{C}_{q_2}$. Then $q_1 = q_2$.*

A similar result holds for the conductivity equation.

Corollary 1.21 *Let $\gamma_j \in C^{4+\alpha}(\overline{\Omega\backslash D})$ $j = 1, 2$ be non vanishing functions. Assume*

$$\Lambda_{\gamma_1}(f) = \Lambda_{\gamma_2}(f) \text{ on } V \quad \forall f \in H^{\frac{1}{2}}(\partial(\overline{\Omega}\backslash D)), \ \text{supp}\, f \subset V$$

Then $\gamma_1 = \gamma_2$.

The two dimensional case has special features since one can construct a much larger set of complex geometrical optics solutions than in higher dimensions. On the other hand, the problem is formally determined in two dimensions and therefore more difficult. The proof of Theorem 1.18 is based on the construction of appropriate complex geometrical optics solutions by Carleman estimates with degenerate weight functions.

Sketch of the Proof. For the partial data problem we need a more general class of CGO solutions than the ones constructed by Bukhgeim, since we would like to have the imaginary part of the phase vanish on Γ. So we consider more general holomorphic functions with non-degenerate critical points as phases. Let the function $\Phi(z) = \varphi(x_1, x_2) + i\psi(x_1, x_2) \in C^2(\overline{\Omega})$ be holomorphic in Ω and $\text{Im}\ \Phi|_{\partial\Omega\backslash\Gamma} = 0$. Notice that this implies $\nabla\varphi \cdot \nu = 0$ on $\partial\Omega\backslash\Gamma$. We denote the set of critical points of Φ by

$$\mathcal{H} = \{z \in \overline{\Omega} | \partial_z \Phi(z) = 0\}.$$

We assume that Φ has a finite number of non-degenerate critical points in $\overline{\Omega}$, that is $\partial_z^2\Phi(z) \neq 0,\ z \in \mathcal{H}$. We denote the critical points by $\mathcal{H} = \{\widetilde{x}_1, ..., \widetilde{x}_\ell\}$ As in the partial data problem considered in Sect. 1.5 we construct appropriate CGO solutions by proving a Carleman estimate.

Carleman estimate Let $u \in H_0^1(\Omega)$, real valued. Then for all large $\tau > 0$:

$$\tau\|ue^{\tau\varphi}\|^2_{L^2(\Omega)} + \|ue^{\tau\varphi}\|^2_{H^1(\Omega)} + \left\|\frac{\partial u}{\partial \nu}e^{\tau\varphi}\right\|^2_{L^2(\partial\Omega\setminus\Gamma)} + \tau^2\left\|\left|\frac{\partial\Phi}{\partial z}\right|ue^{\tau\varphi}\right\|^2_{L^2(\Omega)}$$
$$\leq C\left(\|(\Delta u)e^{\tau\varphi}\|^2_{L^2(\Omega)} + \tau\int\limits_\Gamma\left|\frac{\partial u}{\partial \nu}\right|^2 e^{2\tau\varphi}d\sigma\right)$$

with σ the standard measure on $\partial\Omega$. The Carleman estimate implies the existence of a solution to the boundary value problem for the Schrödinger equation

$$(\Delta - q)u = f \text{ in } \Omega; \quad u|_{\partial\Omega\setminus\Gamma} = g \tag{73}$$

and that it satisfies an estimate. More precisely we have

Proposition 1.22 *Let $q \in L^\infty(\Omega)$. There exists $\tau_0 > 0$ such that for all $|\tau| > \tau_0$ there exists a solution of (72) such that*

$$\|ue^{-\tau\varphi}\|_{L^2(\Omega)} \leq C\big(\|fe^{-\tau\varphi}\|_{L^2(\Omega)}/\tau + \|ge^{-\tau\varphi}\|_{L^2(\partial\Omega\setminus\Gamma)}\big).$$

We next find CGO solutions of

$$(\Delta - q)u = 0 \quad \text{in } \Omega; \quad u|_{\partial\Omega\setminus\widetilde{\Gamma}} = 0 \tag{74}$$

of the form

$$u(x) = e^{\tau\Phi(z)}(a(z) + a_0(z)/\tau) + e^{\tau\overline{\Phi(z)}}\overline{(a(z) + a_1(z)/\tau)} + e^{\tau\varphi}u_1 + e^{\tau\varphi}u_2. \tag{75}$$

The functions $a, a_0, a_1 \in C^2(\overline{\Omega})$ are holomorphic in Ω and $\operatorname{Re} a|_{\partial\Omega\setminus\Gamma} = 0$. Moreover

$$\|u_j\|_{L^2(\Omega)} = o\left(\frac{1}{\tau}\right), \quad \tau \to \infty, j = 1, 2. \tag{76}$$

Now we take two potentials q_1 and q_2 satisfying the hypothesis of Theorem 1.18. We take for the potential q_1 a solution u of the corresponding Schrödinger equation of the form (75) and for the Schrödinger equation associated to q_2 a solution v of the form

$$v(x) = e^{-\tau\Phi(z)}(a(z) + b_0(z)/\tau) + e^{-\tau\overline{\Phi(z)}}\overline{(a(z) + b_1(z)/\tau)} + e^{-\tau\varphi}v_1 + e^{-\tau\varphi}v_2 \tag{77}$$

with v_1, v_2 satisfying the same decay for large τ as u_1, u_2. Using arguments similar to those of Sect. 1.4.1 we get

$$\int_\Omega (q_1 - q_2)uv dx = 0. \tag{78}$$

Substituting (74) and (77) into this identity and applying stationary phase we conclude

Proposition 1.23 *Let $\{\widetilde{x}_1, \ldots, \widetilde{x}_\ell\}$ be the set of critical points of the function $Im\,\Phi$. Then for any potentials q_1, q_2 satisfying the hypotheses of Theorem* 1.18 *and for any holomorphic function a, we have*

$$2\sum_{k=1}^{\ell} \frac{\pi((q_1 - q_2)|a|^2)(\widetilde{x}_k) Re\, e^{2i\tau Im\,\Phi(\widetilde{x}_k)}}{|(det\, Im\,\Phi'')(\widetilde{x}_k)|^{\frac{1}{2}}} = 0, \quad \tau > 0.$$

We can choose Φ such that

$$\text{Im}\,\Phi(\widetilde{x}_k) \neq \text{Im}\,\Phi(\widetilde{x}_j), \; j \neq k.$$

Let $a(\widetilde{x}_k) \neq 0$. Then Proposition 1.24 implies

$$q_1(\widetilde{x}_k) = q_2(\widetilde{x}_k).$$

We then show that the non-degenerate critical points of Φ can be chosen to be a dense set concluding the sketch of the proof of the theorem. We refer to [30] for an analysis of a discrete version of the partial data problem in 2D. The case of Riemannian surfaces was considered in [70]. A improvement on the regularity of the potential is given in [86]. For a study of partial data for a general class of second order elliptic operators see [89]. The case of measurements in disjoint sets is considered in [90].

1.7 Anisotropic conductivities

Anisotropic conductivities depend on direction. Muscle tissue in the human body is an important example of an anisotropic conductor. For instance cardiac muscle has a conductivity of 2.3 mho in the transverse direction and 6.3 in the longitudinal direction. The conductivity in this case is represented by a positive definite, smooth, symmetric matrix $\gamma = (\gamma^{ij}(x))$ on Ω. Under the assumption of no sources or sinks of current in Ω, the potential u in Ω, given a voltage potential f on $\partial\Omega$, solves the Dirichlet problem

$$\begin{cases} \sum_{i,j=1}^{n} \frac{\partial}{\partial x_i}\left(\gamma^{ij}\frac{\partial u}{\partial x_j}\right) = 0 \text{ on } \Omega \\ u|_{\partial\Omega} = f. \end{cases} \tag{79}$$

The DN map is defined by

$$\Lambda_\gamma(f) = \sum_{i,j=1}^{n} \nu^i \gamma^{ij} \frac{\partial u}{\partial x_j}\Big|_{\partial\Omega} \tag{80}$$

where $\nu = (\nu^1, \dots, \nu^n)$ denotes the unit outer normal to $\partial\Omega$ and u is the solution of (79). The inverse problem is whether one can determine γ by knowing Λ_γ. Unfortunately, Λ_γ doesn't determine γ uniquely. This observation is due to L. Tartar (see [113] for an account). Let $\psi : \overline{\Omega} \to \overline{\Omega}$ be a C^∞ diffeomorphism with $\psi|_{\partial\Omega} = Id$ where Id denotes the identity map. We have

$$\Lambda_{\widetilde{\gamma}} = \Lambda_\gamma \tag{81}$$

where

$$\widetilde{\gamma} = \left(\frac{(D\psi)^T \circ \gamma \circ (D\psi)}{|\det D\psi|} \right) \circ \psi^{-1}. \tag{82}$$

Here $D\psi$ denotes the (matrix) differential of ψ, $(D\psi)^T$ its transpose and the composition in (82) is to be interpreted as multiplication of matrices. We have then a large number of conductivities with the same DN map: any change of variables of Ω that leaves the boundary fixed gives rise to a new conductivity with the same electrostatic boundary measurements. The question is then whether this is the only obstruction to unique identifiability of the conductivity. In two dimensions this has been shown for $L^\infty(\Omega)$ conductivities in [12]. This is done by reducing the anisotropic problem to the isotropic one by using isothermal coordinates [2,189] and using Astala and Päivärinta's result in the isotropic case [11]. Earlier results were for C^3 conductivities using the result of Nachman [134] and for Lipschitz conductivities in [186] using the techniques of [33]. An extension of some of these results to quasilinear anisotropic conductivities can be found in [188]. In three dimensions, as was pointed out in [121], this is a problem of geometrical nature and makes sense for general compact Riemannian manifolds with boundary. Let (M, g) be a compact Riemannian manifold with boundary. The Laplace-Beltrami operator associated to the metric g is given in local coordinates by

$$\Delta_g u = \frac{1}{\sqrt{\det g}} \sum_{i,j=1}^{n} \frac{\partial}{\partial x_i} \left(\sqrt{\det g}\, g^{ij} \frac{\partial u}{\partial x_j} \right) \tag{83}$$

where (g^{ij}) is the matrix inverse of the matrix (g_{ij}). Let us consider the Dirichlet problem associated to (83)

$$\Delta_g u = 0 \text{ on } \Omega, \quad u|_{\partial\Omega} = f. \tag{84}$$

We define the DN map in this case by

$$\Lambda_g(f) = \sum_{i,j=1}^{n} \nu^i g^{ij} \frac{\partial u}{\partial x_j} \sqrt{\det g}|_{\partial\Omega} \tag{85}$$

The inverse problem is to recover g from Λ_g. We have

$$\Lambda_{\psi^* g} = \Lambda_g \tag{86}$$

where ψ is any C^∞ diffeomorphism of $\overline{M}$ which is the identity on the boundary. As usual $\psi^* g$ denotes the pull back of the metric g by the diffeomorphism ψ. In the case that M is an open, bounded subset of $\mathbb{R}^n$ with smooth boundary, it is easy to see ([121]) that for $n \geq 3$

$$\Lambda_g = \Lambda_\gamma \tag{87}$$

where

$$(g_{ij}) = (\det \gamma^{kl})^{\frac{1}{n-2}} (\gamma^{ij})^{-1}, \quad (\gamma^{ij}) = (\det g_{kl})^{\frac{1}{2}} (g_{ij})^{-1}. \tag{88}$$

In the two dimensional case there is an additional obstruction since the Laplace-Beltrami operator is conformally invariant. More precisely we have

$$\Delta_{\alpha g} = \frac{1}{\alpha} \Delta_g$$

for any function α, $\alpha \neq 0$. Therefore we have, for $n = 2$,

$$\Lambda_{\alpha(\psi^* g)} = \Lambda_g \tag{89}$$

for any smooth function $\alpha \neq 0$ so that $\alpha|_{\partial M} = 1$. Lassas and Uhlmann ([124]) proved that (86) is the only obstruction to unique identifiability of the conductivity for real-analytic manifolds in dimension $n \geq 3$. In the two dimensional case they showed that (89) is the only obstruction to unique identifiability for smooth Riemannian surfaces. Moreover these results assume that the DN map is measured only on an open subset of the boundary. For another proof see [23]. We state the two basic results. Let Γ be an open subset ∂M. We define for f, supp $f \subseteq \Gamma$

$$\Lambda_{g,\Gamma}(f) = \Lambda_g(f)|_\Gamma.$$

Theorem 1.24 ($n = 2$) *Let (M, g) be a compact Riemannian surface with boundary. Let $\Gamma \subseteq \partial M$ be an open subset. Then $\Lambda_{g,\Gamma}$ determines uniquely the conformal class of (M, g).*

Theorem 1.25 ($n \geq 3$) *Let (M, g) be a real-analytic compact, connected Riemannian manifold with boundary. Let $\Gamma \subseteq \partial M$ be real-analytic and assume that g is real-analytic up to Γ. Then $\Lambda_{g,\Gamma}$ determines uniquely (M, g) up to an isometry.*

Einstein manifolds are real-analytic in the interior and it was conjectured by Lassas and Uhlmann that they were uniquely determined up to isometry by the DN map. This was proven in [68]. Notice that these results don't assume any condition on the topology of the manifold except for connectedness. An earlier result of [121] assumed that (M, g) was strongly convex and simply connected and $\Gamma = \partial M$ in both results. Theorem 1.25 was extended in [125] to non-compact, connected real-analytic manifolds with boundary. These results were extended to differential forms in [116].

1.7.1 The Calderón problem on manifolds

The invariant form on a Riemannian manifold with boundary (M, g) for an isotropic conductivity β is given by

$$\mathrm{div}_g(\beta \nabla_g) u = 0 \tag{90}$$

where div_g (resp. ∇_g) denotes divergence (resp. gradient) with respect to the Riemannian metric g. This includes the case considered by Calderón with g the Euclidean metric, the anisotropic case by taking $g^{ij} = \gamma^{ij}\beta$ and $\beta = \sqrt{\det g}$. It was shown in [186] for bounded domains of Euclidean space in two dimensions that the isometric class of (β, g) is determined uniquely by the DN map associated to (90). In two dimensions, when the metric g is known, it is proven in [77] that one can uniquely determine the conductivity β. Guillarmou and Tzou [69] have shown that a potential is uniquely determined for the Schrödinger equation $\Delta_g - q$, with Δ_g the Laplace-Beltrami operator associated to the metric g, generalizing the result of [77]. In dimension $n \geq 3$ it is an open problem whether one can determine the isotropic conductivity β from the corresponding DN map associated to (90). As before one can consider the more general problem of recovering the potential q from the DN map associated to $\Delta_g - q$. We review below the progress that has been made on this problem based on [52].

1.7.2 Complex geometrical optics on manifolds

We review the recent construction of complex geometrical optics construction for a class of Riemannian manifolds based on [52]. In this paper those Riemannian manifolds which admit limiting Carleman weights, were characterized. All such weights in Euclidean space were listed in Theorem 1.13.

Theorem 1.26 *If (M, g) is an open manifold having a limiting Carleman weight, then some conformal multiple of the metric g admits a parallel unit vector field. For simply connected manifolds, the converse is also true.*

Locally, a manifold admits a parallel unit vector field if and only if it is isometric to the product of an Euclidean interval and another Riemannian manifold. This is an instance of the de Rham decomposition [158]. Thus, if (M, g) has an LCW φ, one can choose local coordinates in such a way that $\phi(x) = x_1$ and

$$g(x_1, x') = c(x) \begin{pmatrix} 1 & 0 \\ 0 & g_0(x') \end{pmatrix},$$

where c is a positive conformal factor. Conversely, any metric of this form admits $\varphi(x) = x_1$ as a limiting weight. In the case $n = 2$, limiting Carleman weights in (M, g) are exactly the harmonic functions with non-vanishing differential. Let us now introduce the class of manifolds which admit limiting Carleman weights and for which one can prove uniqueness results. For this we need the notion of simple manifolds [167].

Definition 1.27 A manifold (M, g) with boundary is *simple* if ∂M is strictly convex, and for any point $x \in M$ the exponential map $\exp_x$ is a diffeomorphism from some closed neighborhood of 0 in $T_x M$ onto M.

Definition 1.28 A compact manifold with boundary (M, g), of dimension $n \geq 3$, is *admissible* if it is conformal to a submanifold with boundary of $\mathbb{R} \times (M_0, g_0)$ where (M_0, g_0) is a compact simple $(n-1)$-dimensional manifold.

Examples of admissible manifolds include the following:

1. Bounded domains in Euclidean space, in the sphere minus a point, or in hyperbolic space. In the last two cases, the manifold is conformal to a domain in Euclidean space via stereographic projection.
2. More generally, any domain in a locally conformally flat manifold is admissible, provided that the domain is appropriately small. Such manifolds include locally symmetric 3-dimensional spaces, which have parallel curvature tensor so their Cotton tensor vanishes [54].
3. Any bounded domain M in $\mathbb{R}^n$, endowed with a metric which in some coordinates has the form

$$g(x_1, x') = c(x)\begin{pmatrix} 1 & 0 \\ 0 & g_0(x') \end{pmatrix},$$

   with $c > 0$ and g_0 simple, is admissible.
4. The class of admissible metrics is stable under C^2-small perturbations of g_0.

The first inverse problem involves the Schrödinger operator

$$\mathcal{L}_{g,q} = \Delta_g - q,$$

where q is a smooth complex valued function on (M, g). We make the standard assumption that 0 is not a Dirichlet eigenvalue of $\mathcal{L}_{g,q}$ in M. Then the Dirichlet problem

$$\begin{cases} \mathcal{L}_{g,q} u = 0 \text{ in } M, \\ u = f \text{ on } \partial M \end{cases}$$

has a unique solution for any $f \in H^{1/2}(\partial M)$, and we may define the DN map

$$\Lambda_{g,q} : f \mapsto \partial_\nu u|_{\partial M}.$$

Given a fixed admissible metric, one can determine the potential q from boundary measurements.

Theorem 1.29 *Let* (M, g) *be admissible, and let* q_1 *and* q_2 *be two smooth functions on* M*. If* $\Lambda_{g,q_1} = \Lambda_{g,q_2}$*, then* $q_1 = q_2$*.*

This result was known previously in dimensions $n \geq 3$ for the Euclidean metric [190] and for the hyperbolic metric [100]. It has been generalized to Maxwell's equations in [106].

2 Travel time tomography and boundary rigidity

2.1 Introduction

The question of determining the sound speed or index of refraction of a medium by measuring the first arrival times of waves arose in geophysics in an attempt to determine the substructure of the Earth by measuring at the surface of the Earth the travel times of seismic waves. An early success of this inverse method was the estimate by Herglotz [78] and Wiechert and Zoeppritz [207] of the diameter of the Earth and the location of the mantle, crust and core. The assumption used in those papers is that the index of refraction (inverse proportional to the speed) depends only on the radius. A more realistic model is to assume that it depends on position. The inverse kinematic problem can be formulated mathematically as determining a Riemannian metric on a bounded domain (the Earth) given by $ds^2 = \frac{1}{c^2(x)}dx^2$, where c is a positive function, from the length of geodesics (travel times) joining points in the boundary. More recently it has been realized, by measuring the travel times of seismic waves, that the inner core of the Earth might exhibit anisotropic behavior, that is the speed of waves depends also on direction there with the fast direction parallel to the Earth's spin axis [44]. Given the complications presented by modeling the Earth as an anisotropic elastic medium we consider a simpler model of anisotropy, namely that the wave speed is given by a symmetric, positive definite matrix $g = (g_{ij})(x)$, that is, a Riemannian metric in mathematical terms. The problem is to determine the metric from the lengths of geodesics joining points in the boundary (the surface of the Earth in the motivating example). It is useful to consider a more general and geometric formulation of the problem. Let (M, g) be a compact Riemannian manifold with boundary ∂M. Let $d_g(x, y)$ denote the geodesic distance between x and y, two points in the boundary. This is defined as the infimum of the length of all sufficiently smooth curves joining the two points. The function d_g measures the first arrival time of waves joining points of the boundary. One of the inverse problems we discuss in this section is whether we can determine the Riemannian metric g knowing $d_g(x, y)$ for any $x \in \partial M$, $y \in \partial M$. This problem also arose in rigidity questions in Riemannian geometry [45,67,129]. The metric g cannot be determined from this information alone. We have $d_{\psi^* g} = d_g$ for any diffeomorphism $\psi : M \to M$ that leaves the boundary pointwise fixed, i.e., $\psi|_{\partial M} = Id$, where Id denotes the identity map and $\psi^* g$ is the pull-back of the metric g. The natural question is whether this is the only obstruction to unique identifiability of the metric. It is easy to see that this is not the case. Namely one can construct a metric g and find a point x_0 in M so that $d_g(x_0, \partial M) > \sup_{x,y \in \partial M} d_g(x, y)$. For such a metric, d_g is independent of a change of g in a neighborhood of x_0. The hemisphere of the round sphere is another example. Therefore it is necessary to impose some a-priori restrictions on the metric. One such restriction is to assume that the Riemannian manifold (M, g) is **simple**, i.e., M is simply-connected, any geodesic has no conjugate points and ∂M is strictly convex. ∂M is strictly convex if the second fundamental form of the boundary is positive definite in every boundary point. R. Michel conjectured in [129] that simple manifolds are boundary distance rigid, that is d_g determines g uniquely up to an isometry which is the identity on the boundary. This is known for simple subspaces of Euclidean space (see [67]), simple subspaces of an open hemisphere in two dimensions (see [130]), simple subspaces of symmetric spaces of constant negative curvature [26], simple two dimensional spaces of negative curvature (see [46, 149]). If one metric is close to the Euclidean metric boundary rigidity was proven in [123] that was improved in [36]. We remark that simplicity of a compact manifold with boundary can be determined from the boundary distance function. Michel's conjecture was proven in generality in [155] in two dimensions and we describe the details of the proof in Sect. 2.2. In the case that both g_1 and g_2 are conformal to the Euclidean metric e (i.e., $(g_k)_{ij} = \alpha_k \delta_{ij}, k = 1, 2$ with δ_{ij} the Kronecker symbol), as mentioned earlier, the problem we are considering here is known in seismology as the inverse kinematic problem. In this case, it has been proven by Mukhometov in two dimensions [131] that if $(M, g_i), i = 1, 2$ is simple and $d_{g_1} = d_{g_2}$, then $g_1 = g_2$. More generally the same method of proof shows that if $(M, g_i), i = 1, 2$, are simple compact Riemannian manifolds with boundary and they are in the same conformal class then the metrics are determined by the boundary distance function. More precisely we have:

Theorem 2.1 *Let $(M, g_i), i = 1, 2$ be simple compact Riemannian manifolds with boundary of dimension $n \geq 2$. Assume $g_1 = \rho g_2$ for a positive, smooth function $\rho, \rho|_{\partial M} = 1$ and $d_{g_1} = d_{g_2}$ then $g_1 = g_2$.*

This result and a stability estimate were proven in [131]. We remark that in this case the diffeomorphism ψ that is present in the general case must be the identity if the metrics are conformal to each other. For related results and generalizations see [25, 27, 45, 60, 133].

In Sect. 2.2 we consider the boundary rigidity in the two dimensional case and in Sect. 2.3 in the higher dimensional case. In Sect. 2.4 we discuss some results on the non-simple case where the measurements are given by the *scattering relation*. Roughly speaking one measures the point of exit and direction of exit of a geodesic for which we know the point of entrance and direction of entrance besides this we also know the travel time, that is the length of that geodesic.

2.2 Boundary rigidity in two dimensions

We will sketch in this section the proof of the following result.

Theorem 2.2 *Let $(M, g_i), i = 1, 2$ be two dimensional simple compact Riemannian manifolds with boundary. Assume*

$$d_{g_1}(x, y) = d_{g_2}(x, y) \quad \forall (x, y) \in \partial M \times \partial M.$$

Then there exists a diffeomorphism $\psi : M \to M$, $\psi|_{\partial M} = Id$, so that

$$g_2 = \psi^* g_1.$$

The proof of Theorem 2.2 involves a connection between the scattering relation and the Dirichlet-to-Neumann map (DN) associated to the Laplace-Beltrami operator discussed in section. In Sect. 2.2.1 we define the scattering relation which quantizes the scattering operator. In Sect. 2.2.2 we discuss the geodesic X-ray transform and compute the commutator of the fiberwise Hilbert transform and geodesic flow (see Theorem 2.7 and Theorem 2.8). In Sect. 2.2.3 we discuss the main step of the proof of Theorem 2.2 which consists in showing that, under the assumptions of the theorem, we can determine the Dirichlet-to-Neumann map if we know the scattering relation. We also sketch the proof of Theorem 2.9. In Sect. 2.2.4 we use the connection indicated in Sect. 2.2.1, to give a characterization of the range of the geodesic X-ray transform in terms of the scattering relation and we give Fredholm type inversion formulas for the geodesic X-ray transform acting on scalar functions and vector fields.

2.2.1 The scattering relation

Suppose we have a Riemannian metric in Euclidean space which is the Euclidean metric outside a compact set. The inverse scattering problem for metrics is to determine the Riemannian metric by measuring the scattering operator (see [72]). A similar obstruction to the boundary rigidity problem occurs in this case with the diffeomorphism ψ equal to the identity outside a compact set. It was proven in [72] that from the wave front set of the scattering operator, one can determine, under some non-trapping assumptions on the metric, the **scattering relation** on the boundary of a large ball. This uses high frequency information of the scattering operator. In the semiclassical setting Alexandrova has shown for a large class of operators that the scattering operator associated to potential and metric perturbations of the Euclidean Laplacian is a semiclassical Fourier integral operator quantized by the scattering relation [8]. The scattering relation maps the point and direction of a geodesic entering the manifold to the point and direction of exit of the geodesic. We proceed to define in more detail the scattering relation and its relation with the boundary distance function. Let ν denote the unit-inner normal to ∂M. We denote by $S(M) \to M$ the unit-sphere bundle over M:

$$S(M) = \bigcup_{x \in M} S_x, \quad S_x = \{\xi \in T_x(M) : |\xi|_g = 1\}.$$

$S(M)$ is a (2 dim $M - 1$)-dimensional compact manifold with boundary, which can be written as the union $\partial\Omega(M) = \partial_+ S(M) \cup \partial_- S(M)$

$$\partial_\pm S(M) = \{(x, \xi) \in \partial S(M), \ \pm(\nu(x), \xi) \geq 0\}.$$

The manifold of inner vectors $\partial_+ S(M)$ and outer vectors $\partial_- S(M)$ intersect at the set of tangent vectors

$$\partial_0 S(M) = \{(x, \xi) \in \partial S(M), \quad (\nu(x), \xi) = 0\}.$$

Let (M, g) be an n-dimensional compact manifold with boundary. We say that (M, g) is **non-trapping** if each maximal geodesic is finite. Let (M, g) be non-trapping and the boundary ∂M is strictly convex. Denote by $\tau(x, \xi)$ the length of the geodesic $\gamma(x, \xi, t), t \geq 0$, starting at the point x in the direction $\xi \in S_x$. This function is smooth on $S(M)\backslash\partial_0 S(M)$. The function $\tau^0 = \tau|_{\partial S(M)}$ is equal to zero on $\partial_- S(M)$ and is smooth on $\partial_+ S(M)$. Its odd part with respect to ξ

$$\tau_-^0(x, \xi) = \frac{1}{2}\left(\tau^0(x, \xi) - \tau^0(x, -\xi)\right)$$

is a smooth function.

Definition 2.3 Let (M, g) be non-trapping with strictly convex boundary. The scattering relation $\alpha : \partial S(M) \to \partial S(M)$ is defined by

$$\alpha(x, \xi) = (\gamma(x, \xi, 2\tau_-^0(x, \xi)), \dot{\gamma}(x, \xi, 2\tau_-^0(x, \xi))).$$

The scattering relation is a diffeomorphism $\partial S(M) \to \partial S(M)$. Notice that $\alpha|_{\partial_+ S(M)} : \partial_+ S(M) \to \partial_- S(M)$, $\alpha|_{\partial_- S(M)} : \partial_- S(M) \to \partial_+ S(M)$ are diffeomorphisms as well. Obviously, α is an involution, $\alpha^2 = id$ and $\partial_0 S(M)$ is the hypersurface of its fixed points, $\alpha(x, \xi) = (x, \xi)$, $(x, \xi) \in \partial_0 S(M)$. A natural inverse problem is whether the scattering relation determines the metric g up to an isometry which is the identity on the boundary. This information takes into account all the travel times not just the first arrivals. In the case that (M, g) is a simple manifold, and we know the metric at the boundary (and this is determined if d_g is known, see [129], knowing the scattering relation is equivalent to knowing the boundary distance function ([129]) so that we concentrate on studying the scattering relation. We introduce the operators of even and odd continuation with respect to α:

$$A_\pm w(x, \xi) = w(x, \xi), \qquad (x, \xi) \in \partial_+ S(M),$$
$$A_\pm w(x, \xi) = \pm\left(\alpha^* w\right)(x, \xi), \quad (x, \xi) \in \partial_- S(M).$$

The scattering relation preserves the measure $|(\xi, \nu)|d\Sigma$, ($d\Sigma$ is the measure of the boundary $\partial S(M)$ induced by the metric g) and therefore the operators $A_\pm$: $L_\mu^2(\partial_+ S(M)) \to L_{|\mu|}^2(\partial S(M))$ are bounded, where $L_{|\mu|}^2(\partial S(M))$ is the real Hilbert space with scalar product

$$(u, v)_{L_{|\mu|}^2(\partial S(M))} = \int_{\partial S(M)} |\mu| uv d\Sigma, \quad \mu = (\xi, \nu).$$

and $L_\mu^2(\partial_+ S(M))$ is the real Hilbert space with scalar product

$$(u, v)_{L_\mu^2(\partial_+ S(M))} = \int_{\partial_+ S(M)} \mu uv d\Sigma.$$

The adjoint of $A_\pm$ is a bounded operator $A^*_\pm : L^2_{|\mu|}(\partial S(M)) \to L^2_\mu(\partial_+ S(M))$ given by

$$A^*_\pm u = (u \pm u \circ \alpha)|_{\partial_+ S(M)}.$$

2.2.2 The geodesic X-ray transform

The X-ray transform integrates a function along lines. Radon found in 1917 an inversion formula in two dimensions to determine a function knowing the X-ray transform. This formula is non-local in the sense that in order to find the function at a point x one needs to know the integral of the function along lines far from the point. Radon's inversion formula has been implemented numerically using the filtered backprojection algorithm which is used today in CT scans. Another important transform in medical imaging and other applications is the Doppler transform which integrates a vector field along lines. The motivation is ultrasound Doppler tomography. It is known that blood flow is irregular and faster around tumor tissue than in normal tissue and Doppler tomography attempts to reconstruct the blood flow pattern. Mathematically the problem is to what extend a vector field is determined from its integral along lines. In this paper we consider the case of integrating functions and vector fields along geodesics of a Riemannian metric. This arises in geophysics since the ray paths are no longer straight lines. We obtain inversion formulas for the constant curvature case and Fredholm type formulas in general which are *non-local*. We define next the geodesic X-ray transform for any compact Riemannian manifold (M, g) with boundary of any dimension. We embed (M, g) into a compact Riemannian manifold $(\widetilde{M}, g)$ with no boundary. Let φ_t be the geodesic flow on $S(\widetilde{M})$ and $\mathcal{H} = \frac{d}{dt}\varphi_t|_{t=0}$ be the geodesic vector field. Let u^f be the solution of the boundary value problem

$$\mathcal{H}u = -f, \quad u|_{\partial_- S(M)} = 0,$$

which can be written as

$$u^f(x,\xi) = \int_0^{\tau(x,\xi)} f(\varphi_t(x,\xi))dt, \quad (x,\xi) \in S(M).$$

In particular

$$\mathcal{H}\tau = -1.$$

The trace

$$If = u^f|_{\partial_+ S(M)}$$

is called **the geodesic X-ray transform** of the function f. If the manifold (M, g) is non-trapping and has a strictly convex boundary the operator $I : C^\infty(S(M)) \to C^\infty(\partial_+ S(M))$. Clearly a function f is not determined by its geodesic X-ray transform alone, since it depends on more variables than If. We consider the geodesic X-ray transform acting on symmetric tensor fields. We denote by $f_m(x,\xi)$ an homogeneous polynomial of degree m with respect to ξ, induced by the symmetric tensor field f on (M,g) of m degree:

$$f_m(x,\xi) = f_{i_1...i_m}(x)\,\xi^{i_1}...\xi^{i_m}.$$

The operator I_m, defined by

$$I_m f = I f_m$$

is called **the geodesic X-ray transform** of the symmetric tensor field. If the manifold (M,g) is non-trapping and the boundary ∂M is strictly convex $I_m{:}C^\infty(M, S_m(M)) \to C^\infty(\partial_+ S(M))$, where $S_m(M)$ denotes the bundle of symmetric tensors over (M,g). It is known that any symmetric smooth enough tensor field f may be decomposed in a potential and solenoidal part [167]:

$$f = dp + h, \quad p|_{\partial M} = 0, \quad \delta h = 0, \tag{91}$$

where δ denotes the divergence and $d = \sigma\nabla$ is the symmetric part of covariant derivative. It is easy to see that the geodesic X-ray transform of the potential part dp is zero. We denote by $C^\infty_{sol}(M, S_m(M))$ the space of smooth solenoidal symmetric tensor fields. The case of the geodesic X-ray transform acting on functions independent of ξ and the geodesic X-ray transform acting on vector fields which, following the notation above, are denoted by I_0 and I_1 respectively. It is known that I_0 is injective on simple manifolds [131,132,161] and that I_1 is injective acting on solenoidal vector fields on simple manifolds [10] and for tensors of order two in [168]. Recently in [151] it has been proven injectivity of I_m for all m for simple two-dimensional manifolds. It was shown in [170] that I_2 is injective for surfaces with no focal points. We mention also that the transform I_2 arises in the linearization of the boundary rigidity problem (see [167]). We define $\psi : S(M) \to \partial_+ S(M)$ by

$$\psi(x,\xi) = \varphi_{-\tau(x,-\xi)}(x,\xi),\ (x,\xi) \in S(M).$$

So, φ is a retract which maps vector (x,ξ) along the geodesic $\gamma(x,\xi,t)$ in the back direction into an incoming vector. The solution of the boundary value problem for the transport equation

$$\mathcal{H}u = 0, \quad u|_{\partial_+ S(M)} = w$$

can be written in the form

$$u = w_\psi = w \circ \psi.$$

The adjoint of the operator I_m is the bounded operator $I_m^*:L_\mu^2(\partial_+ S(M)) \to L^2(M, S_m(M))$ which is given by

$$\left(I_m^* w\right)^{i_1 \dots i_m}(x) = \int\limits_{S_x} w_\psi(x, \xi)\xi^{i_1} \dots \xi^{i_m} dS_x.$$

The Hilbert space $L^2(M, S_m(M))$ may be considered as subspace of $L^2(S(M))$ of homogeneous polynomials with respect to ξ of degree m. The field $I_m^* w$ is solenoidal in the sense of the theory of distributions. Notice, that the adjoint of the bounded operator $I : L^2(S(M)) \to L_\mu^2(\partial_+ S(M))$ is given by

$$I^* w = w_\psi.$$

We also remark that by the fundamental theorem of calculus we have

$$I\mathcal{H}f = (f \circ \alpha - f)|_{\partial_+ S(M)} = -A_-^* f^0, \; f^0 = f|_{\partial S(M)}. \tag{92}$$

The space $C_\alpha^\infty(\partial_+ S(M))$ is defined by

$$C_\alpha^\infty(\partial_+ S(M)) = \{w \in C^\infty(\partial_+ S(M)) : w_\psi \in C^\infty(S(M))\}.$$

In [155] the following characterization of the space of smooth solutions of the transport equation was given

Lemma 2.4

$$C_\alpha^\infty(\partial_+ S(M)) = \{w \in C^\infty(\partial_+ S(M)) : A_+ w \in C^\infty(\partial S(M))\}.$$

In the scalar case the following result holds on the solvability of I_0^* [155,156].

Theorem 2.5 *Let (M, g) be a simple, compact Riemannian manifold with boundary. Then the operator $I_0^* : C_\alpha^\infty(\partial_+ S(M)) \to C^\infty(M)$ is onto.*

The analog result for vector fields was proven in [50].

Theorem 2.6 *Let (M, g) be a simple, compact Riemannian manifold with boundary. Then for any field $v \in C_{sol}^\infty(M, T(M))$ there exists a function $w \in C_\alpha^\infty(\partial_+ S(M))$ such that*

$$v = I_1^* w.$$

Now we define the Hilbert transform in the ξ variable:

$$Hu(x, \xi) = \frac{1}{2\pi} \int\limits_{S_x} \frac{1 + (\xi, \eta)}{(\xi_\perp, \eta)} u(x, \eta) dS_x(\eta), \quad \xi \in S_x,$$

where the integral is understood as a principle value integral. Here $\perp$ means a 90° degree rotation. In coordinates $(\xi_\perp)_i = \varepsilon_{ij}\xi^j$, where

$$\varepsilon = \sqrt{\det g}\begin{pmatrix} 0 & 1 \\ -1 & 0 \end{pmatrix}.$$

The Hilbert transform H transforms even (respectively odd) functions with respect to ξ to even (respectively odd) ones. If H_+ (respectively H_-) is the even (respectively odd) part of the operator H:

$$H_+u(x,\xi) = \frac{1}{2\pi}\int\limits_{S_x} \frac{(\xi,\eta)}{(\xi_\perp,\eta)}u(x,\eta)dS_x(\eta),$$

$$H_-u(x,\xi) = \frac{1}{2\pi}\int\limits_{S_x} \frac{1}{(\xi_\perp,\eta)}u(x,\eta)dS_x(\eta)$$

and u_+, u_- are the even and odd parts of the function u, then $H_+u = Hu_+, H_-u = Hu_-$.

We introduce the notation $\mathcal{H}_\perp = (\xi_\perp, \nabla) = -(\xi, \nabla_\perp)$, where $\nabla_\perp = \varepsilon\nabla$ and ∇ is the covariant derivative with respect to the metric g. The following commutator formula for the geodesic vector field and the Hilbert transform, is a crucial ingredient in the proofs of the main theorems surveyed in this paper (see [155]).

Theorem 2.7 *Let (M, g) be a two dimensional Riemannian manifold. For any smooth function u on $S(M)$ we have the identity*

$$[H, \mathcal{H}]u = \mathcal{H}_\perp u_0 + (\mathcal{H}_\perp u)_0$$

where

$$u_0(x) = \frac{1}{2\pi}\int\limits_{S_x} u(x,\xi)dS_x(\xi)$$

is the average value.

We define

$$P_- = A_-^* H_- A_+, \quad P_+ = A_-^* H_+ A_+.$$

If the manifold (M, g) is simple, the following factorizations hold:

Theorem 2.8

$$P_- = -\frac{1}{2\pi} I\delta_\perp I_1^*, \quad P_+ = -\frac{1}{2\pi} I_1 \nabla_\perp I_0^*. \tag{93}$$

2.2.3 The scattering relation and the Dirichlet-to-Neumann Map

The DN map for the Laplace-Beltrami operator was defined in Sect. 1.7. The connection in two dimensions between the DN map and the scattering relation is given by

Theorem 2.9 *Let $(M, g_i), i = 1, 2,$ be compact, simple two dimensional Riemannian manifolds with boundary. Assume that $\alpha_{g_1} = \alpha_{g_2}$. Then $\Lambda_{g_1} = \Lambda_{g_2}$.*

The proof of Theorem 2.2 is reduced then to the proof of Theorem 2.9. In fact from Theorem 2.9 and Theorem 1.24 we obtain that we can determine the conformal class of the metric up to an isometry which is the identity on the boundary. Now by Theorem 2.1 we have that the conformal factor must be one proving that the metrics are isometric via a diffeomorphism which is the identity at the boundary. In other words $d_{g_1} = d_{g_2}$ implies that $\alpha_{g_1} = \alpha_{g_2}$. By Theorem 2.9 $\Lambda_{g_1} = \Lambda_{g_2}$. By Theorem 1.24, there exists a diffeomorphism $\psi : M \longrightarrow M$, $\psi|_{\partial M} =$ Identity and a function $\beta \neq 0, \beta|_{\partial M} =$ identity such that $g_1 = \beta\psi^* g_2$. By Mukhometov's theorem $\beta = 1$ showing that $g_1 = \psi^* g_2$ proving Theorem 2.2. Before starting the proof of Theorem 2.9 we recall that Michel [130] has proven that for two dimensional Riemannian manifolds with strictly convex boundary one can determine from the boundary distance function, up to the natural obstruction, all the derivatives of the metric at the boundary. This result was generalized to any dimensions in [123]. The proof of Theorem 2.9 consists in showing that from the scattering relation we can determine the traces at the boundary of conjugate harmonic functions, which is equivalent information to knowing the DN map associated to the Laplace-Beltrami operator.

Sketch of the proof of Theorem 2.9 Let (h, h_*) be a pair of conjugate harmonic functions on M,

$$\nabla h = \nabla_\perp h_*, \quad \nabla h_* = -\nabla_\perp h.$$

Notice, that $\delta\nabla = \Delta$ is the Laplace-Beltrami operator and $\delta\nabla_\perp = 0$. Let $I_0^* w = h$. Since $I_1 \mathcal{H}_\perp h = I_1 \mathcal{H} h_* = -A_-^* h_*^0$, where $h_*^0 = h_*|_{\partial M}$, we obtain from the second identity (93)

$$2\pi A_-^* H_+ A_+ w = -A_-^* h_*^0. \tag{94}$$

The following theorem gives the key to obtain the DN map from the scattering relation.

Theorem 2.10 *Let M be a 2-dimensional simple manifold. Let $w \in C_\alpha^\infty(\partial_+ S(M))$ and h_* the harmonic continuation of function h_*^0. Then the equation (94) holds iff the functions $h = I_0^* w$ and h_* are conjugate harmonic functions.*

In summary we have the following procedure to obtain the DN map from the scattering relation. For a given smooth function h_*^0 on ∂M we find a solution $w \in C_\alpha^\infty(\partial_+ S(M))$ of the equation (94). Then the functions $h^0 = 2\pi(A_+ w)_0$ (notice, that $2\pi(A_+ w)_0 = I_0^* w|_{\partial M}$) and h_*^0 are the traces of conjugate harmonic functions. It is easy to see that this gives the DN map.

2.2.4 Range and inversion of the geodesic X-ray transform

Let $T(M)$ be the tangent bundle of M. We denote by δ the divergence operator $\delta : C^{\infty}(M, TM)) \to C^{\infty}(M)$. In local coordinates this is given by $\delta u = g^{kj}\nabla_k u_j$ using Einstein's summation convention. We define the operator $\delta_{\perp} : C^{\infty}(M, T(M)) \to C^{\infty}(M)$ by

$$\delta_{\perp} u = -\delta u_{\perp}.$$

Then

$$\delta_{\perp}\nabla_{\perp} f = \delta\nabla f = \Delta f, \quad \delta_{\perp}\nabla f = -\delta\nabla_{\perp} f = 0.$$

We now give the characterization of the range of I_0 and I_1 in terms of the scattering relation only. We have that these are the projections of the operators P_-, P_+ respectively. For the details see [157].

Theorem 2.11 *Let (M, g) be simple two dimensional compact Riemannian manifold with boundary. Then i) The maps*

$$\delta_{\perp} I_1^* : C_{\alpha}^{\infty}(\partial_+ S(M)) \to C^{\infty}(M),$$
$$\nabla_{\perp} I_0^* : C_{\alpha}^{\infty}(\partial_+ S(M)) \to C_{sol}^{\infty}(M, T(M))$$

are onto. ii). A function $u \in C^{\infty}(\partial_+ S(M))$ belong to Range I_0 iff $u = P_- w$, $w \in C_{\alpha}^{\infty}(\partial_+ S(M))$. iii). A function $u \in C^{\infty}(\partial_+ S(M))$ belong to Range I_1 iff $u = P_+ w$, $w \in C_{\alpha}^{\infty}(\partial_+ S(M))$.

Proposition 2.12 *The operator $W : C_0^{\infty}(M) \to C^{\infty}(M)$, defined by*

$$Wf = (\mathcal{H}_{\perp} u^f)_0$$

can be extended to a smoothing operator $W : L^2(M) \to C^{\infty}(M)$.

We remark that in the case of constant Gaussian curvature $W = 0$ and this does not depend on whether the metric has conjugate points so that the inversion formulas of Theorem 2.13 hold for all two dimensional manifolds with boundary with constant curvature. The inversion formulas are (see [157])

Theorem 2.13 *Let (M, g) be a two-dimensional simple manifold. Then we have*

$$f + W^2 f = \frac{1}{2\pi}\delta_{\perp} I_1^* w, \quad w = \frac{1}{2}\alpha^* H(I_0 f)^- |_{\partial_+ S(M)}, \quad f \in L^2(M),$$
$$h + (W^*)^2 h = \frac{1}{2\pi} I_0^* w, \quad w = \frac{1}{2}\alpha^* H(I_1 \mathcal{H}_{\perp} h)^+ |_{\partial_+ S(M)}, \; h \in H_0^1(M),$$

where $W, W^* : L^2(M) \to C^{\infty}(M)$. In the case of a manifold of constant curvature $W = 0$, $W^* = 0$. Here W^* denotes the adjoint of W.

2.2.5 Remarks

The Hilbert transform for 2-dimensional Riemannian manifolds is the map that relates the restrictions on the boundary of conjugate harmonic functions. In this sense the Hilbert transform, up to a constant, is just the Dirichlet-to-Neumann (DN) map. In Sect. 2.2 we fixed a point x and started with the microlocal Hilbert transform on the circle S_x in the tangent space with Euclidean metric and we ended with the global Hilbert transform (the DN map). The scattering relation and the boundary distance function are determined by the singularities of the DN map associated to the wave equation for the Laplace-Beltrami operator, the so-called hyperbolic (or dynamic) Dirichlet-to-Neumann map [201]. We have found, in two dimensions, a connection between the scattering relation and the *elliptic* Dirichlet-to-Neumann map which led to a solution of the boundary rigidity problem in two dimensions. Is there a similar connection in higher dimensions?

2.3 Boundary rigidity and tensor tomography in dimensions $n \geq 3$

For an earlier review see [177]. In [178], it was proven a local result for metrics in a small neighborhood of the Euclidean one. This result was used in [123] to prove a semiglobal solvability result assuming that one metric is close to the Euclidean and the other has bounded curvature. As it was mentioned earlier it is known [167], that a linearization of the boundary rigidity problem near a simple metric g is given by the following integral geometry problem: recover a symmetric tensor of order 2, which in any coordinates is given by $f = (f_{ij})$, by the geodesic X-ray transform

$$I_g f(\gamma) = \int f_{ij}(\gamma(t))\dot{\gamma}^i(t)\dot{\gamma}^j(t)\,\mathrm{d}t$$

known for all geodesics γ in M. In this section we denote by I_g the geodesic X-ray transform of tensors of order two. It can be easily seen that $I_g dv = 0$ for any vector field v with $v|_{\partial M} = 0$, where dv denotes the symmetric differential

$$[dv]_{ij} = \frac{1}{2}\left(\nabla_i v_j + \nabla_j v_i\right), \tag{95}$$

and $\nabla_k v$ denote the covariant derivatives of the vector field v. This is the linear version of the fact that d_g does not change on $(\partial M)^2 := \partial M \times \partial M$ under an action of a diffeomorphism as above. The natural formulation of the linearized problem is therefore that $I_g f = 0$ implies $f = dv$ with v vanishing on the boundary. We will refer to this property as *s-injectivity* of I_g. More precisely, we have.

Definition 2.14 We say that I_g is *s-injective* in M, if $I_g f = 0$ and $f \in L^2(M)$ imply $f = dv$ with some vector field $v \in H_0^1(M)$.

Any symmetric tensor $f \in L^2(M)$ admits an orthogonal decomposition $f = f^s + dv$ into a *solenoidal* and *potential* parts with $v \in H_0^1(M)$, and f^s divergence free, i.e., $\delta f^s = 0$, where δ is the adjoint operator to $-d$ given by $[\delta f]_i = g^{jk}\nabla_k f_{ij}$ [167]. Therefore, I_g is s-injective, if it is injective on the space of solenoidal tensors. The inversion of I_g is a problem of independent interest in integral geometry, also called *tensor tomography*. We first survey the recent results on this problem. S-injectivity, respectively injectivity for 1-tensors (1-forms) and functions is known, see [167] for references. S-injectivity of I_g was proved in [154] for metrics with negative curvature, in [167] for metrics with small curvature and in [170] for Riemannian surfaces with no focal points. A conditional and non-sharp stability estimate for metrics with small curvature is also established in [167]. In [179], stability estimates for s-injective metrics [see (99) below] were shown and sharp estimates about the recovery of a 1-form $f = f_j dx^j$ and a function f from the associated $I_g f$ which is defined by

$$I_g f(\gamma) = \int f_i(\gamma(t))\dot{\gamma}^i(t)\,dt.$$

The stability estimates proven in [179] were used to prove local uniqueness for the boundary rigidity problem near any simple metric g with s-injective I_g. Similarly to [195], we say that f is analytic in the set K (not necessarily open), if it is real analytic in some neighborhood of K. The results that follow in this section are based on [181]. The first main result we discuss is about s-injectivity for simple analytic metrics.

Theorem 2.15 *Let g be a simple, real analytic metric in M. Then I_g is s-injective.*

Sketch of the proof Note that a simple metric g in M can be extended to a simple metric in some M_1 with $M \Subset M_1$. A simple manifold is diffeomorphic to a (strictly convex) domain $\Omega \subset \mathbf{R}^n$ with the Euclidean coordinates x in a neighborhood of Ω and a metric $g(x)$ there. For this reason, it is enough to prove the results of this section for domains Ω in $\mathbf{R}^n$ provided with a Riemannian metric g. The proof of Theorem 2.15 is based on the following. For smooth metrics, the normal operator $N_g = I_g^* I_g$ is a pseudodifferential operator with a non-trivial null space which is given by

$$(N_g f)_{kl}(x) = \frac{2}{\sqrt{\det g}} \int \frac{f^{ij}(y)}{d_g(x,y)^{n-1}} \frac{\partial d_g}{\partial y^i}\frac{d_g}{\partial y^j}\frac{\partial d_g}{\partial x^k}\frac{\partial d_g}{\partial x^l} \det \frac{\partial^2(d_g^2/2)}{\partial x \partial y}\,dy, \quad x \in \Omega. \tag{96}$$

In the case that the metric g is real-analytic, N_g is an analytic pseudodifferential operator with a non-trivial kernel. We construct an analytic parametrix, using the analytic pseudodifferential calculus in [195], that allows us to reconstruct the solenoidal part of a tensor field from its geodesic X-ray transform, up to a term that is analytic near Ω. If $I_g f = 0$, we show that for some v vanishing on $\partial\Omega$, $\tilde{f} := f - dv$ must be flat at $\partial\Omega$ and analytic in $\bar{\Omega}$, hence $\tilde{f} = 0$. This is similar to the known argument that an analytic elliptic pseudodifferential operator resolves the analytic singularities, hence cannot have compactly supported functions in its kernel. In our case we have a non-trivial kernel, and complications due to the presence of a boundary, in particular a lost of one derivative. For more details see [181]. □

As shown in [179], the s-injectivity of I_g for analytic simple g implies a stability estimate for I_g. In next theorem we show something more, namely that we have a stability estimate for g in a neighborhood of each analytic metric, which leads to stability estimates for generic metrics. As above, let $M_1 \Supset M$ be a compact manifold which is a neighborhood of M and g extends as a simple metric there. We always assume that our tensors are extended as zero outside M, which may create jumps at ∂M. Define the *normal operator* $N_g = I_g^* I_g$, where I_g^* denotes the operator adjoint to I_g with respect to an appropriate measure. We showed in [179] that N_g is a pseudo-differential operator in M_1 of order -1. We introduce the norm $\| \cdot \|_{\tilde{H}^2(M_1)}$ of $N_g f$ in $M_1 \supset M$ in the following way. Choose $\chi \in C_0^\infty$ equal to 1 near ∂M and supported in a small neighborhood of ∂M and let $\chi = \sum_{j=1}^J \chi_j$ be a partition of χ such that for each j, on supp χ_j we have coordinates (x'_j, x^n_j), with x^n_j a normal coordinate. Set

$$\|f\|^2_{\tilde{H}^1} = \int \sum_{j=1}^{J} \chi_j \left(\sum_{i=1}^{n-1} |\partial_{x^i_j} f|^2 + |x^n_j \partial_{x^n_j} f|^2 + |f|^2 \right) \mathrm{d}x, \tag{97}$$

$$\|N_g f\|_{\tilde{H}^2(M_1)} = \sum_{i=1}^{n} \|\partial_{x^i} N_g f\|_{\tilde{H}^1} + \|N_g f\|_{H^1(M_1)}. \tag{98}$$

In other words, in addition to derivatives up to order 1, $\|N_g f\|_{\tilde{H}^2(M_1)}$ includes also second derivatives near ∂M but they are realized as first derivatives of $\nabla N_g f$ tangent to ∂M. The reason to use the $\tilde{H}^2(M_1)$ norm, instead of the stronger $H^2(M_1)$ norm, is that this allows us to work with $f \in H^1(M)$, not only with $f \in H^1_0(M)$, since for such f, extended as 0 outside M, we still have that $N_g f \in \tilde{H}^2(M_1)$, see [179]. On the other hand, $f \in H^1(M)$ implies $N_g f \in \tilde{H}^2(M_1)$ despite the possible jump of f at ∂M. Our stability estimate for the linearized inverse problem is as follows:

Theorem 2.16 *There exists k_0 such that for each $k \geq k_0$, the set $\mathcal{G}^k(M)$ of simple $C^k(M)$ metrics in M for which I_g is s-injective is open and dense in the $C^k(M)$ topology. Moreover, for any $g \in \mathcal{G}^k(M)$,*

$$\left\| f^s \right\|_{L^2(M)} \leq C \|N_g f\|_{\tilde{H}^2(M_1)}, \quad \forall f \in H^1(M), \tag{99}$$

with a constant $C > 0$ that can be chosen locally uniform in $\mathcal{G}^k(M)$ in the $C^k(M)$ topology.

Of course, $\mathcal{G}^k(M)$ includes all real analytic simple metrics in M, according to Theorem 2.15.

Sketch of the proof The proof of the basic estimate (99) is based on the following ideas. For g of finite smoothness, one can still construct a parametrix Q_g of N_g as above that allows us to reconstruct f^s from $N_g f$ up to smoothing operator terms. This is done in a way similar to that in [179] in two steps: first we invert N_g modulo smoothing operators in a neighborhood M_1 of M, and that gives us $f^s_{M_1}$, i.e., the solenoidal projection of f but associated to the manifold M_1. Next, we compare $f^s_{M_1}$

and f^s and show that one can get the latter from the former by an operator that loses one derivative. This is the same construction as in the proof of Theorem 2.15 above but the metric is only C^k, $k \gg 1$. After applying the parametrix Q_g, the equation for recovering f^s from $N_g f$ is reduced to solving the Fredholm equation

$$(\mathcal{S}_g + K_g)f = Q_g N_g f, \quad f \in \mathcal{S}_g L^2(M) \tag{100}$$

where $\mathcal{S}_g$ is the projection to solenoidal tensors, similarly we denote by $\mathcal{P}_g$ the projection onto potential tensors. Here, K_g is a compact operator on $\mathcal{S}_g L^2(M)$. We can write this as an equation in the whole $L^2(M)$ by adding $\mathcal{P}_g f$ to both sides above to get

$$(I + K_g)f = (Q_g N_g + \mathcal{P}_g)f. \tag{101}$$

Then the solenoidal projection of the solution of (101) solves (100). A finite rank modification of K_g above can guarantee that $I + K_g$ has a trivial kernel, and therefore is invertible, if and only if N_g is s-injective. The problem then reduces to that of invertibility of $I + K_g$. The operators above depend continuously on $g \in C^k$, $k \gg 1$. Since for g analytic, $I + K_g$ is invertible by Theorem 2.15, it would still be invertible in a neighborhood of any analytic g, and estimate (99) is true with a locally uniform constant. Analytic (simple) metrics are dense in the set of all simple metrics, and this completes the sketch of the proof of Theorem 2.16. For more details see [181]. □

The analysis of I_g can also be carried out for symmetric tensors of any order, see e.g., [167, 169]. Since we are motivated by the boundary rigidity problem, and to simplify the exposition, we study only tensors of order 2. Theorem 2.16 and especially estimate (99) allow us to prove the following local generic uniqueness result for the non-linear boundary rigidity problem.

Theorem 2.17 *Let k_0 and $\mathcal{G}^k(M)$ be as in Theorem 2.16. There exists $k \geq k_0$, such that for any $g_0 \in \mathcal{G}^k$, there is $\varepsilon > 0$, such that for any two metrics g_1, g_2 with $\|g_m - g_0\|_{C^k(M)} \leq \varepsilon$, $m = 1, 2$, we have the following:*

$$d_{g_1} = d_{g_2} \text{ on } (\partial M)^2 \text{ implies } g_2 = \psi_* g_1 \tag{102}$$

with some $C^{k+1}(M)$-diffeomorphism $\psi : M \to M$ fixing the boundary.

Sketch of the proof We prove Theorem 2.17 by linearizing and using Theorem 2.16, and especially (99), see also [179]. This requires first to pass to special semigeodesic coordinates related to each metric in which $g_{in} = \delta_{in}, \forall i$. We denote the corresponding pull-backs by g_1, g_2 again. Then we show that if g_1 and g_2 have the same distance on the boundary, then $g_1 = g_2$ on the boundary with all derivatives. As a result, for $f := g_1 - g_2$ we get that $f \in C_0^l(\bar{\Omega})$ with $l \gg 1$, if $k \gg 1$; and $f_{in} = 0, \forall i$. Then we linearize to get

$$\|N_{g_1} f\|_{L^\infty(\Omega_1)} \leq C\|f\|_{C^1}^2,$$

where $\Omega_1 \supset \bar{\Omega}$ is as above. Combine this with (99) and interpolation estimates, to get $\forall \mu < 1$,

$$\|f^s\|_{L^2} \le C\|f\|_{L^2}^{1+\mu}.$$

One can show that tensors satisfying $f_{in} = 0$ also satisfy $\|f\|_{L^2} \le C\|f^s\|_{H^2}$, and using this, and interpolation again, we get

$$\|f\|_{L^2} \le C\|f\|_{L^2}^{1+\mu'}, \quad \mu' > 0.$$

This implies $f = 0$ for $\|f\| \ll 1$. Note that the condition $f \in C_0^l(\bar{\Omega})$ is used to make sure that f, extended as zero in $\Omega_1 \backslash \Omega$, is in $H_0^l(\Omega)$, and then use this fact in the interpolation estimates. Again, for more details see [181]. □

Finally, in [181] it is proven a conditional stability estimate of Hölder type. A similar estimate near the Euclidean metric was proven in [206] based on the approach in [178].

Theorem 2.18 *Let k_0 and $\mathcal{G}^k(M)$ be as in Theorem 2.16. Then for any $\mu < 1$, there exits $k \ge k_0$ such that for any $g_0 \in \mathcal{G}^k$, there is an $\varepsilon_0 > 0$ and $C > 0$ with the property that for any two metrics g_1, g_2 with $\|g_m - g_0\|_{C(M)} \le \varepsilon_0$, and $\|g_m\|_{C^k(M)} \le A$, $m = 1, 2$, with some $A > 0$, we have the following stability estimate*

$$\|g_2 - \psi_* g_1\|_{C^2(M)} \le C(A)\|d_{g_1} - d_{g_2}\|_{C(\partial M \times \partial M)}^{\mu}$$

with some diffeomorphism $\psi : M \to M$ fixing the boundary.

Sketch of the proof To prove Theorem 2.18, we basically follow the uniqueness proof sketched above by showing that each step is stable. The analysis is more delicate near pairs of points too close to each other. An important ingredient of the proof is stability at the boundary, that is also of independent interest: □

Theorem 2.19 *Let g_0 and g_1 be two simple metrics in Ω, and $\Gamma \subset\subset \Gamma' \subset \partial\Omega$ be two sufficiently small open subsets of the boundary. Then for some diffeomorphism ψ fixing the boundary,*

$$\left\| \partial_{x^n}^k (\psi_* g_1 - g_0) \right\|_{C^m(\bar{\Gamma})} \le C_{k,m} \left\| d_{g_1}^2 - d_{g_0}^2 \right\|_{C^{m+2k+2}\left(\overline{\Gamma' \times \Gamma'}\right)},$$

where $C_{k,m}$ depends only on Ω and on an upper bound of g_0, g_1 in $C^{m+2k+5}(\bar{\Omega})$.

Theorem 2.18 can be used to obtain stability near generic simple metrics for the inverse problem of recovering g from the hyperbolic *Dirichlet-to-Neumann map* Λ_g^h. It is known that g can be recovered uniquely from Λ_g, up to a diffeomorphism as above, see e.g. [24]. This result however relies on a unique continuation theorem by Tataru [194] and it is unlikely to provide Hölder type of stability estimate as above. By using the fact that d_g is related to the leading singularities in the kernel of Λ_g^h, it was proven a Hölder stability estimate under the assumptions above, relating g and Λ_g. We refer to [180] for details.

2.4 Lens rigidity

For non-simple manifolds in particular, if we have conjugate points or the boundary is not strictly convex, we need to look at the behavior of all the geodesics and the scattering relation encodes this information. We proceed to define in more detail the scattering relation for non-convex manifolds and the lens rigidity problem and state our results. We note that we will also consider the case of incomplete data, that is when we don't have information about all the geodesics entering the manifold. More details can be found in [182,183].

The scattering relation

$$\Sigma : \partial_+ SM \to \overline{\partial_- SM} \tag{103}$$

is defined by $\Sigma(x, \xi) = (y, \eta) = \Phi^{\ell}(x, \xi)$, where Φ^t is the geodesic flow, and $\ell > 0$ is the first moment, at which the unit speed geodesic through (x, ξ) hits ∂M again. If such an ℓ does not exists, we formally set $\ell = \infty$ and we call the corresponding initial condition and the corresponding geodesic *trapping*. This defines also $\ell(x, \xi)$ as a function $\ell : \partial_+ SM \to [0, \infty]$. Note that Σ and ℓ are not necessarily continuous. This coincides with the scattering relation α defined in Sect. 2.2 for strictly convex manifolds.

It is convenient to think of Σ and ℓ as defined on the whole ∂SM with $\Sigma = Id$ and $\ell = 0$ on $\overline{\partial_- SM}$. We parametrize the scattering relation in a way that makes it independent of pulling it back by diffeomorphisms fixing ∂M pointwise. Let $\kappa_\pm$: $\partial_\pm SM \to B(\partial M)$ be the orthogonal projection onto the (open) unit ball tangent bundle that extends continuously to the closure of $\partial_\pm SM$. Then $\kappa_\pm$ are homeomorphisms, and we set

$$\sigma = \kappa_+ \circ \Sigma \circ \kappa_-^{-1} : \overline{B(\partial M)} \longrightarrow \overline{B(\partial M)}. \tag{104}$$

According to our convention, $\sigma = Id$ on $\partial(\overline{B(\partial M)}) = S(\partial M)$. We equip $\overline{B(\partial M)}$ with the relative topology induced by $T(\partial M)$, where neighborhoods of boundary points (those in $S(\partial M)$) are given by half-neighborhoods, i.e., by neighborhoods in $T(\partial M)$ intersected with $\overline{B(\partial M)}$. It is possible to define σ in a way that does not require knowledge of $g|_{T(\partial M)}$ by thinking of any boundary vector ξ as characterized by its angle with ∂M and the direction of its tangential projection. Let $\mathcal{D}$ be an open subset of $\overline{B(\partial M)}$. A priori, the latter depends on $g|_{T(\partial M)}$. By the remark above, we can think of it as independent of $g|_{T(\partial M)}$ however. The *lens rigidity* problem we study is the following: *Given M and $\mathcal{D}$, do σ and ℓ, restricted to $\mathcal{D}$, determine g uniquely, up to a pull back of a diffeomorphism that is identity on ∂M?* The answer to this question, even when $\mathcal{D} = B(\partial M)$, is negative, see [48]. The known counter-examples are trapping manifolds. The boundary rigidity problem and the lens rigidity one are equivalent for simple metrics.

2.5 Main assumptions

Definition 2.20 We say that $\mathcal{D}$ is **complete** for the metric g, if for any $(z,\zeta) \in T^*M$ there exists a maximal in M, finite length unit speed geodesic $\gamma : [0,l] \to M$ through z, normal to ζ, such that

$$\{(\gamma(t),\dot\gamma(t));\ 0 \le t \le l\} \cap S(\partial M) \subset \mathcal{D}, \tag{105}$$

$$\text{there are no conjugate points on } \gamma. \tag{106}$$

We call the C^k metric g **regular**, if a complete set $\mathcal{D}$ exists, i.e., if $\overline{B(\partial M)}$ is complete.

If $z \in \partial M$ and ζ is conormal to ∂M, then γ may reduce to one point.

Topological Condition (T): Any path in M connecting two boundary points is homotopic to a polygon $c_1 \cup \gamma_1 \cup c_2 \cup \gamma_2 \cup \cdots \cup \gamma_k \cup c_{k+1}$ with the properties that for any j, (i) c_j is a path on ∂M; (ii) $\gamma_j : [0,l_j] \to M$ is a geodesic lying in M^{int} with the exception of its endpoints and is transversal to ∂M at both ends; moreover, $\kappa_-(\gamma_j(0),\dot\gamma_j(0)) \in \mathcal{D}$; Notice that (T) is an open condition w.r.t. g, i.e., it is preserved under small C^2 perturbations of g. To define the $C^K(M)$ norm below in a unique way, we choose and fix a finite atlas on M.

2.5.1 Results about the linear problem

We refer to [182] for more details about the results in this section. It turns out that a linearization of the lens rigidity problem is again the problem of s-injectivity of the ray transform I. Here and below we sometimes drop the subscript g. Given $\mathcal{D}$ as above, we denote by $I_{\mathcal{D}}$ (or $I_{g,\mathcal{D}}$) the ray transform I restricted to the maximal geodesics issued from $(x,\xi) \in \kappa_-^{-1}(\mathcal{D})$. The first result of this section generalizes Theorem 2.15.

Theorem 2.21 *Let g be an analytic, regular metric on M. Let $\mathcal{D}$ be complete and open. Then $I_{\mathcal{D}}$ is s-injective.*

Sketch of the proof Since we know integrals over a subset of geodesics only, this creates difficulties with cut-offs in the phase variable that cannot be analytic. For this reason, the proof of Theorem 2.21 is different from that of Theorem 2.15. Let g be an analytic regular metrics in M, and let $M_1 \Supset M$ be the manifold where g is extended analytically. There is an analytic atlas in M, and ∂M can be assumed to be analytic, too. In other words, now $(M,\partial M,g)$ is a real analytic manifold with boundary. We denote by $\mathcal{A}(M)$ (respectively $\mathcal{A}(M_1)$) the set of analytic functions on M (respectively

M_1). Next, $f^s_{M_1}$ denotes the solenoidal part of the tensor f, extended as zero to M_1, in the manifold M_1. The main step is to show that $I_{\mathcal{D}} f = 0$ implies $f^s \in \mathcal{A}(M)$. In order to do that one shows that $f^s_{M_1} \in \mathcal{A}(M_1)$. Let us first notice, that in $M_1 \backslash M$, $f^s_{M_1} = -dv_{M_1}$, where v_{M_1} satisfies $\delta d v_{M_1} = 0$ in $M_1 \backslash M$, $v|_{\partial M_1} = 0$ since $f = 0$ in $M_1 \backslash M$. Therefore, v_{M_1} is analytic up to ∂M_1. Therefore, we only need to show that $f^s_{M_1}$ is analytic in the interior of M_1. Below, $\mathrm{WF_A}(f)$ stands for the analytic wave front set of f, see [173,195]. The crucial point is the following microlocal analytic regularity result. □

Proposition 2.22 *Let γ_0 be a fixed maximal geodesic in M with endpoints on ∂M, without conjugate points, and let $I_g f(\gamma) = 0$ for $\gamma \in$ neigh(γ_0). Let g be analytic in* neigh(γ_0). *Then*

$$N^*\gamma_0 \cap \mathrm{WF_A}(f^s_{M_1}) = \emptyset. \tag{107}$$

Sketch of the proof Set $f = f^s_{M_1}$. Let U_ε be a tubular neighborhood of γ_0, and $x = (x', x^n)$ be semigeodesic coordinates in it such that $x' = 0$ on γ_0. Fix $x_0 \in \gamma_0 \cap M$. We can assume that $x_0 = 0$ and $g_{ij}(0) = \delta_{ij}$. Then we can assume that $U_\varepsilon = \{-l_1 - \varepsilon < x^n < l_2 + \varepsilon,\ |x'| < \varepsilon\}$ with the part of γ_0 corresponding to $x^n \notin [-l_1, l_2]$ outside M. Fix $\xi^0 = ((\xi^0)', 0)$ with $\xi^0_n = 0$. We will show that

$$(0, \xi^0) \notin \mathrm{WF_A}(f). \tag{108}$$

We choose a local chart for the geodesics close to γ_0. Set first $Z = \{x^n = 0;\ |x'| < 7\varepsilon/8\}$, and denote the x' variable on Z by z'. Then z', θ' (with $|\theta'| \ll 1$) are local coordinates in neigh(γ_0) determined by $(z', \theta') \to \gamma_{(z',0),(\theta',1)}$ where the latter denotes the geodesic through the point $(z', 0)$ in the direction $(\theta', 1)$. Let $\chi_N(z')$ be a smooth cut-off function equal to 1 for $|z'| \le 3\varepsilon/4$ and supported in Z, also satisfying $|\partial^\alpha \chi_N| \le (CN)^{|\alpha|}$, $|\alpha| \le N$. Set $\theta = (\theta', 1)$, $|\theta'| \ll 1$, and multiply

$$I f\left(\gamma_{(z',0),\theta}\right) = 0$$

by $\chi_N(z') e^{\mathrm{i}\lambda z' \cdot \xi'}$, where $\lambda > 0$, ξ' is in a complex neighborhood of $(\xi^0)'$, and integrate w.r.t. z' to get

$$\iint e^{\mathrm{i}\lambda z' \cdot \xi'} \chi_N(z') f_{ij}\left(\gamma_{(z',0),\theta}(t)\right) \dot{\gamma}^i_{(z',0),\theta}(t) \dot{\gamma}^j_{(z',0),\theta}(t)\, \mathrm{d}t\, \mathrm{d}z' = 0. \tag{109}$$

Set $x = \gamma_{(z',0),\theta}(t)$. If $\theta' = 0$, we have $x = (z', t)$. By a perturbation argument, for θ' fixed and small enough, (t, z') are analytic local coordinates, depending analytically on θ'. In particular, $x = (z' + t\theta', t) + O(|\theta'|)$ but this expansion is not enough for the analysis below. Performing a change of variables in (109), we get

$$\int e^{\mathrm{i}\lambda z'(x,\theta') \cdot \xi'} a_N(x, \theta') f_{ij}(x) b^i(x, \theta') b^j(x, \theta')\, \mathrm{d}x = 0 \tag{110}$$

for $|\theta'| \ll 1$, $\forall \lambda$, $\forall \xi'$, where, for $|\theta'| \ll 1$, the function $(x, \theta') \mapsto a_N$ is positive for x in a neighborhood of γ_0, vanishing for $x \notin U_\varepsilon$, and satisfies the same estimate as

χ_N. The vector field b is analytic, and $b(0,\theta') = \theta$, $a_N(0,\theta') = 1$. To clarify the approach, note that if g is Euclidean in neigh(γ_0), then (110) reduces to

$$\int e^{\mathrm{i}\lambda(\xi',-\theta'\cdot\xi')\cdot x}\chi f_{ij}(x)\theta^i\theta^j\,\mathrm{d}x = 0,$$

where $\chi = \chi(x' - x^n\theta')$. Then $\xi = (\xi', -\theta'\cdot\xi')$ is perpendicular to $\theta = (\theta', 1)$. This implies that

$$\int e^{\mathrm{i}\lambda\xi\cdot x}\chi f_{ij}(x)\theta^i(\xi)\theta^j(\xi)\,\mathrm{d}x = 0 \tag{111}$$

for any function $\theta(\xi)$ defined near ξ^0, such that $\theta(\xi)\cdot\xi = 0$. This has been noticed and used before if g is close to the Euclidean metric (with $\chi = 1$), see e.g., [178]. We will assume that $\theta(\xi)$ is analytic. A simple argument (see e.g. [167,178]) shows that a constant symmetric tensor f_{ij} is uniquely determined by the numbers $f_{ij}\theta^i\theta^j$ for finitely many θ's (actually, for $N' = (n+1)n/2$ θ's); and in any open set on the unit sphere, there are such θ's. On the other hand, f is solenoidal. To simplify the argument, assume for a moment that f vanishes on ∂M. Then $\xi^i\widehat{f_{ij}}(\xi) = 0$. Therefore, combining this with (111), we need to choose $N = n(n-1)/2$ vectors $\theta(\xi)$, perpendicular to ξ, that would uniquely determine the tensor $\widehat{f}$ on the plane perpendicular to ξ. To this end, it is enough to know that this choice can be made for $\xi = \xi^0$, then it would be true for $\xi \in$ neigh(ξ^0). This way, $\xi^i\widehat{f_{ij}}(\xi) = 0$ and the N Eqs. (111) with the so chosen $\theta_p(\xi)$, $p = 1,\dots,N$, form a system with a tensor-valued symbol elliptic near $\xi = \xi^0$. The C^∞ ΨDOcalculus easily implies the statement of the lemma in the C^∞ category, and the complex stationary phase method below, or the analytic ΨDOcalculus in [195] with appropriate cut-offs in ξ, implies the lemma in this special case (g locally Euclidean). The general case is considered in [182], and is based on an application of a complex stationary phase argument [173] to (110) as in [107]. □

Proposition 2.22 makes it possible to prove that $f^s \in \mathcal{A}(M)$. We combine this with a boundary determination theorem for tensors, a linear version of Theorem 2.28 below, to conclude that then $f = 0$.

Next, we formulate a stability estimate in the spirit of Theorem 2.16. We need first to parametrize (a complete subset of) the geodesics issued from $\mathcal{D}$ in a different way that would make them a manifold. The parametrization provided by $\mathcal{D}$ is inconvenient near the directions tangent to ∂M. Let H_m be a finite collection of smooth hypersurfaces in M_1^{int}. Let $\mathcal{H}_m$ be an open subset of $\{(z,\theta) \in SM_1;\ z \in H_m, \theta \notin T_z H_m\}$, and let $\pm l_m^{\pm}(z,\theta) \geq 0$ be two continuous functions. Let $\Gamma(\mathcal{H}_m)$ be the set of geodesics

$$\Gamma(\mathcal{H}_m) = \left\{\gamma_{z,\theta}(t);\ l_m^-(z,\theta) \leq t \leq l_m^+(z,\theta),\ (z,\theta) \in \mathcal{H}_m\right\}, \tag{112}$$

that, depending on the context, is considered either as a family of curves, or as a point set. We also assume that each $\gamma \in \Gamma(\mathcal{H}_m)$ is a simple geodesic (no conjugate points). If g is simple, then one can take a single $H = \partial M_1$ with $l^- = 0$ and an appropriate $l^+(z,\theta)$. If g is regular only, and Γ is any complete set of geodesics, then any small enough neighborhood of a simple geodesic in Γ has the properties listed in the paragraph above and by a compactness argument one can choose a finite complete set of such $\Gamma(\mathcal{H}_m)$'s, that is included in the original Γ. Given $\mathcal{H} = \{\mathcal{H}_m\}$ as above, we consider an open set $\mathcal{H}' = \{\mathcal{H}'_m\}$, such that $\mathcal{H}'_m \Subset \mathcal{H}_m$, and let $\Gamma(\mathcal{H}'_m)$ be the associated set of geodesics defined as in (112), with the same $l^\pm_m$. Set $\Gamma(\mathcal{H}) = \cup\Gamma(\mathcal{H}_m)$, $\Gamma(\mathcal{H}') = \cup\Gamma(\mathcal{H}'_m)$. The restriction $\gamma \in \Gamma(\mathcal{H}'_m) \subset \Gamma(\mathcal{H}_m)$ can be modeled by introducing a weight function α_m in $\mathcal{H}_m$, such that $\alpha_m = 1$ on $\mathcal{H}'_m$, and $\alpha_m = 0$ otherwise. More generally, we allow α_m to be smooth but still supported in $\mathcal{H}_m$. We then write $\alpha = \{\alpha_m\}$, and we say that $\alpha \in C^k(\mathcal{H})$, if $\alpha_m \in C^k(\mathcal{H}_m)$, $\forall m$. We consider $I_{\alpha_m} = \alpha_m I$, or more precisely, in the coordinates $(z,\theta) \in \mathcal{H}_m$,

$$I_{\alpha_m} f = \alpha_m(z,\theta) \int\limits_0^{l_m(z,\theta)} \left\langle f(\gamma_{z,\theta}), \dot\gamma^2_{z,\theta}\right\rangle \mathrm{d}t, \quad (z,\theta) \in \mathcal{H}_m. \tag{113}$$

Next, we set

$$I_\alpha = \{I_{\alpha_m}\}, \quad N_{\alpha_m} = I^*_{\alpha_m} I_{\alpha_m} = I^*|\alpha_m|^2 I, \quad N_\alpha = \sum N_{\alpha_m}, \tag{114}$$

where the adjoint is taken w.r.t. the measure $\mathrm{d}\mu := |\langle \nu(z), \theta\rangle|\,\mathrm{d}S_z\,\mathrm{d}\theta$ on $\mathcal{H}_m$, $\mathrm{d}S_z\,\mathrm{d}\theta$ being the induced measure on SM, and $\nu(z)$ being a unit normal to H_m. S-injectivity of N_α is equivalent to s-injectivity for I_α, which in turn is equivalent to s-injectivity of I restricted to $\operatorname{supp}\alpha$.

Theorem 2.23 *(a) Let $g = g_0 \in C^k$, $k \gg 1$ be regular, and let $\mathcal{H}' \Subset \mathcal{H}$ be as above with $\Gamma(\mathcal{H}')$ complete. Fix $\alpha = \{\alpha_m\} \in C^\infty$ with $\mathcal{H}'_m \subset \operatorname{supp}\alpha_m \subset \mathcal{H}_m$. Then if I_α is s-injective, we have*

$$\|f^s\|_{L^2(M)} \leq C\|N_\alpha f\|_{\tilde H^2(M_1)}. \tag{115}$$

(b) Assume that $\alpha = \alpha_g$ in (a) depends on $g \in C^k$, so that $C^k(M_1) \ni g \to C^l(\mathcal{H}) \ni \alpha_g$ is continuous with $l \gg 1$, $k \gg 1$. Assume that $I_{g_0,\alpha_{g_0}}$ is s-injective. Then estimate (115) *remains true for g in a small enough neighborhood of g_0 in $C^k(M_1)$ with a uniform constant $C > 0$.*

The theorem above allows us to formulate a generic result:

Theorem 2.24 *Let $\mathcal{G} \subset C^k(M)$ be an open set of regular Riemannian metrics on M such that (T) is satisfied for each one of them. Let the set $\mathcal{D}' \subset \partial SM$ be open and complete for each $g \in \mathcal{G}$. Then there exists an open and dense subset $\mathcal{G}_s$ of $\mathcal{G}$ such that $I_{g,\mathcal{D}'}$ is s-injective for any $g \in \mathcal{G}_s$.*

Of course, the set $\mathcal{G}_s$ includes all real analytic metrics in $\mathcal{G}$.

Corollary 2.25 *Let $\mathcal{R}(M)$ be the set of all regular C^k metrics on M satisfying (T) equipped with the $C^k(M_1)$ topology. Then for $k \gg 1$, the subset of metrics for which the X-ray transform I over all simple geodesics through all points in M is s-injective, is open and dense in $\mathcal{R}(M)$.*

2.6 Results about the non-linear lens rigidity problem

Using the results above, we prove the following about the lens rigidity problem on manifolds satisfying the assumptions in Sect. 2.5. More details can be found in [183]. Theorem 2.26 below says, loosely speaking, that for the classes of manifolds and metrics we study, the uniqueness question for the non-linear lens rigidity problem can be answered locally by linearization. This is a non-trivial implicit function type of theorem however because our success heavily depends on the a priori stability estimate that the s-injectivity of $I_{\mathcal{D}}$ implies; see Theorem 2.23; and the latter is based on the hypoelliptic properties of $I_{\mathcal{D}}$. We work with two metrics g and $\widehat{g}$; and will denote objects related to $\widehat{g}$ by $\widehat{\sigma}$, $\widehat{\ell}$, etc.

Theorem 2.26 *Let (M, g_0) satisfy the topological assumption (T), with $g_0 \in C^k(M)$ a regular Riemannian metric with $k \gg 1$. Let $\mathcal{D}$ be open and complete for g_0, and assume that there exists $\mathcal{D}' \Subset \mathcal{D}$ so that $I_{g_0,\mathcal{D}'}$ is s-injective. Then there exists $\varepsilon > 0$, such that for any two metrics g, $\widehat{g}$ satisfying*

$$\|g - g_0\|_{C^k(M)} + \|\widehat{g} - g_0\|_{C^k(M)} \le \varepsilon, \tag{116}$$

the relations

$$\sigma = \widehat{\sigma}, \quad \ell = \widehat{\ell} \quad \text{on } \mathcal{D}$$

imply that there is a C^{k+1} diffeomorphism $\psi : M \to M$ fixing the boundary such that

$$\widehat{g} = \psi^* g.$$

By Theorem 2.24, the requirement that $I_{g_0,\mathcal{D}'}$ is s-injective is a generic one for g_0. Therefore, Theorems 2.26 and 2.24 combined imply that there is local uniqueness, up to isometry, near a generic set of regular metrics.

Corollary 2.27 *Let $\mathcal{D}' \Subset \mathcal{D}$, $\mathcal{G}$, $\mathcal{G}_s$ be as in Theorem 2.24. Then the conclusion of Theorem 2.26 holds for any $g_0 \in \mathcal{G}_s$.*

2.6.1 Boundary determination of the jet of g

The first step of the proof of Theorem 2.26 is to determine all derivatives of g on ∂M. The following theorem is interesting by itself. Notice that g below does not need to be analytic or generic.

Theorem 2.28 *Let (M, g) be a compact Riemannian manifold with boundary. Let $(x_0, \xi_0) \in S(\partial M)$ be such that the maximal geodesic γ_{x_0,ξ_0} through it is of finite length, and assume that x_0 is not conjugate to any point in $\gamma_{x_0,\xi_0} \cap \partial M$. If σ and ℓ are known on some neighborhood of (x_0, ξ_0), then the jet of g at x_0 in boundary normal coordinates is determined uniquely.*

Sketch of the proof of Theorem 2.28 To make the arguments below more transparent, assume that the geodesic γ_0 issued from (x_0, ξ_0) hits ∂M for the first time transversally at $\gamma_0(l_0) = y_0, l_0 > 0$. Then y_0 is the only point on ∂M reachable from (x_0, ξ_0), and x_0, y_0 are not conjugate points on γ_0 by assumption. Assume also that γ_0 is tangent of finite order at x_0. Then there is a half neighborhood V of x_0 on ∂M visible from y_0. The latter is not always true if γ_0 is tangent to ∂M of infinite order at x_0. Choose local boundary normal coordinates near x_0 and y_0, and let g_0 be the Euclidean metric in each of them w.r.t. to the so chosen coordinates. We can then consider a representation of Σ, denoted by $\Sigma^\sharp$ below, defined locally on $\mathbb{R}^{n-1} \times S^{n-1}$, with values on another copy of the same space. If $(x, \theta) \in \mathbb{R}^{n-1} \times S^{n-1}$, then the associated vector at $x \in \partial M$ is $\xi = \theta/|\theta|_g$; and $\Sigma^\sharp(x, \theta) = \Sigma(x, \xi)$. The same applies to the second component of $\Sigma^\sharp(x, \theta)$. Namely, if $(y, \eta) = \Sigma(x, \xi)$, then we set $\omega = \eta/|\eta|_{g_0}$, then $\Sigma^\sharp : (x, \theta) \mapsto (y, \omega)$. Similarly, we set $\ell^\sharp(x, \theta) = \ell(x, \xi)$. Let also θ_0 and ω_0 correspond to ξ_0 and η_0, respectively, where $\Sigma(x_0, \xi_0) = (y_0, \eta_0)$. Set $\tau(x) := \tau(x, y_0)$, where τ is the smooth travel time function localized near $x = x_0$ such that $\tau(x_0, y_0) = l_0$. Then τ is well defined in a small neighborhood of x_0 by the implicit function theorem and the assumption that x_0 and y_0 are not conjugate on γ_0. In the normal boundary coordinates $x = (x', x^n)$ near x_0, $g_{in} = \delta_{in}, \forall i$. Since x_0 and y_0 are not conjugate, for $\eta \in S_{y_0}M$ close enough to η_0, the map $\eta \mapsto x \in \partial M$ is a local diffeomorphism as long as the geodesic connecting x and y_0 is not tangent to ∂M at x. Moreover, that map is known, being the inverse of Σ. Similarly, the map $S^{n-1} \ni \omega \mapsto x$ is a local diffeomorphism and is also known. Then we know $(x, -\theta) = \Sigma^\sharp(y_0, -\omega)$, and we know $\ell^\sharp(y_0, -\omega) = \ell^\sharp(x, \theta) = \tau(x)$. Then we can recover $\operatorname{grad}' \tau = -\theta'/|\theta|_g$, where the prime stands for tangential projection as usual. Taking the limit $\omega \to \omega_0$, we recover $|\theta_0|_g^2 = g_{\alpha\beta}\theta_0^\alpha\theta_0^\beta$. We use again the fact that a symmetric $n \times n$ tensor f_{ij} can be recovered by knowledge of $f_{ij}p_k^i p_k^j$ for $N = n(n+1)/2$ "generic" vectors $p_k, k = 1, \dots, N$; and such N vectors exist in any open set on S^{n-1}, see e.g. [183]. Thus choosing appropriate $n(n-1)/2$ perturbations of θ_0's, we recover $g(x_0)$. Thus, we recover g in a neighborhood of x_0 as well; we can assume that V covers that neighborhood. Note that we know all tangential derivatives of g in $V \ni x_0$. Then τ solves the eikonal equation

$$g^{\alpha\beta}\tau_{x^\alpha}\tau_{x^\beta} + \tau_{x^n}^2 = 1. \tag{117}$$

Next, in V, we know τ_{x^α}, $\alpha \leq n-1$, we know g, therefore by (117), we get $\tau_{x^n}^2$. It is easy to see that $\tau_{x^n} \leq 0$ on the visible part, so we recover τ_{x^n} there. We therefore know the tangential derivatives of τ_{x^n} on ∂M near x_0. Differentiate (117) w.r.t. x^n at $x = x_0$ to get

$$\left[\frac{dg^{\alpha\beta}}{dx^n}\tau_{x^\alpha}\tau_{x^\beta} + 2g^{\alpha\beta}\tau_{x^\alpha x^n}\tau_{x^\beta} + 2\tau_{x^n x^n}\tau_{x^n}\right]\Bigg|_{x=x_0} = 0. \tag{118}$$

Since γ_0 is tangent to ∂M at x_0, we have $\tau_{x^n}(x_0) = 0$ by (117). The third term in the l.h.s. of (118) therefore vanishes. Therefore the only unknown term in (118) is $\gamma^{\alpha\beta} := dg^{\alpha\beta}/dx^n$ at $x = x_0$. Since $\tau_{x^\alpha}(x_0) = -\xi_0$, using the fact that $\operatorname{grad}\tau(x_0) = -\xi_0$ again, we get that we have to determine $\gamma^{\alpha\beta}$ from $\gamma_{\alpha\beta}\xi_0^\alpha\xi_0^\beta$. This is possible if as above, we repeat the construction and replace ξ_0 by a finite number of vectors, close enough to ξ_0. So we get an explicit formula for $\partial g/\partial x^i|_{\partial M}$ in fact. Next, for $x \in V$ but not on ∂V, we can recover $\tau_{x^n x^n}(x)$ by (118) because $\tau_{x^n}(x) < 0$. By continuity, we recover $\tau_{x^n x^n}(x_0)$, therefore we know $\tau_{x^n x^n}$ near x_0, and all tangential derivatives of the latter. We differentiate (118) w.r.t. x^n again, and as above, recover $d^2g/d(x^n)^2|_{\partial M}$ near x_0. Then we recover $d^3\tau/d(x^n)^3$, etc. In the general case, we repeat those arguments with ξ_0 replaced by $\xi_0 + \varepsilon\nu$, where ν is the interior unit normal, and take the limit $\varepsilon \to 0$. □

Sketch of the proof of Theorem 2.26 We first find suitable metric $\widehat{g}_1$ isometric to $\widehat{g}$, and then we show that $\widehat{g}_1 = g$. First, we can always assume that g and $\widehat{g}$ have the same boundary normal coordinates near ∂M. By [47], there is a metric h isometric to $\widehat{g}$ so that h is solenoidal w.r.t. g. Moreover, $h = \widehat{g} + O(\varepsilon)$. By a standard argument, by a diffeomorphism that identifies normal coordinates near ∂M for h and g, and is identity away from some neighborhood of the boundary, we find a third $\widehat{g}_1$ isometric to h (and therefore to $\widehat{g}$), so that $\widehat{g}_1 = \widehat{g}$ near ∂M, and $\widehat{g}_1 = h$ away from some neighborhood of ∂M (and there is a region that $\widehat{g}_1$ is neither). Then $\widehat{g}_1 - h$ is as small as $g - h$, more precisely,

$$\|\widehat{g}_1 - h\|_{C^{k-3}} \le C\|g - h\|_{C^{k-1}}, \quad k \gg 1. \tag{119}$$

Set

$$f = h - g, \quad \widetilde{f} = \widehat{g}_1 - g. \tag{120}$$

Estimate (119) implies

$$\|\widetilde{f} - f\|_{C^{l-3}} \le C\|f\|_{C^{l-1}}, \quad \forall l \le k. \tag{121}$$

By (116), (121),

$$\|f\|_{C^{k-1}} \le C\varepsilon, \quad \|\widetilde{f}\|_{C^{k-3}} \le C\varepsilon. \tag{122}$$

By Theorem 2.28,

$$\partial^\alpha \widetilde{f} = 0 \quad \text{on } \partial M \text{ for } |\alpha| \le k - 5. \tag{123}$$

It is known [167] that $2dv$ is the linearization of $\psi_\tau^* g$ at $\tau = 0$, where ψ_τ is a smooth family of diffeomorphisms, and $v = d\psi_\tau/d\tau$ at $\tau = 0$. Next proposition is therefore a version of Taylor's expansion: □

Proposition 2.29 *Let $\widehat{g}$ and g be in C^k, $k \ge 2$ and isometric, i.e.,*

$$\widehat{g} = \psi^* g$$

for some diffeomorphism ψ fixing ∂M. Set $f = \widehat{g} - g$. Then there exists v vanishing on ∂M, so that

$$f = 2dv + f_2,$$

and for g belonging to any bounded set U in C^k, there exists $C(U) > 0$, such that

$$\|f_2\|_{C^{k-2}} \le C(U)\|\psi - Id\|^2_{C^{k-1}}, \quad \|v\|_{C^{k-1}} \le C(U)\|\psi - Id\|_{C^{k-1}}.$$

We will sketch now the rest of the proof of Theorem 2.26. We apply Proposition 2.29 to h and $\widehat{g}_1$ to get

$$\widetilde{f} = f + 2dv + f_2, \quad \|f_2\|_{C^{l-3}} \le C\|f\|^2_{C^{l-1}}, \quad \forall l \le k. \tag{124}$$

In other words, $\widetilde{f}^s = f$ up to $O(\|f\|^2)$. We can assume that g is extended smoothly on $M_1 \Supset M$. Next, with g extended as above, we extend $\widehat{g}_1$ so that $\widehat{g}_1 = g$ outside M. This can be done in a smooth way by Theorem 2.28. The next step is to reparametrize the scattering relation. We show that one can extend the maximal geodesics of g, respectively $\widehat{g}_1$, outside M (where $g = \widehat{g}_1$), and since the two metrics have the same scattering relation and travel times, they will still have the same scattering relation and travel times if we locally push ∂M a bit outside M. Then we can arrange that the new pieces of ∂M are transversal to the geodesics close to a fixed one, which provides a smooth parametrization. By a compactness argument, one can do this near finitely many geodesics issued from point on $\mathcal{D}$, and still have a complete set. This puts as in the situation of Theorem 2.23, where the set of geodesics is parametrized by $\alpha = \{\alpha_j\}$. Next, we linearize the energy functional near each geodesic (in our set of data) related to g. Using the assumption that g and $\widehat{g}_1$ have the same scattering relation and travel times, we deduct

$$\|N_{\alpha_j}\widetilde{f}\|_{L^\infty} \le C\|\widetilde{f}\|^2_{C^1}, \quad \forall j. \tag{125}$$

Using interpolation inequalities, and the fact that the extension of $\widetilde{f}$ outside M is smooth enough across ∂M as a consequence of the boundary recovery, we get by (125), and (121),

$$\|N_\alpha \widetilde{f}\|_{\widetilde{H}^2(M_1)} \le C\|\widetilde{f}\|^{3/2}_{C^3} \le C'\|f\|^{3/2}_{C^5}. \tag{126}$$

Since $I_{g_0,\mathcal{D}'}$ is s-injective, so is N_α, related to g_0, by the support properties of α. Now, since g is close enough to g_0 with s-injective N_α by (116), N_α (the one related to g) is s-injective as well by Theorem 2.23. Therefore, by (126) and (115),

$$\|f^s\|_{L^2(M)} \le C\|N_\alpha \widetilde{f}\|_{\widetilde{H}^2} \le C'\|f\|^{3/2}_{C^5}. \tag{127}$$

A decisive moment of the proof is that by Proposition 2.29, see (124), $\widetilde{f^s} = f + f_2^s$, the latter being the solenoidal projection of f_2. Therefore,

$$\|\widetilde{f^s}\|_{L^2(M)} \geq \|f\|_{L^2(M)} - C\|f\|_{C^2}^2.$$

Together with (127), this yields

$$\|f\|_{L^2(M)} \leq C\left(\|f\|_{C^2}^2 + \|f\|_{C^5}^{3/2}\right) \leq C'\|f\|_{C^5}^{3/2}$$

because the C^5 norm of f is uniformly bounded when $\varepsilon \leq 1$. Using interpolation again, we easily deduct $\|f\|_{L^2(M)} \geq 1/C$ if $f \neq 0$. This contradicts (122) if $\varepsilon \ll 1$. Now, $f = 0$ implies $h = g$, therefore, g and $\widehat{g}$ are isometric. This concludes the sketch proof of Theorem 2.26.

3 Invisibility for electrostatics

We discuss here only invisibility results for electrostatics. For similar results for electromagnetic waves, acoustic waves, quantum waves, etc., see the review papers [61,62] and the references given there. The fact that the boundary measurements do not change, when a conductivity is pushed forward by a smooth diffeomorphism leaving the boundary fixed (see Sect. 1.7), can already be considered as a weak form of invisibility. Different media appear to be the same, and the apparent location of objects can change. However, this does not yet constitute real invisibility, as nothing has been hidden from view. In invisibility cloaking the aim is to hide an object inside a domain by surrounding it with a material so that even the presence of this object can not be detected by measurements on the domain's boundary. This means that all boundary measurements for the domain with this cloaked object included would be the same as if the domain were filled with a homogeneous, isotropic material. Theoretical models for this have been found by applying diffeomorphisms having singularities. These were first introduced in the framework of electrostatics, yielding counterexamples to the anisotropic Calderón problem in the form of singular, anisotropic conductivities in $\mathbb{R}^n$, $n \geq 3$, indistinguishable from a constant isotropic conductivity in that they have the same Dirichlet-to-Neumann map [64,65]. The same construction was rediscovered for electromagnetism in [153], with the intention of actually building such a device with appropriately designed metamaterials; a modified version of this was then experimentally demonstrated in [171]. (See also [126] for a somewhat different approach to cloaking in the high frequency limit.) The first constructions in this direction were based on blowing up the metric around a point [125]. In this construction, let (M, g) be a compact 2-dimensional manifold with non-empty boundary, let $x_0 \in M$ and consider the manifold

$$\widetilde{M} = M \backslash \{x_0\}$$

with the metric

$$\widetilde{g}_{ij}(x) = \frac{1}{d_M(x, x_0)^2} g_{ij}(x),$$

where $d_M(x, x_0)$ is the distance between x and x_0 on (M, g). Then $(\widetilde{M}, \widetilde{g})$ is a complete, non-compact 2-dimensional Riemannian manifold with the boundary $\partial\widetilde{M} = \partial M$. Essentially, the point x_0 has been "pulled to infinity". On the manifolds M and $\widetilde{M}$ we consider the boundary value problems

$$\begin{cases} \Delta_g u = 0 \text{ in } M, \\ u = f \quad \text{on } \partial M, \end{cases} \quad \text{and} \quad \begin{cases} \Delta_{\widetilde{g}} \widetilde{u} = 0 & \text{in } \widetilde{M}, \\ \widetilde{u} = f & \text{on } \partial\widetilde{M}, \\ \widetilde{u} \in L^\infty(\widetilde{M}). \end{cases}$$

These boundary value problems are uniquely solvable and define the DN maps

$$\Lambda_{M,g} f = \partial_\nu u|_{\partial M}, \quad \Lambda_{\widetilde{M},\widetilde{g}} f = \partial_\nu \widetilde{u}|_{\partial\widetilde{M}}$$

where ∂_ν denotes the corresponding conormal derivatives. Since, in the two dimensional case, functions which are harmonic with respect to the metric g stay harmonic with respect to any metric which is conformal to g, one can see that $\Lambda_{M,g} = \Lambda_{\widetilde{M},\widetilde{g}}$. This can be seen using e.g. Brownian motion or capacity arguments. Thus, the boundary measurements for (M, g) and $(\widetilde{M}, \widetilde{g})$ coincide. This gives a counter example for the inverse electrostatic problem on Riemannian surfaces – even the topology of possibly non-compact Riemannian surfaces can not be determined using boundary measurements (see Fig. 1). The above example can be thought as a "hole" in a Riemann surface that does not change the boundary measurements. Roughly speaking, mapping the manifold $\widetilde{M}$ smoothly to the set $M\backslash\overline{B}_M(x_0, \rho)$, where $B_M(x_0, \rho)$ is a metric ball of M, and by putting an object in the obtained hole $\overline{B}_M(x_0, \rho)$, one could hide it from detection at the boundary. This observation was used in [64,65], where "undetectability" results were introduced in three dimensions, using degenerations of Riemannian metrics, whose singular limits can be considered as coming directly from singular changes of variables. The degeneration of the metric (see Fig. 2) can be obtained by considering surfaces (or manifolds in the higher dimensional cases) with a thin "neck" that is pinched. At the limit the manifold contains a pocket about which the boundary measurements do not give any information. If the collapsing of the manifold is done in an appropriate way, we have, in the limit, a singular Riemannian manifold which is indistinguishable in boundary measurements from a flat surface. Then the conductivity which corresponds to this metric is also singular at the pinched points, cf. the first formula in (130). The electrostatic measurements on the boundary for this singular conductivity will be the same as for the original regular conductivity corresponding to the metric g. To give a precise, and concrete, realization of this idea, let $B(0, R) \subset \mathbb{R}^3$ denote the open ball with center 0 and radius R. We use in the sequel the set $N = B(0, 2)$, the region at the boundary of which the electrostatic measurements will be made, decomposed into two parts, $N_1 = B(0, 2)\backslash\overline{B}(0, 1)$ and

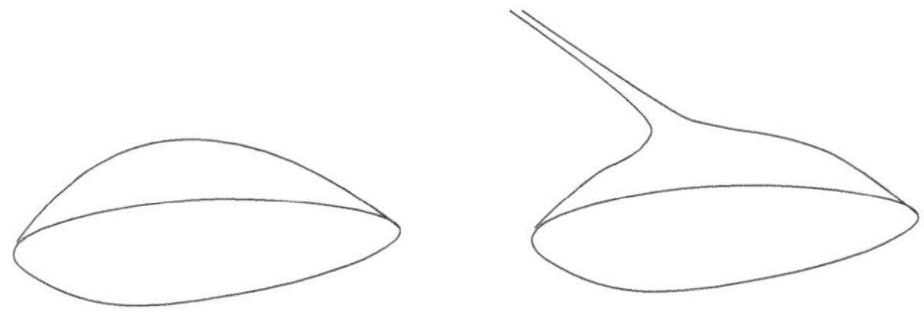

Fig. 1 Blowing up a metric at a point, after [125]. The electrostatic boundary measurements on the boundary of the surfaces, one compact and the other noncompact but complete, coincide

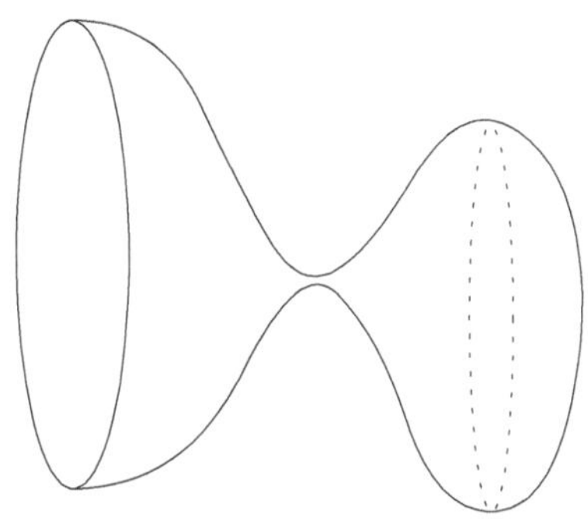

Fig. 2 A typical member of a family of manifolds developing a singularity as the width of the neck connecting the two parts goes to zero

$N_2 = B(0,1)$. We call the interface $\Sigma = \partial N_2$ between N_1 and N_2 the *cloaking surface*. We also use a "copy" of the ball $B(0,2)$, with the notation $M_1 = B(0,2)$, another ball $M_2 = B(0,1)$, and the disjoint union M of M_1 and M_2. (We will see the reason for distinguishing between N and M.) Let $g_{jk} = \delta_{jk}$ be the Euclidian metrics in M_1 and M_2 and let $\gamma = 1$ be the corresponding isotropic homogeneous conductivity. We define a singular transformation

$$F_1 : M_1 \backslash \{0\} \to N_1, \quad F_1(x) = \left(\frac{|x|}{2} + 1\right)\frac{x}{|x|}, \quad 0 < |x| \le 2, \tag{128}$$

(see Fig. 3).

We also consider a regular transformation (diffeomorphism) $F_2 : M_2 \to N_2$, which for simplicity we take to be the identity map $F_2 = Id$. Considering the maps F_1 and F_2 together, $F = (F_1, F_2)$, we define a map $F : M\backslash\{0\} = (M_1\backslash\{0\}) \cup M_2 \to N\backslash\Sigma$. The push-forward $\widetilde{g} = F_* g$ of the metric g in M by F is the metric in N given by

$$(F_* g)_{jk}(y) = \left. \sum_{p,q=1}^{n} \frac{\partial F^p}{\partial x^j}(x) \frac{\partial F^q}{\partial x^k}(x) g_{pq}(x) \right|_{x=F^{-1}(y)}. \tag{129}$$

This metric gives rise to a conductivity $\widetilde{\sigma}$ in N which is singular in N_1,

$$\widetilde{\sigma} = \begin{cases} |\widetilde{g}|^{1/2} \widetilde{g}^{jk} & \text{for } x \in N_1, \\ \delta^{jk} & \text{for } x \in N_2. \end{cases} \tag{130}$$

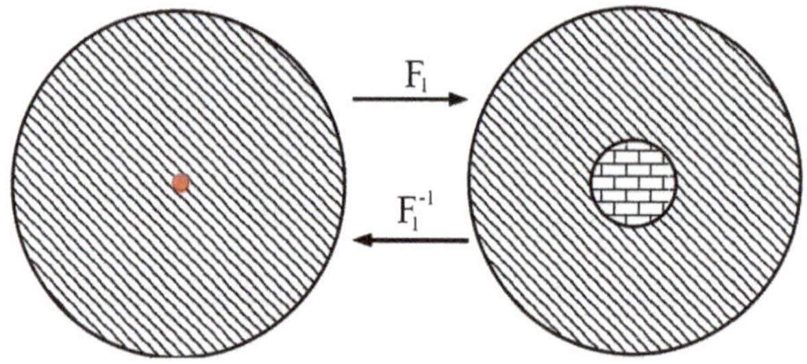

Fig. 3 Map $F_1 : B(0, 2)\backslash\{0\} \to B(0, 2)\backslash\overline{B}(0, 1)$

Thus, F forms an invisibility construction that we call "blowing up a point". Denoting by $(r, \phi, \theta) \mapsto (r \sin\theta \cos\phi, r \sin\theta \sin\phi, r \cos\theta)$ the spherical coordinates, we have

$$\widetilde{\sigma} = \begin{pmatrix} 2(r-1)^2 \sin\theta & 0 & 0 \\ 0 & 2\sin\theta & 0 \\ 0 & 0 & 2(\sin\theta)^{-1} \end{pmatrix}, \quad 1 < |x| \leq 2. \tag{131}$$

Note that the anisotropic conductivity $\widetilde{\sigma}$ is singular degenerate on Σ in the sense that it is not bounded from below by any positive multiple of I. (See [112] for a similar calculation.) The Euclidian conductivity δ^{jk} in N_2 (130) could be replaced by any smooth conductivity bounded from below and above by positive constants. This would correspond to cloaking of a general object with non-homogeneous, anisotropic conductivity. Here, we use the Euclidian metric just for simplicity. Consider now the *Cauchy data* of all solutions in the Sobolev space $H^1(N)$ of the conductivity equation corresponding to $\widetilde{\sigma}$, that is,

$$C_1(\widetilde{\sigma}) = \{(u|_{\partial N}, \nu \cdot p\widetilde{\sigma}\nabla u|_{\partial N}) : u \in H^1(N), \nabla \cdot p\widetilde{\sigma}\nabla u = 0\},$$

where ν is the Euclidian unit normal vector of ∂N.

Theorem 3.1 ([65]) *The Cauchy data of all H^1-solutions for the conductivities $\widetilde{\sigma}$ and γ on N coincide, that is, $C_1(\widetilde{\sigma}) = C_1(\gamma)$.*

This means that all boundary measurements for the homogeneous conductivity $\gamma = 1$ and the degenerated conductivity $\widetilde{\sigma}$ are the same. The result above was proven in [64,65] for the case of dimension $n \geq 3$. The same basic construction works in the two dimensional case [112]. Figure 4 portrays an analytically obtained solution on a disc with conductivity $\widetilde{\sigma}$. As seen in the figure, no currents appear near the center of the disc, so that if the conductivity is changed near the center, the measurements on the boundary ∂N do not change. The above invisibility result is valid for a more general class of singular cloaking transformations. A general class, sufficing at least for electrostatics, is given by the following result from [65]:

Theorem 3.2 *Let $\Omega \subset \mathbb{R}^n$, $n \geq 3$, and $g = (g_{ij})$ a smooth metric on Ω bounded from above and below by positive constants. Let $D \subset\subset \Omega$ be such that there is a C^∞-diffeomorphism $F : \Omega\backslash\{y\} \to \Omega\backslash\overline{D}$ satisfying $F|_{\partial\Omega} = Id$ and such that*

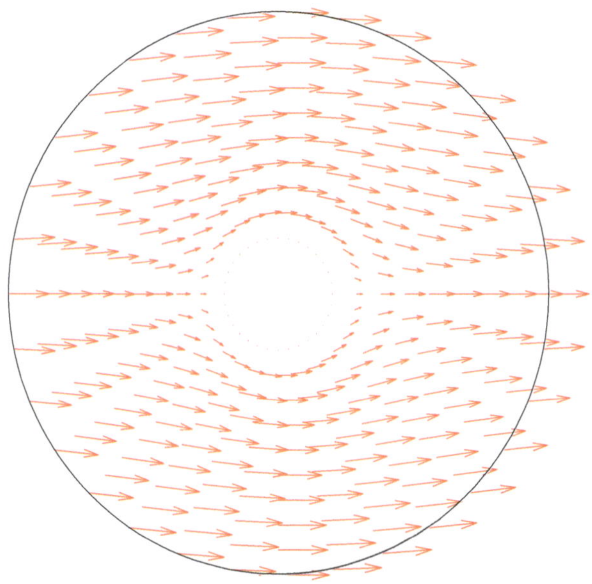

Fig. 4 Analytic solutions for the currents

$$dF(x) \geq c_0 I, \quad det(dF(x)) \geq c_1 \, dist_{\mathbb{R}^n}(x, y)^{-1} \tag{132}$$

where dF is the Jacobian matrix in Euclidian coordinates on $\mathbb{R}^n$ and $c_0, c_1 > 0$. Let $\widehat{g}$ be a metric in Ω which coincides with $\widetilde{g} = F_* g$ in $\Omega \backslash \overline{D}$ and is an arbitrary regular positive definite metric in D^{int}. Finally, let σ and $\widehat{\sigma}$ be the conductivities corresponding to g and $\widehat{g}$, cf. (88)*. Then,*

$$C_1(\widehat{\sigma}) = C_1(\sigma).$$

The key to the proof of Theorem 3.2 is a removable singularities theorem that implies that solutions of the conductivity equation in $\Omega \backslash \overline{D}$ pull back by this singular transformation to solutions of the conductivity equation in the whole Ω. Returning to the case $\Omega = N$ and the conductivity given by (130), similar types of results are valid also for a more general class of solutions. Consider an unbounded quadratic form, A in $L^2(N, |\widetilde{g}|^{1/2} dx)$,

$$A_{\widetilde{\sigma}}[u, v] = \int_N \widetilde{\sigma} \nabla u \cdot p \nabla v \, dx$$

defined for $u, v \in \mathcal{D}(A_{\widetilde{\sigma}}) = C_0^\infty(N)$. Let $\overline{A_{\widetilde{\sigma}}}$ be the closure of this quadratic form and say that

$$\nabla \cdot p \widetilde{\sigma} \nabla u = 0 \quad \text{in } N$$

is satisfied in the finite energy sense if there is $u_0 \in H^1(N)$ supported in N_1 such that $u - u_0 \in \mathcal{D}(\overline{A}_{\widetilde{\sigma}})$ and

$$\overline{A}_{\widetilde{\sigma}}[u-u_0, v] = -\int_N \widetilde{\sigma}\nabla u_0 \cdot p\nabla v\, dx, \quad \text{for all } v \in \mathcal{D}(\overline{A}_{\widetilde{\sigma}}).$$

Then the Cauchy data set of the finite energy solutions, denoted by

$$C_{f.e.}(\widetilde{\sigma}) = \Big\{(u|_{\partial N}, \nu \cdot p\widetilde{\sigma}\nabla u|_{\partial N}) \,|u \text{ is a finite energy solution of } \nabla \cdot p\widetilde{\sigma}\nabla u = 0\Big\},$$

coincides with the Cauchy data $C_{f.e.}(\gamma)$ corresponding to the homogeneous conductivity $\gamma = 1$, that is,

$$C_{f.e.}(\widetilde{\sigma}) = C_{f.e.}(\gamma). \tag{133}$$

Kohn, Shen, Vogelius and Weinstein [112] in an interesting article have considered the case when instead of blowing up a point one stretches a small ball into the cloaked region. In this case the conductivity is non-singular and one gets “almost” invisibility with a precise estimate in terms of the radius of the small ball.

3.1 Quantum shielding

In [63], using CGO solutions, uniqueness was proven for the Calderón problem for Schrödinger operators having a more singular class of potentials, namely potentials conormal to submanifolds of $\mathbb{R}^n$, $n \geq 3$. However, for more singular potentials, there are counterexamples to uniqueness. It was constructed in [63] a class of potentials that shield any information about the values of a potential on a region D contained in a domain Ω from measurements of solutions at $\partial\Omega$. In other words, the boundary information obtained outside the shielded region is independent of $q|_D$. On $\Omega \setminus D$, these potentials behave like $q(x) \sim -Cd(x, \partial D)^{-2-\epsilon}$ where d denotes the distance to ∂D and C is a positive constant. In D, Schrödinger’s cat could live forever. From the point of view of quantum mechanics, q represents a potential barrier so steep that no tunneling can occur. From the point of view of optics and acoustics, no sound waves or electromagnetic waves will penetrate, or emanate from, D. However, this construction should be thought of as shielding, not cloaking, since the potential barrier that shields $q|_D$ from boundary observation is itself detectable .

References

1. Ablowitz, M., Yaacov, D.B., Fokas, A.: On the inverse scattering transform for the Kadomtsev–Petviashvili equation. Stud. Appl. Math. **69**, 135–143 (1983)
2. Ahlfors, L.: Quasiconformal Mappings. Van Nostrand, Princeton (1966)
3. Albin, P., Guillarmou, C., Tzou, L., Uhlmann, G.: Inverse boundary problems for systems in two dimensions. Annales Institut Henri Poincaré **14**, 1151–1571 (2013)
4. Alessandrini, G.: Stable determination of conductivity by boundary measurements. App. Anal. **27**, 153–172 (1988)
5. Alessandrini, G.: Singular solutions of elliptic equations and the determination of conductivity by boundary measurements. J. Differ. Equ. **84**, 252–272 (1990)
6. Alessandrini, G.: Open issues of stability for the inverse conductivity problem. J. Inverse Ill-Posed Probl. **15**, 451–460 (2007)
7. Alessandrini, G., Vessella, S.: Lipschitz stability for the inverse conductivity problem. Adv. Appl. Math. **35**, 207–241 (2005)
8. Alexandrova, I.: Structure of the semi-classical amplitude for general scattering relations. Comm. PDE **30**, 1505–1535 (2005)
9. Ammari, H., Uhlmann, G.: Reconstruction of the potential from partial Cauchy data for the Schrödinger equation. Indiana Univ. Math. J. **53**, 169–183 (2004)
10. Anikonov, YuE: Some Methods for the Study of Multidimensional Inverse Problems. Nauka. Sibirsk Otdel, Novosibirsk (1978)
11. Astala, K., Päivärinta, L.: Calderón's inverse conductivity problem in the plane. Ann. Math. **163**, 265–299 (2006)
12. Astala, K., Lassas, M., Päiväirinta, L.: Calderón's inverse problem for anisotropic conductivity in the plane. Comm. Partial Differ. Equ. **30**, 207–224 (2005)
13. Bal, G.: Hybrid inverse problems and internal functionals. In: Uhlmann, G. (ed.) Chapter in Inside Out II, MSRI Publications, vol. 60, pp. 271–323. Cambridge University Press, Cambridge (2012)
14. Bal, G., Langmore, I., Monard, F.: Inverse transport with isotropic sources and angularly averaged measurements. Inverse Probl. Imaging **2**, 23–42 (2008)
15. Bal, G., Ren, K., Uhlmann, G., Zhou, T.: Quantitative thermo-acoustics and related problems. Inverse Probl. **27**, 055007 (2011)
16. Bal, G., Uhlmann, G.: Inverse diffusion theory of photoacoustics. Inverse Probl. **26**, 085010 (2010)
17. Bal, G., Uhlmann, G.: Reconstructions for some coupled-physics inverse problems. Appl. Math. Lett. **25**, 1030–1033 (2012)
18. Bal, G., Uhlmann, G.: Reconstructions of coefficients in scalar second-order elliptic equations from knowledge of their solutions. Comm. Pure Appl. Math. **66**, 1629–1652
19. Barber, D., Brown, B.: Applied potential tomography. J. Phys. E **17**, 723–733 (1984)
20. Barceló, T., Faraco, D., Ruiz, A.: Stability of Calderón's inverse problem in the plane. Journal des Mathématiques Pures et Appliquées **88**, 522–556 (2007)
21. Beals, R., Coifman, R.: Transformation spectrales et equation d'evolution non lineares. Seminaire Goulaouic-Meyer-Schwarz, exp. 21 (1981–1982)
22. Beals, R., Coifman, R.: Multidimensional inverse scattering and nonlinear PDE. Proc. Symp. Pure Math., vol. 43, pp. 45–70. American Mathematical Society, Providence (1985).
23. Belishev, M.I.: The Calderón problem for two-dimensional manifolds by the BC-method. SIAM J. Math. Anal. **35**, 172–182 (2003)
24. Belishev, M., Kurylev, Y.: To the reconstruction of a Riemannian manifold via its spectral data (BC-method). Comm. Partial Differ. Equ. **17**, 767–804 (1992)
25. Bernstein, I.N., Gerver, M.L.: Conditions on distinguishability of metrics by hodographs. In: Methods and Algorithms of Interpretation of Seismological Information, Computerized Seismology, vol. 13, pp. 50–73. Nauka, Moscow (1980) (in Russian)
26. Besson, G., Courtois, G., Gallot, S.: Entropies et rigidités des espaces localement symétriques de courbure strictment négative. Geom. Funct. Anal. **5**, 731–799 (1995)
27. Beylkin, G.: Stability and uniqueness of the solution of the inverse kinematic problem in the multidimensional case. J. Soviet Math. **21**, 251–254 (1983)
28. Blasten, E.: Stability and uniqueness for the inverse problem of the Schrödinger equation with potentials in $W^{p,\epsilon}$.
29. Borcea, L.: Electrical impedance tomography. Inverse Probl. **18**, R99–R136 (2002)

30. Borcea, L., Druskin, V., Guevara Vasquez, F., Mamonov, A.V.: Resistor network approaches to electrical impedance tomography. In: Uhlmann, G. (ed.) Inside Out II. MSRI Publications, vol. 60, pp. 55–118 (2012).
31. Brown, R.: Recovering the conductivity at the boundary from the Dirichlet to Neumann map: a pointwise result. J. Inverse Ill-Posed Probl. **9**, 567–574 (2001)
32. Brown, R., Torres, R.: Uniqueness in the inverse conductivity problem for conductivities with $3/2$ derivatives in L^p, $p > 2n$. J. Fourier Analysis Appl. **9**, 1049–1056 (2003)
33. Brown, R., Uhlmann, G.: Uniqueness in the inverse conductivity problem with less regular conductivities in two dimensions. Comm. PDE **22**, 1009–10027 (1997)
34. Bukhgeim, A.: Recovering the potential from Cauchy data in two dimensions. J. Inverse Ill-Posed Probl. **16**, 19–34 (2008)
35. Bukhgeim, A., Uhlmann, G.: Recovering a potential from partial Cauchy data. Comm. PDE **27**, 653–668 (2002)
36. Burago, D., Ivanov, S.: Boundary rigidity and filling volume minimality for metrics close to a Euclidean metric. Annals of Math. **171**, 1183–1211 (2010)
37. Calderón, A.P.: On an inverse boundary value problem. In: Seminar on Numerical Analysis and its Applications to Continuum Physics (Rio de Janeiro, 1980), pp. 65–73. Soc. Brasil. Mat., Rio de Janeiro (1980)
38. Calderón, A.P.: Reminiscencias de mi vida matemática, Discurso de investidura de "Doctor Honoris Causa", pp. 117–125. Publicaciones UAM, Universidad Autónoma de Madrid (1997)
39. Calderón, A.P.: Boundary value problems for elliptic equations. Outlines of the joint Soviet-American symposium on partial differential equations, pp. 303–304, Novisibirsk (1963)
40. Caro, P., Ola, P., Salo, M.: Inverse boundary value problem for Maxwell equations with local data. Comm. PDE **34**, 1425–1464 (2009)
41. Chanillo, S.: A problem in electrical prospection and a n-dimensional Borg-Levinson theorem. Proc. AMS **108**, 761–767 (1990)
42. Chen, J., Yang, Y.: Quantitative photo-acoustic tomography with partial data. Inverse Probl. **28**, 115014 (2012)
43. Cheney, M., Isaacson, D., Newell, J.C.: Electrical impedance tomography. SIAM Rev. **41**, 85–101 (1999)
44. Creager, K.C.: Anisotropy of the inner core from differential travel times of the phases PKP and PKIPK. Nature **356**, 309–314 (1992)
45. Croke, C.: Rigidity and the distance between boundary points. J. Differ. Geom. **33**, 445–464 (1991)
46. Croke, C.: Rigidity for surfaces of non-positive curvature. Comment. Math. Helv. **65**, 150–169 (1990)
47. Croke, C., Dairbekov, D., Sharafutdinov, V.: Local boundary rigidity of a compact Riemannian manifold with curvature bounded above. Trans. Amer. Math. Soc. **352**(9), 3937–3956 (2000)
48. Croke, C., Kleiner, B.: Conjugacy and rigidity for manifolds with a parallel vector field. J. Differ. Geom. **39**, 659–680 (1994)
49. Dos Santos Ferreira, D., Kenig, C.E., Sjöstrand, J., Uhlmann, G.: On the linearized local Calderón problem. G. Math. Res. Lett. **16**, 955–970 (2009)
50. Dairbekov, N., Uhlmann, G.: Reconstructing the metric and magnetic field from the scattering relation. Inverse Probl. Imaging **4**, 397–409 (2010)
51. Ferreira, Dos Santos: D., Kenig, C.E., Sjöstrand, J., Uhlmann, G.: Determining a magnetic Schrödinger operator from partial Cauchy data. Comm. Math. Phys. **271**, 467–488 (2007)
52. Dos Santos Ferreira, D., Kenig, C.E., Salo, M., Uhlmann, G.: Limiting Carleman weights and anisotropic inverse problems. Inventiones Math. **178**, 119–171 (2009)
53. Duistermaat, J.J., Hörmander, L.: Fourier integral operators II. Acta Math. **128**, 183–269 (1972)
54. Eisenhart, L.: Riemannian geometry, 2nd printing. Princeton University Press, Princeton (1949)
55. Eskin, G., Ralston, J.: On the inverse boundary value problem for linear isotropic elasticity. Inverse Probl. **18**, 907–921 (2002)
56. Faddeev D.: Growing solutions of the Schrödinger equation. Dokl. Akad. Nauk SSSR **165**, 514–517 (1965) (translation in, Sov. Phys. Dokl. 10, 1033)
57. Francini, E.: Recovering a complex coefficient in a planar domain from the Dirichlet-to-Neumann map. Inverse Probl. **16**, 107–119 (2000)
58. Fridman, B., Kuchment, P., Ma, D., Papanicolaou, V.G.: Solution of the linearized inverse conductivity problem in a half space via integral geometry. Voronezh Winter Mathematical Schools, pp. 85–95.

Amer. Math. Soc. Transl. Ser. 2, 184, pp. 85–95. American Mathematical Society, Providence, RI (1998)
59. Gilbarg, D., Trudinger, N.: Elliptic Partial Differential Equations. Interscience Publishers, New York (1964)
60. Gerver, M.L., Nadirashvili, N.S.: An isometricity conditions for Riemannian metrics in a disk. Soviet Math. Dokl. **29**, 199–203 (1984)
61. Greenleaf, A., Kurylev, Y., Lassas, M., Uhlmann, G.: Cloaking devices, electromagnetic wormholes, and transformation optics. SIAM Rev. **51**, 3–33 (2009)
62. Greenleaf, A., Kurylev, Y., Lassas, M., Uhlmann, G.: Invisibility and inverse problems. Bull. Amer. Math. Soc. (N.S.) **46**, 55–97 (2009)
63. Greenleaf, A., Lassas, M., Uhlmann, G.: The Calderón problem for conormal potentials, I: Global uniqueness and reconstruction. Comm. Pure Appl. Math **56**, 328–352 (2003)
64. Greenleaf, A., Lassas, M., Uhlmann, G.: Anisotropic conductivities that cannot be detected in EIT. Physiolog. Meas. (special issue on Impedance Tomography) **24**, 413–420 (2003)
65. Greenleaf, A., Lassas, M., Uhlmann, G.: On nonuniqueness for Calderón's inverse problem. Math. Res. Lett. **10**, 685–693 (2003)
66. Greenleaf, A., Uhlmann, G.: Local uniqueness for the Dirichlet-to-Neumann map via the two-plane transform. Duke Math. J. **108**, 599–617 (2001)
67. Gromov, M.: Filling Riemannian manifolds. J. Differ. Geom. **18**(1), 1–148 (1983)
68. Inverse problems for Einstein manifolds: Guillarmou, C., Sá Barreto, A. Inverse Probl. Imag. **3**, 1–15 (2009)
69. Guillarmou, C., Tzou, L.: Calderón inverse problem on Riemann surfaces. In: Proceedings of CMA, vol. 44, pp. 129–142. Volume for the AMSI/ANU workshop on Spectral Theory and Harmonic, Analysis (2009)
70. Guillarmou, C., Tzou, L.: Calderón inverse problem with partial data on Riemann surfaces. Duke Math. J. **158**, 83–120 (2011)
71. Guillarmou, C., Tzou, L.: Identification of a connection from Cauchy data space on a Riemann surface with boundary. Geom. Funct. Anal. (GAFA) **21**, 393–418 (2011)
72. Guillemin, V.: Sojourn times and asymptotic properties of the scattering matrix. In: Proceedings of the Oji Seminar on Algebraic Analysis and the RIMS Symposium on Algebraic Analysis (Kyoto Univ., Kyoto, 1976). Publ. Res. Inst. Math. Sci., vol. 12, supplement, pp. 69–88 (1976/77).
73. Hähner, P.: A periodic Faddeev-type solution operator. J. Differ. Equ. **128**, 300–308 (1996)
74. Hanke, M., Brühl, M.: Recent progress in electrical impedance tomography. Special Sect. Imaging Inverse Probl. **19**, S65–S90 (2003)
75. Haberman, B., Tataru, D.: Uniqueness in Calderón's problem with Lipschitz conductivities. Duke Math. J. **162**, 497–516 (2013)
76. Heck, H., Wang, J.-N.: Stability estimates for the inverse boundary value problem by partial Cauchy data. Inverse Probl. **22**, 1787–1796 (2006)
77. Henkin, G., Michel, V.: Inverse conductivity problem on Riemann surfaces. J. Geom. Anal. **18**, 1033–1052 (2008)
78. Herglotz, G.: Uber die elastizitaet derErde bei beruecksichtigung ihrer variablen dichte. Zeitschr. fur Math. Phys. **52**, 275–299 (1905)
79. Holder, D.: Electrical Impedance Tomography. Institute of Physics Publishing, Bristol and Philadelphia (2005)
80. Holder, D., Isaacson, D., Müller, J., Siltanen, S. (eds.) Physiol. Meas. 25(1) (2003)
81. Hörmander, L.: The Analysis of Linear Partial Differential Operators, vol. I. Springer, Berlin (1983)
82. Ide, T., Isozaki, H., Nakata, S., Siltanen, S., Uhlmann, G.: Probing for electrical inclusions with complex spherical waves. Comm. Pure Appl. Math. **60**, 1415–1442 (2007)
83. Ikehata, M.: The enclosure method and its applications. In: Analytic extension formulas and their applications, Chapter 7. (Fukuoka, 1999/Kyoto, 2000). Int. Soc. Anal. Appl. Comput., vol. 9, pp. 87–103. Kluwer Academic Publishers, Boston (2001)
84. Ikehata, M.: How to draw a picture of an unknown inclusion from boundary measurements. Two mathematical inversion algorithms. J. Inverse Ill-Posed Probl. **7**, 255–271 (1999)
85. Ikehata, M., Siltanen, S.: Numerical method for finding the convex hull of an inclusion in conductivity from boundary measurements. Inverse Probl. **16**, 273–296 (2000)
86. Imanuvilov, O., Yamamoto, M.: Inverse boundary value for Schrödinger equation in two dimensions, arXiv.

87. Imanuvilov, O., Uhlmann, G., Yamamoto, M.: The Calderón problem with partial data in two dimensions. J. AMS **23**, 655–691 (2010)
88. Imanuvilov, O., Uhlmann, G., Yamamoto, M.: On determination of second order operators from partial Cauchy data. Proc. Natl. Acad. Sci. **108**, 467–472 (2011)
89. Imanuvilov, O., Uhlmann, G., Yamamoto, M.: Partial data for general second order elliptic operators in two dimensions. Publ. Res. Inst. Math. Sci. **48**, 971–1055 (2012)
90. Imanuvilov, O., Uhlmann, G., Yamamoto, M.: Inverse boundary problem with Cauchy data on disjoint sets. Inverse Probl. **27**, 085007 (2011)
91. Imanuvilov, O., Uhlmann, G., Yamamoto, M.: On reconstruction of Lamé coefficients from partial Cauchy data in three dimensions. Inverse Probl. **28**, 125002 (2012)
92. Isaacson, D., Isaacson, E.: Comment on Calderón's paper: "On an inverse boundary value problem". Math. Comput. **52**, 553–559 (1989)
93. Isaacson, D., Müller, J.L., Newell, J.C., Siltanen, S.: Reconstructions of chest phantoms by the d-bar method for electrical impedance tomography. IEEE Trans. Med. Imaging **23**, 821–828 (2004)
94. Isaacson, D., Newell, J.C., Goble, J.C., Cheney, M.: Thoracic impedance images during ventilation. Ann. Conf. IEEE Eng. Med. Biol. Soc. **12**, 106–107 (1990)
95. Isakov, V.: On uniqueness in the inverse conductivity problem with local data. Inverse Probl. Imaging **1**, 95–105 (2007)
96. Isakov, V.: On uniqueness in inverse problems for semilinear parabolic equations. Arch. Rat. Mech. Anal. **124**, 1–12 (1993)
97. Isakov, V.: Completeness of products of solutions and some inverse problems for PDE. J. Differ. Equ. **92**, 305–317 (1991)
98. Isakov, V., Nachman, A.: Global uniqueness for a two-dimensional semilinear elliptic inverse problem. Trans. AMS **347**, 3375–3390 (1995)
99. Isakov, V., Sylvester, J.: Global uniqueness for a semilinear elliptic inverse problem. Comm. Pure Appl. Math. **47**, 1403–1410 (1994)
100. Isozaki, H.: Inverse spectral problems on hyperbolic manifolds and their applications to inverse boundary value problems in Euclidean space. Amer. J. Math. **126**, 1261–1313 (2004)
101. Isozaki, H., Uhlmann, G.: Hyperbolic geometric and the local Dirichlet-to-Neumann map. Adv. Math. **188**, 294–314 (2004)
102. Jordana, J., Gasulla, J.M., Paola's-Areny, R.: Electrical resistance tomography to detect leaks from buried pipes. Meas. Sci. Technol. **12**, 1061–1068 (2001)
103. Jossinet, J.: The impedivity of freshly excised human breast tissue. Physiol. Meas. **19**, 61–75 (1998)
104. Kang, H., Uhlmann, G.: Inverse problems for the Pauli Hamiltonian in two dimensions. J. Fourier Anal. Appl. **10**, 201–215 (2004)
105. Kashiwara, M.: On the structure of hyperfunctions. Sagaku no Ayumi **15**, 19–72 (1970). (in Japanese)
106. Kenig, C., Salo, M., Uhlmann, G.: Inverse Problems for the Anisotropic Maxwell's Equations. Duke Math. J. **157**, 369–419 (2011)
107. Kenig, C., Sjöstrand, J., Uhlmann, G.: The Calderón problem with partial data. Ann. Math. **165**, 567–591 (2007)
108. Knudsen, K.: The Calderón problem with partial data for less smooth conductivities. Comm. Partial Differ. Equ. **31**, 57–71 (2006)
109. Knudsen, K., Salo, M.: Determining nonsmooth first order terms from partial boundary measurements. Inverse Probl. Imaging **1**, 349–369 (2007)
110. Kocyigit, I.: Acoustic-electric tomography and CGO solutions with internal data. Inverse Probl. **28**, 125004 (2012)
111. Kolehmainen, V., Lassas, M., Ola, P.: Inverse conductivity problem with an imperfectly known boundary. SIAM J. Appl. Math. **66**, 365–383 (2005)
112. Kohn, R., Shen, H., Vogelius, M., Weinstein, M.: Cloaking via change of variables in electrical impedance tomography. Inverse Probl. **24**, 015016 (2008)
113. Kohn, R., Vogelius, M.: Identification of an unknown conductivity by means of measurements at the boundary. Inverse Probl. SIAM-AMS Proc. 14 (1984)
114. Kohn, R., Vogelius, M.: Determining conductivity by boundary measurements. Comm. Pure Appl. Math. **37**, 289–298 (1984)
115. Kohn, R., Vogelius, M.: Determining conductivity by boundary measurements II. Interior results. Comm. Pure Appl. Math. **38**, 643–667 (1985)

116. Krupchyk, K., Lassas, M., Uhlmann, G.: Inverse problems for differential forms on Riemannian manifolds with boundary". Comm. PDE. **36**, 1475–1509 (2011)
117. Krupchyk, K., Lassas, M., Uhlmann, G.: Inverse problems with partial data for the magnetic Schrödinger operator in an infinite slab and on a bounded domain Comm. Math. Phys. **312**, 87–126 (2012)
118. Krupchyk, K., Lassas, M., Uhlmann, G.: Inverse boundary value problems for the polyharmonic operator. J. Funct. Anal. **262**, 1781–1801 (2012)
119. Krupchyk, K., Lassas, M., Uhlmann, G: Determining a first order perturbation of the biharmonic operator by partial boundary measurements. Trans. AMS **366**, 95–112 (2014)
120. Krupchyk, K., Uhlmann, G.: Determining a magnetic Schrödinger operator with a bounded magnetic potential from boundary measurements. Comm. Math. Phys. (2014, to appear)
121. Lee, J., Uhlmann, G.: Determining anisotropic real-analytic conductivities by boundary measurements. Comm. Pure Appl. Math. **42**, 1097–1112 (1989)
122. Li, X., Uhlmann, G.: Inverse problems on a slab. Inverse Probl. Imaging **4**, 449–462 (2010)
123. Lassas, M., Sharafutdinov, V., Uhlmann, G.: Semiglobal boundary rigidity for Riemannian metrics. Math. Annalen **325**, 767–793 (2003)
124. Lassas, M., Uhlmann, G.: Determining a Riemannian manifold from boundary measurements. Ann. Sci. École Norm. Sup. **34**, 771–787 (2001)
125. Lassas, M., Taylor, M., Uhlmann, G.: The Dirichlet-to-Neumann map for complete Riemannian manifolds with boundary. Comm. Geom. Anal. **11**, 207–222 (2003)
126. Leonhardt, U.: Optical Conformal Mapping. Science **312**, 1777–1780 (2006)
127. Mandache, N.: Exponential instability in an inverse problem for the Schrödinger equation. Inverse Probl. **17**, 1435–1444 (2001)
128. Melrose, R.B.: Geometric Scattering Theory. Cambridge University Press, Cambridge (1995)
129. Michel, R.: Sur la rigidité imposée par la longueur des géodésiques. Invent. Math. **65**, 71–83 (1981)
130. Michel, R.: Restriction de la distance géodésique a un arc et rigidité. Bull. Soc. Math. France **122**, 435–442 (1994)
131. Mukhometov, R.G.: The reconstruction problem of a two-dimensional Riemannian metric, and integral geometry (Russian). Dokl. Akad. Nauk SSSR **232**(1), 32–35 (1977)
132. Mukhometov, R.G.: On one problem of reconstruction of Riemannian metric (Russian). Siberian Math. J. **22**(3), 119–135 (1981)
133. Mukhometov, R.G., Romanov, V.G.: On the problem of finding an isotropic Riemannian metric in an n-dimensional space (Russian). Dokl. Akad. Nauk SSSR **243**(1), 41–44 (1978)
134. Nachman, A.: Global uniqueness for a two-dimensional inverse boundary value problem. Ann. Math. **143**, 71–96 (1996)
135. Nachman, A.: Reconstructions from boundary measurements. Ann. Math. **128**, 531–576 (1988)
136. Nachman, A., Ablowitz, N.: A multidimensional inverse scattering method. Stud. Appl. Math. **71**, 243–250 (1984)
137. Nachman, A., Street, B.: Reconstruction in the Calderón problem with partial data. Comm. PDE **35**, 375–390 (2010)
138. Nagayasu, S., Uhlmann, G., Wang, J.-N.: Depth dependent stability estimate in electrical impedance tomography. Inverse Probl. **25**, 075001 (2009)
139. Nagayasu, S., Uhlmann, G., Wang, J.-N.: Reconstruction of penetrable obstacles in acoustics. SIAM J. Math. Anal. **43**, 189–211 (2011)
140. Nagayasu, S., Uhlmann, G., Wang, J.-N.: Increasing stability for the acoustic equation. Inverse Probl. **29**, 229–247 (2013)
141. Nakamura, G., Tanuma, K.: Local determination of conductivity at the boundary from the Dirichlet-to-Neumann map. Inverse Probl. **17**, 405–419 (2001)
142. Nakamura, G., Uhlmann, G.: Global uniqueness for an inverse boundary value problem arising in elasticity, Invent. Math., 118, : 457–474. Erratum. Invent. Math. **152**(2003), 205–207 (1994)
143. Nakamura, G., Uhlmann, G.: Inverse problems at the boundary for an elastic medium. SIAM J. Math. Anal. **26**, 263–279 (1995)
144. Nakamura, G., Sun, Z., Uhlmann, G.: Global identifiability for an inverse problem for the Schrödinger equation in a magnetic field. Math. Annalen **303**, 377–388 (1995)
145. Novikov, R.G.: Multidimensional inverse spectral problems for the equation $-\Delta\psi + (v(x) - Eu(x))\psi = 0$, Funktsionalny Analizi Ego Prilozheniya 22, 11–12 (1988). Translation in Functional Analysis and its Applications 22, 263–272 (1988)

146. Novikov, R.G., Henkin, G.M.: The $\bar{\partial}$-equation in the multidimensional inverse scattering problem. Russ. Math. Surv. **42**, 109–180 (1987)
147. Ola, P., Päivärinta, L., Somersalo, E.: An inverse boundary value problem in electrodynamics. Duke Math. J. **70**, 617–653 (1993)
148. Ola, P., Somersalo, E.: Electromagnetic inverse problems and generalized Sommerfeld potentials. SIAM J. Appl. Math. **56**, 1129–1145 (1996)
149. Otal, J.P.: Sur les longuer des géodésiques d'une métrique a courbure négative dans le disque. Comment. Math. Helv. **65**, 334–347 (1990)
150. Paternain, G., Salo, M., Uhlmann, G.: The attenuated ray transform for connections and Higgs fields. Geom. Funct. Anal. (GAFA) **22**, 1460–1489 (2012)
151. Paternain, G., Salo, M., Uhlmann, G.: Tensor tomography on surfaces. Inventiones Math. **193**, 20518–20530 (2013)
152. Päivärinta, L., Panchenko, A., Uhlmann, G.: Complex geometrical optics for Lipschitz conductivities. Revista Matematica Iberoamericana **19**, 57–72 (2003)
153. Pendry, J.B., Schurig, D., Smith, D.R.: Controlling electromagnetic fields. Science **312**, 1780–1782 (2006)
154. Pestov, L., Sharafutdinov, V.A.: Integral geometry of tensor fields on a manifold of negative curvature. Siberian Math. J. **29**, 427–441 (1988)
155. Pestov, L., Uhlmann, G.: Two dimensional simple Riemannian manifolds with boundary are boundary distance rigid. Ann. Math. **161**, 1089–1106 (2005)
156. Pestov, L., Uhlmann, G.: The boundary distance function and the Dirichlet-to-Neumann map. Math. Res. Lett. **11**, 285–298 (2004)
157. Pestov, P., Uhlmann, G.: Characterization of the range and inversion formulas for the geodesic X-ray transform. Int. Math. Res. Notices **80**, 4331–4347 (2004)
158. Petersen, P.: Riemannian Geometry. Springer, Berlin (1998)
159. Ramm, A.G.: Recovery of the potential from fixed energy scattering data. Inverse Probl. **4**, 877–886 (1988)
160. Rondi, L.: A remark on a paper by G. Alessandrini and S. Vessella: "Lipschitz stability for the inverse conductivity problem" [Adv. Appl. Math. 35, 207–241 (2005)]. Adv. Appl. Math. 36, 67–69 (2006).
161. Romanov, V.G.: Inverse Problems of Mathematical Physics. VNU Science Press, Utrech, the Netherlands (1987)
162. Salo, M.: Semiclassical pseudodifferential calculus and the reconstruction of a magnetic field. Comm. PDE **31**, 1639–1666 (2006)
163. Salo, M., Inverse problems for nonsmooth first order perturbations of the Laplacian. Ann. Acad. Sci. Fenn. Math. Diss. 139 (2004)
164. Salo, M., Tzou, L.: Inverse problems with partial data for a Dirac system: a Carleman estimate approach. Adv. Math. **225**, 487–513 (2010)
165. Salo, M., Wang, J.-N.: Complex spherical waves and inverse problems in unbounded domains. Inverse Probl. **22**, 2299–2309 (2006)
166. Santosa, F., Vogelius, M.: A backprojection algorithm for electrical impedance imaging. SIAM J. Appl. Math. **50**, 216–243 (1990)
167. Sharafutdinov, V.: Integral geometry of tensor fields. In: Inverse and Ill-Posed Problems Series, VSP, Utrecht (1994)
168. Sharafutdinov, V.A.: Variations of Dirichlet-to-Neumann map and deformation boundary rigidity of simple 2-manifolds. J. Geom. Anal. **17**, 147–187 (2007)
169. Sharafutdinov, V., Skokan, M., Uhlmann, G.: Regularity of ghosts in tensor tomography. J. Geom. Anal. **15**, 517–560 (2005)
170. Sharafutdinov, V., Uhlmann, G.: On deformation boundary rigidity and spectral rigidity for Riemannian surfaces with no focal points. J. Differ. Geom. **56**, 93–110 (2001)
171. Schurig, D., Mock, J., Justice, B., Cummer, S., Pendry, J., Starr, A., Smith, D.: Metamaterial electromagnetic cloak at microwave frequencies. Science **314**, 977–980 (2006)
172. Siltanen, S., Müller, J.L., Isaacson, D.: A direct reconstruction algorithm for electrical impedance tomography. IEEE Trans. Med. Imaging **21**, 555–559 (2002)
173. Sjöstrand, J., Singularités analytiques microlocales, Astérisque (1985)
174. Sjöstrand, J.: Remark on extensions of the Watermelon theorem. Math. Res. Lett. **1**, 309–317 (1994)
175. Somersalo, E., Isaacson, D., Cheney, M.: A linearized inverse boundary value problem for Maxwell's equations. J. Comput. Appl. Math. **42**, 123–136 (1992)

176. Stefanov, P., Uhlmann, G.L.: Multi-wave methods via ultrasound. In: Uhlmann, G. (ed.) Inverse Problems and Applications, Inside Out II, MSRI Publications, vol. 60, pp. 271–323. Cambridge University Press, Cambridge (2012)
177. Stefanov, P., Uhlmann, G.: Recent progress on the boundary rigidity problem. Electr. Res. Announc. Amer. Math. Soc. **11**, 64–70 (2005)
178. Stefanov, P., Uhlmann, G.: Rigidity for metrics with the same lengths of geodesics. Math. Res. Lett. **5**, 83–96 (1998)
179. Stefanov, P., Uhlmann, G.: Stability estimates for the X-ray transform of tensor fields and boundary rigidity. Duke Math. J. **123**, 445–467 (2004)
180. Stefanov, P., Uhlmann, G.: Stable determination of generic simple metrics from the hyperbolic Dirichlet-to-Neumann map. Int. Math. Res. Notices **17**, 1047–1061 (2005)
181. Stefanov, P., Uhlmann, G.: Boundary rigidity and stability for generic simple metrics. J. Amer. Math. Soc. **18**, 975–1003 (2005)
182. Stefanov, P., Uhlmann, G.: Integral geometry of tensor fields on a class of non-simple Riemannian manifolds. Am. J. Math. **130**, 239–268 (2008)
183. Stefanov, P., Uhlmann, G.: Local lens rigidity with incomplete data for a class of non-simple Riemannian manifolds. J. Differ. Geom. **82**, 383–409 (2009)
184. Sun, Z.: On a quasilinear boundary value problem. Math. Z. **221**, 293–305 (1996)
185. Sun, Z.: Conjectures in inverse boundary value problems for quasilinear elliptic equations. Cubo **7**, 65–73 (2005)
186. Sun, Z., Uhlmann, G.: Anisotropic inverse problems in two dimensions. Inverse Probl. **19**, 1001–1010 (2003)
187. Sun, Z., Uhlmann, G.: Generic uniqueness for an inverse boundary value problem. Duke Math. J. **62**, 131–155 (1991)
188. Sun, Z., Uhlmann, G.: Inverse problems in quasilinear anisotropic media. Amer. J. Math. **119**, 771–797 (1997)
189. Sylvester, J.: An anisotropic inverse boundary value problem. Comm. Pure Appl. Math. **43**, 201–232 (1990)
190. Sylvester, J., Uhlmann, G.: A global uniqueness theorem for an inverse boundary value problem. Ann. Math. **125**, 153–169 (1987)
191. Sylvester, J., Uhlmann, G.: A uniqueness theorem for an inverse boundary value problem in electrical prospection. Comm. Pure Appl. Math. **39**, 92–112 (1986)
192. Sylvester, J., Uhlmann, G.: Inverse boundary value problems at the boundary - continuous dependence. Comm. Pure Appl. Math. **41**, 197–221 (1988)
193. Sylvester, J., Uhlmann, G.: Inverse problems in anisotropic media. Contemp. Math. **122**, 105–117 (1991)
194. Tataru, D.: Unique continuation for solutions to PDE's; between Hörmander's theorem and Holmgren's theorem. Comm. PDE **20**, 855–884 (1995)
195. Treves, F.: Introduction to pseudodifferential and fourier integral operators, vol. 1. In: Pseudodifferential Operators. The University Series in Mathematics, Plenum Press, New York (1980)
196. Tsai, T.Y.: The Schrödinger equation in the plane. Inverse Probl. **9**, 763–787 (1993)
197. Tolmasky, C.: Exponentially growing solutions for nonsmooth first-order perturbations of the Laplacian. SIAM J. Math. Anal. **29**, 116–133 (1998)
198. Tzou, L.: Stability estimates for coefficients of magnetic Schrödinger equation from full and partial measurements. Comm. PDE **33**, 161–184 (2008)
199. Uhlmann, G., Inverse boundary value problems for partial differential equations. Documenta Math. Extra Volume ICM 98, III 77–86 (1998)
200. Uhlmann, G.: Inverse boundary value problems and applications. Astérisque **207**, 153–211 (1992)
201. Uhlmann, G.: Developments in inverse problems since Calderón's foundational paper. In: Christ, M., Kenig, C., Sadosky, C. (eds.) Harmonic Analysis and Partial Differential Equations, Chapter 19, pp. 295–345. University of Chicago Press, Chicago (1999)
202. Uhlmann, G.: Scattering by a metric. In: Pike, R., Sabatier, P. (eds.) Encyclopedia on Scattering, Chap. 6.1.5, pp. 1668–1677. Academic Press, New York (2002)
203. Uhlmann, G., Vasy, A.: Low-energy inverse problems in three-body scattering. Inverse Probl. **18**, 719–736 (2002)
204. Uhlmann, G., Wang, J.-N.: Complex spherical waves for the elasticity system and probing of inclusions. SIAM J. Math. Anal. **38**, 1967–1980 (2007)

205. Uhlmann, G., Wang, J.-N.: Reconstruction of discontinuities in systems. SIAM J. Appl. Math. **28**, 1026–1044 (2008)
206. Wang, J.-N.: Stability for the reconstruction of a Riemannian metric by boundary measurements. Inverse Probl. **15**, 1177–1192 (1999)
207. Wiechert, E., Zoeppritz, K.: Uber erdbebenwellen. Nachr. Koenigl. Geselschaft Wiss Goettingen **4**, 415–549 (1907)
208. Zhdanov, M.S., Keller, G.V.: The geoelectrical methods in geophysical exploration. In: Methods in Geochemistry and Geophysics, vol. 31. Elsevier, Amsterdam (1994)
209. Zhou, T.: Reconstructing electromagnetic obstacles by the enclosure method. Inverse Probl. Imaging **4**, 547–569 (2010)
210. Zou, Y., Guo, Z.: A review of electrical impedance techniques for breast cancer detection. Med. Eng. Phys. **25**, 79–90 (2003)

7

A note on global regularity in optimal transportion

Neil S. Trudinger

Abstract We indicate how recent work of Figalli–Kim–McCann and Vetois can be used to improve previous results of Trudinger and Wang on classical solvability of the second boundary value problem for Monge–Ampère type partial differential equations arising in optimal transportation together with the global regularity of the associated optimal mappings.

Keywords Optimal transportation · Monge–Ampère equations · Second boundary value problem

Mathematics Subject Classification (1991) Primary 35J66 · 35J96; Secondary 49N60

1 Introduction

In this short note we use recent work of Figalli–Kim–McCann [4] and Vetois [24] on continuous differentiability and strict convexity of potentials in optimal transportation to improve previous results on global regularity of optimal mappings in [20]. The corresponding partial differential equations are Monge–Ampère type equations which have the general form:

Communicated by S.K. Jain.

Research supported by Australian Research Council Grant.

N. S. Trudinger (✉)
Centre for Mathematics and Its Applications, Australian National University,
Canberra, ACT 0200, Australia
e-mail: neil.trudinger@anu.edu.au

$$\det[D^2u - A(\cdot, u, Du)] = B(\cdot, u, Du), \tag{1.1}$$

where A and B are given $n \times n$ matrix and scalar valued function defined on $\Omega \times \mathbb{R} \times \mathbb{R}^n$, where Ω is a domain in Euclidean n-space, $\mathbb{R}^n$. We use (x, z, p) to denote points in $\Omega \times \mathbb{R} \times \mathbb{R}^n$ so that $A(x, z, p) \in \mathbb{R}^n \times \mathbb{R}^n$, $B(x, z, p) \in \mathbb{R}$. Equation (1.1) is elliptic with respect to a solution $u \in C^2(\Omega)$ whenever

$$D^2u - A(\cdot, u, Du) > 0, \tag{1.2}$$

whence also $B > 0$.

We assume that A is generated by a *cost function* $c \in C^4(\bar{\Omega} \times \bar{\Omega}^*)$ satisfying the following conditions, where $\Omega^* \subset \mathbb{R}^n$ is a target domain:

A1: The mappings $c_x(x, \cdot), c_y(\cdot, y)$ are one-to-one for each $x \in \Omega$, $y \in \Omega^*$;
A2: $\det c_{x,y} \neq 0$ in $\bar{\Omega} \times \bar{\Omega}^*$.

Conditions A1, A2 imply the existence of a C^3 mapping Y given by $Y(x, p) = c_x(x, \cdot)^{-1}(p)$ for $x \in \Omega$ and $p \in c_x(x, \cdot)(\Omega^*)$, with $Y_p = [c_{x,y}]^{-1}$. Moreover, for positive densities $f \in C^0(\Omega)$, $g \in C^0(\Omega^*)$, the *prescribed Jacobian equation*,

$$|\det DY(\cdot, Du)| = f/g^\circ Y \tag{1.3}$$

can be written in the form (1.1) with

$$A(\cdot, z, p) = c_{xx}(\cdot, Y(\cdot, p)), \quad B(x, z, p) = |\det c_{x,y}(\cdot, Y)| f/g^\circ Y, \tag{1.4}$$

(independent of z), for elliptic solutions u.

The *second boundary value problem* for the prescribed Jacobian equation (1.3) is to prescribe the image,

$$Tu(\Omega) := Y(\cdot, Du)(\Omega) = \Omega^*, \tag{1.5}$$

and a necessary condition for the existence of an elliptic solution u, for which the mapping Tu is a diffeomorphism, is the *mass balance* condition,

$$\int_\Omega f = \int_{\Omega^*} g. \tag{1.6}$$

For the standard Monge–Ampère equation, $c(x, y) = x.y$, $Y = p$, $A = 0$, with $Tu = Du$, the classical solvability of the second boundary value problem for smooth densities and domains, under the mass balance condition (1.6), was proved by Delanöe, ($n = 2$), [3], Caffarelli [2] and Urbas [22], under the hypothesis that both domains, Ω and Ω^*, are uniformly convex, (in the sense that the principal curvatures of their boundaries are bounded from below by a positive constant). As already pointed out in [10,11], (when Ω^* is a ball), (1.5) implies a nonlinear *oblique* boundary condition. A weaker interpretation of the boundary condition (1.5) arises through optimal transportation, in which case Caffarelli [1] proved that the convexity of the target Ω^* suffices for local smoothness of solutions.

Interest in the general case was stimulated in the last decade through its application to regularity in optimal transportation. In the wake of earlier work on reflector design [27], a condition for local regularity called A3 was found by Ma, Trudinger and Wang in [15] and its degenerate form called A3w was introduced and used for global regularity in [20]. The latter was shown to be sharp by Loeper in [14], utilising the proof in [15] of the necessity of target c-convexity. We may write these conditions in the form:

A3 (A3w)

$$A_{ij}^{kl}\xi_i\xi_j\eta_k\eta_l := (D_{p_k p_l}A_{ij})\xi_i\xi_j\eta_k\eta_l >, (\geq)\ 0,$$

for all $x, p \in \Omega \times \mathbb{R}^n$ such that $Y(x, p) \in \Omega^*$ and $\xi, \eta \in \mathbb{R}^n$ such that $\xi.\eta = 0$. Using the notation,

$$c_{ij\cdots,kl\cdots} = \frac{\partial}{\partial x_i}\frac{\partial}{\partial x_j}\cdots\frac{\partial}{\partial y_k}\frac{\partial}{\partial y_l}\cdots c$$

and writing $[c^{i,j}]$ for the inverse of $c_{x,y} = [c_{i,j}]$, we also have

$$A_{ij}^{kl} = \left(c_{ij,pq} - c^{r,s}c_{ij,s}c_{r,pq}\right)c^{p,k}c^{q,l}, \tag{1.7}$$

in accordance with the original formulation in [15]. Note that from (1.7), we see that conditions A3 and A3w are symmetric in x and y, while the formulation in terms of x and p shows they are invariant under coordinate transformations in x, for fixed y.

For global regularity, we also need uniform convexity conditions on our domains, with respect to c, as also formulated in [20], namely we assume that Ω, Ω^* are respectively uniformly c-convex, c^*-convex with respect to each other in the sense that the images $c_y(\cdot, y)(\Omega)$, $c_x(x, \cdot)(\Omega^*)$ are uniformly convex in $\mathbb{R}^n$ for all $y \in \overline{\Omega}^*$, $x \in \overline{\Omega}$.

Theorem 1.1 *Let $c \in C^4(\overline{\Omega} \times \overline{\Omega}^*)$ be a cost function satisfying conditions A1, A2 and A3w in C^4 domains Ω, Ω^* which are mutually uniformly c-convex, c^*-convex with respect to each other. Suppose $f \in C^2(\overline{\Omega})$, $g \in C^2(\overline{\Omega})$ are positive densities satisfying the mass balance condition* (1.6). *Then there exists an elliptic solution $u \in C^3(\overline{\Omega})$ of the second boundary value problem* (1.3), (1.5), *with Tu a diffeomorphism from $\overline{\Omega}$ to $\overline{\Omega}^*$, which is unique up to additive constants.*

The corresponding Monge–Kantorovich problem in optimal transportation is to find a measure preserving mapping $T_0 : \Omega \to \Omega^*$ which maximizes the cost functional

$$C(T) = \int_{\Omega} f(c(\cdot, T)dx \tag{1.8}$$

among all measure preserving mappings T from Ω to Ω^*. A mapping $T : \Omega \to \Omega^*$ is called *measure preserving*, with respect to densities f and g, if it is Borel measurable and for any Borel set $E \subset \Omega^*$,

$$\int_{T^{-1}(E)} f = \int_E g. \tag{1.9}$$

The reader is referred to the expositions [16,23,25,26] for further information about optimal transportation.

Corollary 1.2 *Under the hypotheses of Theorem 1, there exists a unique diffeomorphism $T = Y(\cdot, Du), \in [C^2(\overline{\Omega})]^n$ maximizing the functional* (1.8)*, where u is an elliptic solution of the second boundary value problem* (1.3), (1.5).

The solution u of (1.3), (1.5) is called a potential. From elliptic regularity theory it follows that if $c, \Omega, \Omega^*, f, g$ are C^∞ smooth, then the resultant potentials u and optimal mapping T are also. Note that we have followed the same sign conventions as in [20] but usually the cost functions and potentials are the negatives of those here and the optimal transportation problem is written equivalently as a minimization problem.

These results are proved in [20] under a further hypothesis, namely either the non degenerate condition A3 holds or a global barrier type condition is satisfied in Ω, (called c-boundedness in [17,20]), or the matrix function A depends only on p. Such restrictions arise from the second derivative bounds in Section 3 of [20] and do not affect the main examples as presented there in Section 8. The last mentioned alternative sufficient condition follows from a duality argument in Section 3 of [20] which does not seem to extend to general dependence of A on (x, p), (as envisaged in an earlier version of [20]). An alternative sub-solution condition for these second derivative bounds is also given in the recent paper [6]. In the next section we indicate the necessary modifications to obtain the full generality.

We remark also that the results and proofs in [20] were extended to the more general prescribed Jacobian equations in [18] and to parabolic equations in [9].

2 Second derivative bounds

Theorem 1.1 follows from a modification to the main second derivative estimate in [20], (also labelled Theorem 1.1). To facilitate the application of the method of continuity from [20], we need to consider more general equations than (1.3), namely we assume still that A is given by (1.4) but consider a general $B > 0, \in C^2(\overline{\Omega} \times \mathbb{R} \times \mathbb{R}^n)$. We also allow for a domain variation.

Theorem 2.1 *Let the matrix function A and domains Ω and Ω^* satisfy the same conditions as in Theorem* 1.1 *with the scalar function B as above and let Ω_t and Ω_t^* denote C^4 subdomains of Ω and Ω^*, which are also uniformly c and c^*-convex with respect to each other. Then any elliptic solution $u \in C^3(\overline{\Omega}_t)$ of* (1.1), (1.5)*, for which the mapping Tu is a diffeomorphism from Ω_t to Ω_t^*, satisfies the a priori estimate*

$$|D^2u| \le C, \tag{2.1}$$

where C depends on $c, \Omega_t, \Omega_t^*, B$ and $sup_{\Omega_t}|u|$.

More specifically, with c, Ω and Ω^* fixed, the constant C depends on the C^2 norms and minima of B over compact subsets of $\Omega \times \mathbb{R} \times \mathbb{R}^n$ together with the C^4 smoothness of the domains Ω_t and Ω_t^* and their uniform convexity constants. We also observe the simple gradient estimate

$$|Du| = |c_x(\cdot, Tu)| \leq sup_{\Omega\times\Omega^*}|c_x|. \tag{2.2}$$

Proof From [20] it suffices to derive a global bound for D^2u in terms of its boundary trace as that is subsequently estimated in Section 4 of [20], using the obliqueness estimate from Sect. 2. First we note from [20] that the uniform c-convexity of Ω implies the existence of a C^2 defining function ϕ for Ω satisfying

$$\left[D_{ij}\phi - c^{l,k}c_{ij,l}(\cdot, Tu)D_k\phi\right]\xi_i\xi_j \geq \delta_0|\xi|^2, \tag{2.3}$$

in a fixed neighbourhood $\mathcal{N}$ of $\partial\Omega$, for all $\xi \in \mathbb{R}^n$ and for a fixed constant $\delta_0 > 0$. Noting also the formula,

$$D_{p_k}A_{ij}(\cdot, Du) = c^{l,k}c_{ij,l}(x, Tu),$$

we have from the global second derivative estimate, Theorem 3.1 in [20],

$$\sup_{\mathcal{N}}|D^2u| \leq C\left(1 + \sup_{\partial\mathcal{N}}|D^2u|\right), \tag{2.4}$$

with constant C depending on c, Ω, Ω^*, B and $\sup(|u| + |Du|)$. Consequently the global estimation is reduced to an interior estimate in a strictly contained subdomain $\Omega' = \Omega - \overline{\mathcal{N}}$.

Interior second order derivative estimates for solutions of Monge–Ampère type equations, under the weak condition A3w, are derived in [12] and [13]. We need to apply these in c-sections of u whose diameters will be controlled by the strict convexity results in [4] and [24]. First we note that the ellipticity condition (1.2) and the c-convexity of the domain Ω implies that u is c-convex in Ω [21], (see also [20] and [8]), that is for any $x_0 \in \Omega$ and $y_0 = Tu(x_0)$, we have

$$u(x) \geq u(x_0) + c(x, y_0) - c(x_0, y_0) \tag{2.5}$$

for all $x \in \Omega$. The *c-section*, $S_h(x_0)$ for $h > 0$ is then defined by

$$S_h(x_0) = \{x \in \Omega| \quad u(x) < u(x_0) + c(x, y_0) - c(x_0, y_0) + h\}. \tag{2.6}$$

Writing $b = B(\cdot, u, Du)/|detc_{x,y}(\cdot, Tu)|$, it follows since u is c-convex and Tu a diffeomorphism, that u is a potential for the corresponding Monge–Kantorovich problem with densities $f = b$ and $g = 1$. Furthermore there exist positive constants b_0 and b_1, depending on B, $|detc_{x,y}|$ and $\sup(|u| + |Du)$ such that $b_0 \leq b \leq b_1$. We now claim that, from [4] and [24], we have a strict convexity estimate, that is for any $R < \text{dist}(x_0, \partial\Omega_t)$, there exists $h > 0$, depending on c, Ω, Ω^* and b_0, b_1, such that

$$S_h(x_0) \subset B_R(x_0) \tag{2.7}$$

where $B_R = B_R(x_0)$ denotes the ball of radius R and centre at x_0. This follows in a standard way by contradiction for otherwise there would exist a sequence of potentials converging uniformly to a c -convex potential of a limiting problem with densities bounded away from 0 and ∞ which would not be strictly c-convex at x_0, thereby contradicting the strict convexity assertions in [4] and [24]. We also note that a crucial hypotheses in these works is the uniform c-convexity of the target domain which is satisfied by hypothesis in our case. So that there is no confusion, we note also that *uniform* convexity in our terminology, as employed for example in [7], is designated *strong* in [4], (in accordance with usage in convexity theory where *uniform* can mean a uniform modulus of convexity).

To complete the proof of Theorem 2.1, we then note that for sufficiently small R there exists a smooth defining function ϕ satisfying (2.3) in B_R so that we can apply the interior second derivative estimate in [12] in $S_h(x_0)$ to obtain an estimate

$$\sup_{B_\rho} |D^2 u| \leq C, \tag{2.8}$$

for sufficiently small ρ depending on h and $\sup|D(u - c(\cdot, y_0)|$. By using a finite covering the estimate (2.8) is extended to Ω' and we conclude Theorem 2.1, with the asserted constant dependence, through replacement of Ω and Ω^* by Ω_t and Ω_t^* respectively. □

We remark that a stronger result is proved in [4] than that used above, namely that the optimal mappings are strictly convex and Hölder continuous whereas in [24] only strict convexity and continuity is proved, (using an earlier version of [4] which assumes condition A3w without the orthogonality restriction on ξ and η). Note also that in two dimensions the strict convexity was already proved in [5] under a one-sided density bound as an extension of the classical result of Aleksandrov for the Monge–Ampère equation, (see for example [19]). However the argument in [5] also depends on a result established in [4]. We remark also that when the strong condition A3 holds, then the proof of the second derivative estimate is much simpler and already follows from [15]; see also [19].

From the estimates (2.1),(2.2), Theorem 1.1 follows immediately by using Theorem 2.1, (instead of Theorem 1.1 in [20]), in the method of continuity argument presented in Section 5 of [20].

References

1. Caffarelli, L.: The regularity of mappings with a convex potential. J. Am. Math. Soc. **5**, 99–104 (1992)
2. Caffarelli, L.: Boundary regularity of maps with convex potentials II. Ann. Math. **144**, 453–496 (1996)
3. Delanoë, Ph.: Classical solvability in dimension two of the second boundary value problem associated with the Monge-Ampère operator. Ann. Inst. Henri Poincaré Analyse Non Linéaire **8**, 443–457 (1991)
4. Figalli, A., Kim, Y.-H., McCann, R.J.: Hölder continuity and injectivity of optimal maps. Arch. Rat. Mech. Anal. **209**, 747–795 (2013)
5. Figalli, A., Loeper, G.: C^1 regularity of solutions of the Monge-Ampère equation for optimal transport in dimension two. Calc. Var. Partial Differential Equations **35**, 537–550 (2009)
6. Jiang, F., Trudinger, N.S., Yang, X.-P.: On the Dirichlet Problem for Monge-Ampère type equations. Calc. Var. Partial Differ. Equ. (to appear).
7. Gilbarg, D., Trudinger, N.S.: Elliptic partial differential equations of second order. Springer, Berlin (1983)
8. Kim, Y.-H., McCann, R.J.: Continuity, curvature and the general covariance of optimal transportation. J. Eur. Math. Soc. **12**, 1009–1040 (2010)
9. Kitagawa, J.: A parabolic flow towards solutions of the optimal transportation problem on domains with boundary. J. Reine Angew. Math. **672**, 127–160 (2012)
10. Lions, P.-L., Trudinger, N.S., Urbas, J.: Neumann problem for equations of Monge-Ampère type, Comm. Pure Appl. Math. **39**, 539–563 (1985)
11. Lions, P.-L., Trudinger, N.S., Urbas, J.: Neumann problem for equations of Monge-Ampère type. Proc. Centre Math. Anal. **10**, 135–140 (1985)
12. Liu, J., Trudinger, N.: On Pogorelov estimates for Monge-Ampere type equations. Discret. Contin. Dyn. Sys. **28**, 1121–1135 (2010)
13. Liu, J., Trudinger, N.S., Wang, X.-J.: Interior $C^{2,\alpha}$ regularity for potential functions in optimal transportation. Comm. Partial Differ. Equ. **35**, 165–184 (2010)
14. Loeper, G.: On the regularity of solutions of optimal transportation problems. Acta Math. **202**, 241–283 (2009)
15. Ma, X.-N., Trudinger, N.S., Wang, X.-J.: Regularity of potential functions of the optimal transportation problem. Arch. Rat. Mech. Anal. **177**, 151–183 (2005)
16. Rachev, S.T., Ruschendorff, L.: Mass transportation problems. Springer, Berlin (1998)
17. Trudinger, N.S.: Recent developments in elliptic partial differential equations of Monge-Ampère type. Int. Congress Math. III, 291–301 (2006, Eur. Math. Soc., Zürich)
18. Trudinger. N.S.: On the prescribed Jacobian equation, Gakuto Intl. Series, Math. Sci. Appl. 20 (2008), Proc. Intl. Conf. for the 25th Anniversary of Viscosity, Solutions, 243–255
19. Trudinger, N.S., Wang, X.-J.: The Monge-Ampère equation and its geometric applications. In: Handbook of geometric analysis. International Press, USA, pp. 467–524 (2008)
20. Trudinger, N.S., Wang, X.J.: On the second boundary value problem for Monge-Ampère type equations and optimal transportation. Ann. Scuola Norm. Sup. Pisa Cl. Sci. Ser. 5 **8**, 143–174 (2009)
21. Trudinger, N.S., Wang, X.-J.: On strict convexity and continuous differentiability of potential functions in optimal transportation. Arch. Rat. Mech. Anal. **192**, 403–418 (2009)
22. Urbas, J.: On the second boundary value problem for equations of Monge-Ampère type. J. Reine Angew. Math. **487**, 115–124 (1997)
23. Urbas, J.: Mass transfer problems. University of Bonn, Lecture Notes (1998)
24. Vetois, J.: Continuity and injectivity of optimal maps, preprint (2011)
25. Villani, C.: Topics in optimal transportation. Amer. Math. Soc. (2003)
26. Villani, C.: Optimal transportation, old and new. Springer, Berlin (2008)
27. Wang, X.-J.: On the design of a reflector antenna. Inverse Prob. **12**, 351–375 (1996)

Permissions

All chapters in this book were first published in BMS, by Springer; hereby published with permission under the Creative Commons Attribution License or equivalent. Every chapter published in this book has been scrutinized by our experts. Their significance has been extensively debated. The topics covered herein carry significant findings which will fuel the growth of the discipline. They may even be implemented as practical applications or may be referred to as a beginning point for another development.

The contributors of this book come from diverse backgrounds, making this book a truly international effort. This book will bring forth new frontiers with its revolutionizing research information and detailed analysis of the nascent developments around the world.

We would like to thank all the contributing authors for lending their expertise to make the book truly unique. They have played a crucial role in the development of this book. Without their invaluable contributions this book wouldn't have been possible. They have made vital efforts to compile up to date information on the varied aspects of this subject to make this book a valuable addition to the collection of many professionals and students.

This book was conceptualized with the vision of imparting up-to-date information and advanced data in this field. To ensure the same, a matchless editorial board was set up. Every individual on the board went through rigorous rounds of assessment to prove their worth. After which they invested a large part of their time researching and compiling the most relevant data for our readers.

The editorial board has been involved in producing this book since its inception. They have spent rigorous hours researching and exploring the diverse topics which have resulted in the successful publishing of this book. They have passed on their knowledge of decades through this book. To expedite this challenging task, the publisher supported the team at every step. A small team of assistant editors was also appointed to further simplify the editing procedure and attain best results for the readers.

Apart from the editorial board, the designing team has also invested a significant amount of their time in understanding the subject and creating the most relevant covers. They scrutinized every image to scout for the most suitable representation of the subject and create an appropriate cover for the book.

The publishing team has been an ardent support to the editorial, designing and production team. Their endless efforts to recruit the best for this project, has resulted in the accomplishment of this book. They are a veteran in the field of academics and their pool of knowledge is as vast as their experience in printing. Their expertise and guidance has proved useful at every step. Their uncompromising quality standards have made this book an exceptional effort. Their encouragement from time to time has been an inspiration for everyone.

The publisher and the editorial board hope that this book will prove to be a valuable piece of knowledge for researchers, students, practitioners and scholars across the globe.

List of Contributors

András Kroó
Alfréd Rényi Institute of Mathematics, Hungarian Academy of Sciences, Budapest, Hungary

Department of Analysis, Budapest University of Technology and Economics, Budapest, Hungary

J. Mostovoy
Departamento de Matemáticas, CINVESTAV-IPN, Apartado Postal 14-740, 07000 México D.F., Mexico

J. M. Perez-Izquierdo
Departamento de Matemáticas y Computación, Universidad de La Rioja, 26004 Logroño, Spain

I. P. Shestakov
Instituto de Matemática e Estatística, Universidade de São Paulo, Caixa Postal 66281, São Paulo, SP 05311-970, Brazil
Sobolev Institute of Mathematics, Novosibirsk 630090, Russia

Tuomo Kuusi
Aalto University Institute of Mathematics, P.O. Box 11100, 00076 Aalto, Finland

Giuseppe Mingione
Dipartimento di Matematica e Informatica, Università di Parma, Parco Area delle Scienze 53/a, Campus, 43124 Parma, Italy

D. R. Yafaev
IRMAR, Université de Rennes I, Campus de Beaulieu, 35042 Rennes Cedex, France

Francisco Marcellán and M. Francisca Pérez-Valero
Departamento de Matemáticas, Universidad Carlos III de Madrid, Avenida de la Universidad 30, 28911 Leganés, Madrid, Spain

Yamilet Quintana and Alejandro Urieles
Departamento de Matemáticas Puras y Aplicadas, Edificio Matemáticas y Sistemas (MYS), Universidad Simón Bolívar, Apartado Postal: 89000, Caracas 1080 A, Venezuela

Gunther Uhlmann
Department of Mathematics, University of Washington, Seattle, WA 98195, USA

Neil S. Trudinger
Centre for Mathematics and Its Applications, Australian National University, Canberra, ACT 0200, Australia